I. F. Pelyvás   C. Monneret   P. Herczegh

Synthetic Aspects of Aminodeoxy Sugars of Antibiotics

With 9 Figures, 206 Schemes, and 23 Tables

Springer-Verlag
Berlin Heidelberg NewYork
London Paris Tokyo

Dr. István F. Pelyvás

Research Group of Antibiotics of the Hungarian Academy of Sciences,
P.O. Box 70, H-4010 Debrecen/Hungary

Dr. Claude Monneret
Centre National de la Recherche Scientifique, Group No. 484,
4, av. de l'Observatoire, F-75270 Paris/France

Dr. Pál Herczegh

Research Group of Antibiotics of the Hungarian Academy of Sciences,
P.O. Box 70, H-4010 Debrecen/Hungary

Illustrations and cover picture designed by Mrs. Erzsébet L. Magyar

ISBN-13:978-3-642-73401-4 e-ISBN-13:978-3-642-73399-4
DOI: 10.1007/978-3-642-73399-4

Library of Congress Cataloging-in-Publication Data.
Pelyvás, I.F. (István), 1947– . Synthetic aspects of aminodeoxy sugars of antibiotics.
Bibliography: p. Includes index.
1. Aminodeoxy sugar antibiotics — Synthesis. I. Monneret, C. (Claude), 1938– . II. Herczegh, P. (Pál), 1947– . III. Title.
RS431.A58P45 1988 615'.329 88-6476
ISBN-13:978-3-642-73401-4 (U.S.)

2152/3020-543210

To Professor Rezső Bognár

on the occasion of his 75th birthday

Foreword

The synthetic chemistry of carbohydrates has advanced at a scarcely equalled rate in the last 25 years, due to the great interest of biologically active natural products containing sugar moieties. It suffices to note that in the review by J. D. Dutcher appearing in "Advances in Carbohydrate Chemistry" vol. 18, 1963, only the structures of less than ten aminodeoxy sugars were reported. This book deals exclusively with a single class of carbohydrates, namely the aminodeoxy sugars of antibiotics, the most popular of which is probably daunosamine, a compound for which more than 20 different synthetic approaches have been reported in the literature since the publication of its structure in 1964. No compound in the 3-amino-2-deoxy-L-hexose series had been prepared by chemical synthesis when we started our synthetic work in this field in 1972 on the wave of the successful therapeutic applications of adriamycin. The compounds with *xylo* stereochemistry were unknown even in the more easily accessible D-series. The size of this book documents the rapid development of the field. I wish to add that the improvements of chemical methodology reported in the volume outspan the specific field and are of importance in the design of synthetic approaches to other carbohydrate structures. These also include compounds involved in chemical interactions of great biological interest, but hitherto unexplained at the molecular level, such as those related with cell recognition, adhesiveness and differentiation.

It is certainly very appropriate and by no means surprising that the endeavour to write the first comprehensive book discussing the currently available knowledge on the synthesis of aminodeoxy sugars be undertaken by two well-known members of the established Hungarian School of carbohydrate chemistry to whose founder the book is dedicated, and by a knowledgeable French researcher who has contributed greatly to the chemistry of the anthracycline antibiotics in recent times. I agree with the authors that the separate presentation of the different reaction steps occurring in the synthesis represents an original organization of the literature data that will be of considerable help to those planning synthetic schemes in this interesting field. All organic chemists will be grateful to Drs. Pelyvás, Monneret and Herczegh for their outstanding work.

Firenze, April 1988 Federico Maria Arcamone

Preface

In the therapy of infections caused by resistant bacteria and also in the treatment of certain neoplastic diseases, antibiotics containing aminodeoxy sugar building elements have emerged as major chemotherapeutic agents. Including only mono-amino-mono-, di- and trideoxyhexoses, more than forty representatives have been isolated from various antibiotic substances so far. The elaboration of definitive and preparative syntheses for such rare aminodeoxy sugars have offered great challenges for carbohydrate chemists during the past two decades. In many cases related efforts have brought the introduction of novel, ingenious methodologies that made valuable contributions to research on carbohydrates and another organic compounds.

The present work is aimed at summarizing the applied and potentially useful strategies for the synthesis of the specific aminodeoxy sugars occurring in antibiotics with a particular emphasis on 3-amino-polydeoxyhexoses. It provides hitherto unexplored contexts and offers a general outlook on the syntheses of such compounds. The book is concerned with the methodological construction and systemic layout of molecules of the daunosamine-type unbranched- and branched-chain natural aminodeoxy hexoses, and their synthetic derivatives starting with both carbohydrates and non-sugar precursors. The methodologies employed for the preparation of 2-deoxyhexoses and for the introduction of a nitrogen function into such sugars, and also the strategies for the conversion of aminodeoxy hexoses into their polydeoxy analogs are outlined as are total synthesis approaches. Examples of the utilization of these type of carbohydrates as "chiral templates" for transformations into non-sugar organic substances with potential biological activities follow.

The reader may notice that the reaction sequences employed for the preparation of 3-amino-2,3,6-trideoxyhexoses are in many cases divided into individual reaction steps that are, in turn, discussed — for the previously mentioned methodological reasons — in separate chapters. However, with such a treatment of the subject the reader can get a thorough inside view of the major synthetic methods and can also obtain a basis for comparing the applied and potential strategies. This kind of benefit has not yet been offered by the reviews covering these aminosugars. In this respect the volume can help carbohydrate chemists in designing the synthesis of similar molecules because many of the methods discussed might also be adapted to other fields of carbohydrate research.

To ensure the general clarity of the complete synthetic routes to 3-amino-2,3,6-trideoxyhexoses, cross references are given throughout the volume. Closely related methods are discussed in tabular form whenever possible. Information on more than four-hundred of the functionalized derivatives of these aminohexoses that have so far been prepared are included and collected in the Appendix.

The literature is covered through December, 1987. Because of the huge number of contributions to this research cited, we have concentrated our efforts on presenting the details of methods that offer significantly exploitable results that might help in the work of those engaged in related areas of carbohydrate chemistry.

Debrecen, April, 1988

I. F. Pelyvás
C. Monneret
P. Herczegh

Acknowledgements. The authors are indebted to Professors Rezső Bognár and Ferenc Sztaricskai for encouragement and to Mrs. Erzsébet L. Magyar for the extensive art work. Financial support for certain parts of the authors' related research work and for the preparation of the manuscript was obtained from grants TPB KKFA and OTKA-1728 given by the Hungarian Academy of Sciences.

Contents

Abbreviations

Ac	acetyl
AIBN	azobisisobutyronitrile
All	allyl
anhydr.	anhydrous
aq.	aqueous
Bn	benzyl
BOC	*tert*-butyloxycarbonyl
Bu	*n*-butyl
Bu^t	*tert*-butyl
Bz	benzoyl
ca.	circa
CAN	cerium ammonium nitrate
cat.	catalyst, catalytic
Cbz	benzyloxycarbonyl
cc.	concentrated
c-Hex	cyclohexyl
d	day(s)
DAST	diethylaminosulfur trifluoride
DBU	1,5-diazabicyclo-[5,4,0]-undec-5-ene
DCAc	dichloroacetyl
DCC	N,N-dicyclohexylcarbodiimide
DEAD	diethylazodicarboxylate
DHP	dihydropyran
DIBAL	diisobutylaluminum hydride
diphos-4	1,4-bis-(diphenylphosphino)-butane
DMF	N,N-dimethylformamide
DMP	2,2-dimethoxypropane
DMSO	dimethylsulfoxide
DPPA	diphenyl phosphorazidate
EE	ethoxyethyl
EOC	ethoxycarbonyl
Et	ethyl
eq.	equivalent, equivalents
HMPT	hexamethylphosphoric triamide
HPLC	high performance liquid chromatography
Ip	isopropyl
LDA	lithium diisopropylamide

liq.	liquid
L-Selectride	lithium tri-(*sec*-butyl)-borohydride
LTBH	lithium triethylborohydride
Me	methyl
MEM	methoxyethoxymethyl
MOC	methoxycarbonyl
MOM	methoxymethyl
MCPBA	*meta*-chloroperbenzoic acid
Ms	methanesulfonyl
nbd	norbornadiene
NBS	N-bromosuccinimide
NIS	N-iodosuccinimide
NMO	N-methylmorpholine oxide
PCC	pyridinium chlorochromate
Ph	phenyl
Pht	phthalyl
*p*NBz	*para*-nitrobenzoyl
Prop	*n*-propyl
pyr.	pyridine
Ref.	reference
refl.	reflux
RT	room temperature
TBAF	tetrabutylammonium fluoride
TBDMS	*tert*-butyldimethylsilyl
TCAc	trichloroacetyl
TEA	triethylamine
Tf	trifluoromethanesulfonyl
TFAA	trifluoroacetic anhydride
TFAc	trifluoroacetyl
THF	tetrahydrofuran
THP	tetrahydropyranyl
TMNO	trimethylamine oxide
TMS	trimethylsilyl
Tr	trityl
Ts	*para*-toluenesulfonyl

1. Introduction

During the past decade, research work on the structural elucidation of sugar-containing antibiotics significantly contributed to the development of certain areas of carbohydrate chemistry. The structural and synthetic studies on various rare monosaccharide derivatives (deoxy and aminodeoxy sugars and their branched-chain analogs) isolated from metabolic substances and antibiotics produced by numerous microorganisms necessitated the elaboration of new and more sophisticated investigation techniques. New methods of isolation and structural determination were accompanied by the introduction of new and ingenious synthetic strategies and methodology.

Since the mid 1960s, research on deoxy and aminodeoxy sugars has been steadily urged on and stimulated by approaches focussed on the chemical modification and total synthesis of aminocyclitol antibiotics (1, 2) and the anticancer anthracycline glycoside antibiotics (3–8). Some characteristic representatives of the latter group are shown in Fig. 1.

A common structural feature of both class I and class II anthracycline antibiotics **(1–7)** is the 3-amino-2,3,6-trideoxy-L-hexose moiety attached to the substituted naphtacenequinone aglycone at position C-7 through an α-O-glycosidic linkage (8).

In addition to the anthracycline glycosides, 3-amino-2,3,6-trideoxyhexoses are important structural elements of other types of antibiotic substances as well. Thus, five of the eight stereoisomeric 3-amino-2,3,6-trideoxy-D- and L-hexoses (Table 1) have been found in nature as building blocks of antibiotics.

The first representative of such sugars, rhodosamine (**14**, 3-dimethylamino-2,3,6-trideoxy-L-*lyxo*-hexose) (33), was isolated from the anthracycline-type rhodomycins in 1963. One year later daunosamine **(12)**, obtained from daunomycin, was identified (19) as the N-di-demethyl analog of rhodosamine. During the past twenty years these two aminosugars have been found to be components of a large number of anthracycline antibiotics.

Isolated third in 1966, angolosamine (**11**, 3-dimethylamino-2,3,6-trideoy-D-*arabino*-hexose) (14) was obtained from the macrolide-type angolamycin and later from antibiotics with an aromatic polycyclic C-glycoside (anthra[1,2-b]-pyran) structure (13–17). These latter antibiotics contain N,N-dimethylvancosamine **(16)** as well. Both aminohexoses are attached to the aglycone moieties with C-glycosidic bonds. Either of the enantiomeric forms of angolosamine can also be incorporated in to the molecule of lactoquinomycin (18), and a recent paper reported that the antibiotics benzanthrin A and benzanthrin B (18a) contain L-rhodosamine and L-angolosamine, respectively, as O-glycosidic components and that both variants carry D-angolosamine as the C-glycosidic moiety.

Table 1. Naturally occuring 3-amino-2,3,6-trideoxyhexoses

Sugar			Name	Antibiotic	References
L	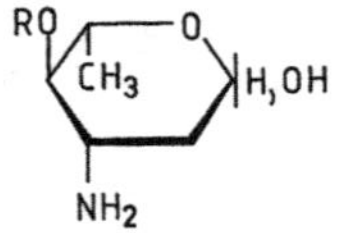	8 R=H 9 $R=CH_3$	L-Acosamine L-Actinosamine	Actinoidin	9–11
Arabino					
D	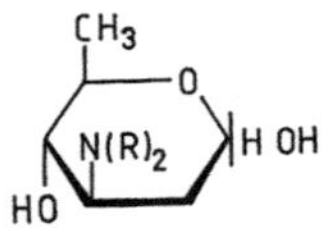	10 R=H 11 $R=CH_3$	D-Acosamine Angolosamine	Sporaviridin Angolamycin, Kidamycin Isokidamycin, Hedamycin Pluramycins	12 13–17
		12 $R=R_1=R_2=H$	L-Daunosamine	Daunomycin Adriamycin Carminomycin and additional anthracyclines	19–21 22 23 24–31
L	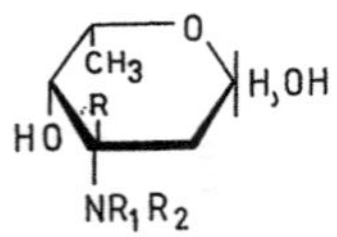	13 $R=R_2=H$; $R_1=CH_3$ 14 $R=H$; $R_1=R_2=CH_3$	N-Methyldaunosamine Rhodosamine	Antibiotic MA 144 L1 Rhodomycins Aclacinomycins and additional anthracyclines	32 33–35 36, 37 38–46
Lyxo		15 $R=CH_3$; $R_1=R_2=H$	Vancosamine	Vancomycin Sporaviridin	47 12
		16 $R=R_1=R_2=CH_3$	N,N-Dimethylvancosamine	Kidamycins Hedamycin Pluramycins	15, 16 16 17
D	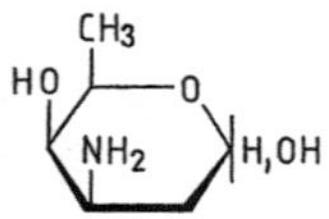	17 D-Daunosamine	Has not been found in Nature		

	Structure	Compound	Name	Antibiotic	Ref.
L	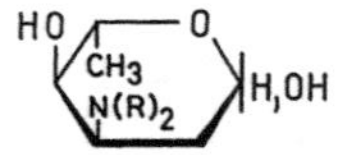	18 R=H	L-Ristosamine	Ristomycin (Ristocetin) Avoparcins	48–50 51 52–53
		19 $R=CH_3$	Megosamine	Megalomycins	54, 55
Ribo					
D	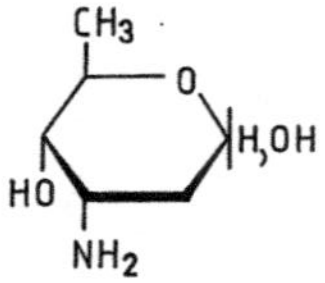	20 D-Ristosamine	Has not been found in Nature		
L	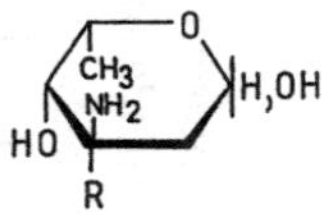	21 $R=CH_3$	3-Epi-vancosamine	Antibiotic A 35512 B	56
Xylo					
D	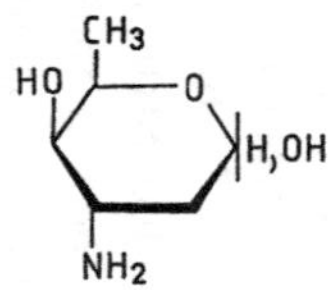	22 Has not been found in Nature			

Class I anthracycline antibiotics

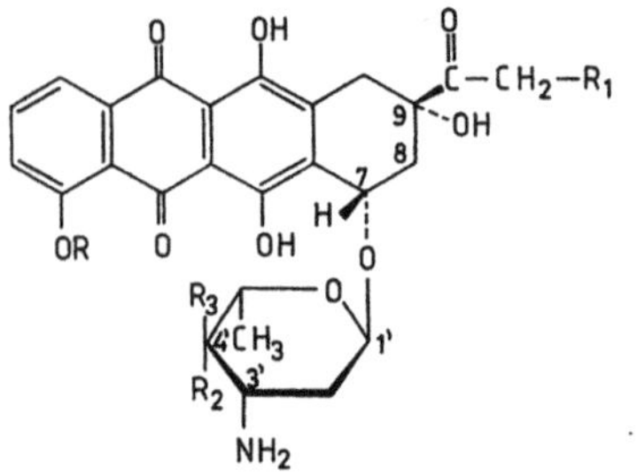

1 R = CH_3; $R_1 = R_3 = H$; $R_2 = OH$ — Daunomycin
2 R = CH_3; $R_1 = R_2 = OH$; $R_3 = H$ — Adriamycin
3 $R = R_1 = R_3 = H$; $R_2 = OH$ — Carminomycin
4 R = CH_3; $R_1 = R_2 = H$ $R_3 = OH$ — 4'-Epidaunomycin (synthetic)
5 R = CH_3; $R_1 = R_3 = OH$; $R_2 = H$ — 4'-Epiadriamycin (synthetic)

Class II anthracycline antibiotics

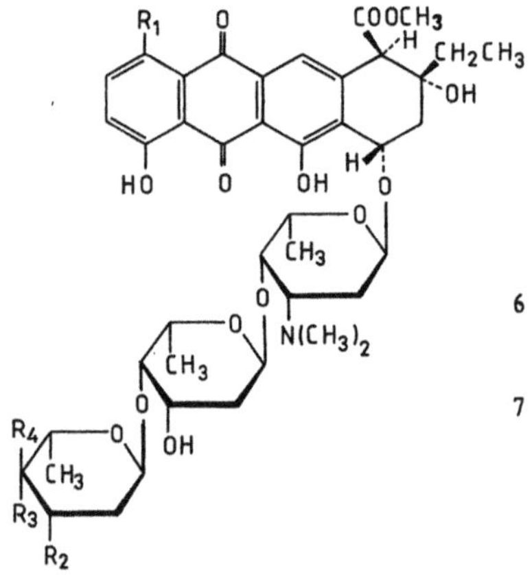

6 $R_1 = R_2 = H$; R_3, $R_4 = O$
Aclacinomycin A

7 $R_1 = R_2 = R_3 = OH$; $R_4 = H$
Marcellomycin

Fig. 1. Characteristic representatives of class I and class II anthracycline glycoside antibiotics

The structure of the aminodeoxy sugar isolated from the macrolide-type megalomycins was identified (57) in 1969 as 3-dimethylamino-2,3,6-trideoxy-D-*arabino*-hexose. However, NMR and X-ray studies performed (54, 55) ten years later proved that the correct structure of this compound (megosamine) is 3-dimethylamino-2,3,6-trideoxy-L-*ribo*-hexopyranose **(19)**.

Vancosamine **(15)**, the 3-C-methyl analog of daunosamine, is present (47) in the molecule of vancomycin (Fig. 2), the first representative of the vancomycin group of antibiotics (58) to be isolated. Structural elucidation (58) of the other members of this family of antibiotics showed that the basic sugar component of each is a 3-amino-2,3,6-trideoxyhexose steroisomer. Thus, ristomycin — identical (49) with ristocetin (51) — contains L-ristosamine (**18**, 3-amino-2,3,6-trideoxy-L-*ribo*-hexose)

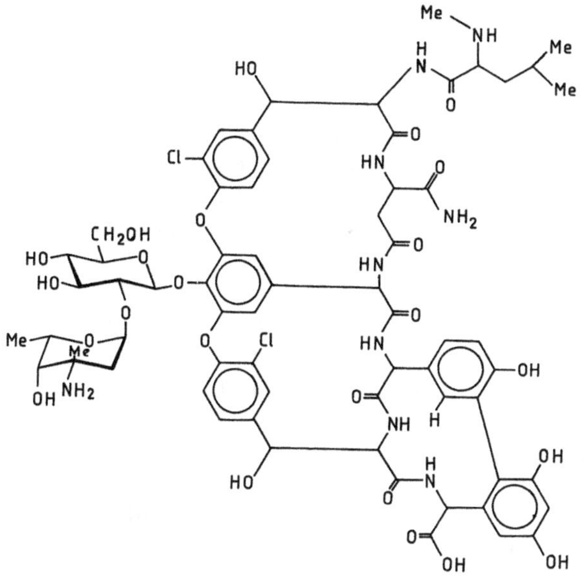

Vancomycin

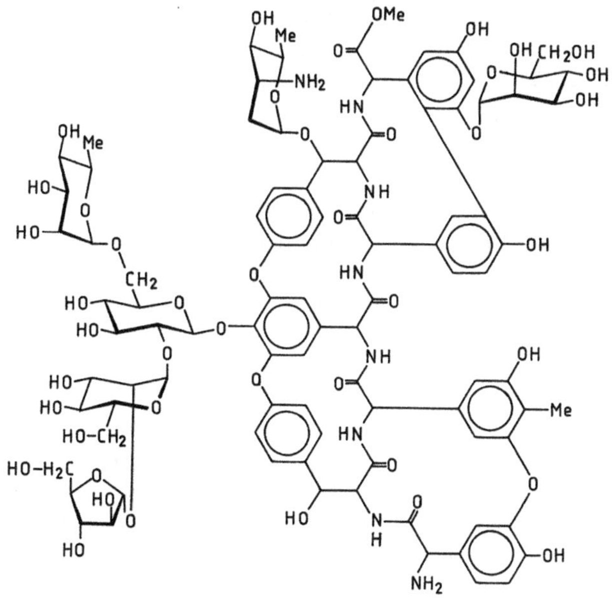

Ristomycin (ristocetin) A

Fig. 2. The structure of antibiotics vancomycin and ristomycin A

(48). This aminosugar was later found also in avoparcins (52, 53). Actinoidins (9–11) contain both L-acosamine (**8**, the corresponding L-*arabino* isomer) and its 4-O-methyl ether (actinosamine, **9**), and 3-epivancosamine **(21)** (56) is a building unit of the most recently isolated representative, antibiotic A 35512 B. So far neither the D-*lyxo* **(17)** and D-*ribo* **(20)**, nor the D-*xylo* **(22)** stereoisomer in the foregoing group of aminosugars has been shown to be a structural element of a natural substance.

It is quite obvious — at least from pharmacological and therapeutical points of view — that among the above-mentioned antibiotics though the anthracycline glycoside-type are the most important. During the past ten years some of these compounds, first daunomycin **(1)**, then adriamycin **(2)**, and more recently carminomycin **(3)**, have emerged as valuable chemotherapeutic agents for the treatment of a broad spectrum of human cancers (5–8). The dose-dependent cardiotoxicity of daunomycin was recognized early. Later it was replaced in clinical trials by adriamycin and carminomycin which are less toxic and more effective against a much broader spectrum of tumors. Aclacinomycin A **(6)** is currently undergoing clinical evaluation as an anticancer agent.

Organic chemists are also attempting to prepare structurally modified analogs of anthracycline antibiotics in the hope of decreasing their toxicity and enhancing their effectiveness against cancer. The original daunosamine moiety of daunomycin and adriamycin has been replaced by other deoxy and aminodeoxy sugars, and the aglycone portions have also been changed. Several hundred new synthetic or semisynthetic anthracycline glycosides have been prepared. Of these derivatives 4′-epidaunomycin **(4)** and 4′-epiadriamycin **(5)** — both containing L-acosamine instead of the daunosamine portion —, 4-demethoxyadriamycin, 4-demethoxydaunomycin, 3′-deamino-3′-(3-cyano-4-morpholinyl)-adriamycin (59), 7-O-(2′-deoxyfucosyl)-daunomycinone, and certain 13-dihydro- and 5-imino analogs reportedly exhibit improved therapeutic indices when compared with those of their parent antibiotics **1–3**.

The experimental efforts to obtain novel semisynthetic anthracyclines modified in the sugar moiety entailed extensive research on aminodeoxy sugars, particularly, the 3-amino-polydeoxyhexoses. This progress is unambigously illustrated by the following data. Prior to the preparation of the first semisynthetic anthracycline antibiotic, only forty-five 3-amino-2,3-dideoxy- and 3-amino-2,3,6-trideoxyhexopyranose derivatives were synthesized. However, between the years 1969 and 1979 an additional 140 novel analogs were produced, and from 1979 to March, 1987 200 more of this type of aminohexose derivative were reported.

During the past decade, excellent reviews covering the syntheses of the daunoamine-type 3-amino-2,3,6-trideoxyhexoses have been published by F. Arcamone (3, 4, 7), H. S. El Khadem et al. (8), J. Yoshimura (60), G. J. McGravey (61), and F. M. Hauser (62). However, the applied and potentially useful strategies of the systemic layout of the molecules of such aminosugars and their possible intermediates have not yet been comprehensively reviewed. In the hope of providing hitherto unexplored contexts and a more general outlook for colleagues working on related research, this volume is concerned with the methodical construction of 3-amino-2,3-di- and -2,3,6-trideoxyhexoses and their branched-chain and unsaturated analogs derived from both carbohydrate and non-sugar precursors.

2. Synthesis of 3-Amino-2,3,6-Trideoxyhexoses from Carbohydrates

Most of the carbohydrate-based synthesis strategies for the preparation of 3-amino-2,3-dideoxy- and 3-amino-2,3,6-trideoxyhexose stereoisomers have involved 2-deoxyhexose derivatives — readily available from natural sugars — as principal intermediates. The more expensive 2,6-dideoxy analogs are also frequently used starting materials, although recently elaborated synthetic methodologies allow the convenient generation of the 6-deoxy moiety of aminodeoxyhexoses with the required stereochemistry. Several examples of the elegant simultaneous introduction of the C-3 amino and C-2 deoxy functionalities into hexose derivatives via various intermediates or the subsequent deoxygenation of substituted 3-aminodeoxyhexoses at position C-2 have also been reported.

In order to explore the methodical construction of 3-amino-2,3,6-trideoxyhexoses from carbohydrates, the most important methods for the preparation of 2-deoxyhexose derivatives are first discussed. Strategies for the introduction of the C-3 amino- or C-3 methyl substituents or both into functionalized carbohydrates are then reviewed, and this is followed by a survey of the synthetic routes for transforming 3-amino-3-deoxy- or -2,3-dideoxyhexopyranosides into their corresponding 6-deoxy analogs and another derivatives.

2.1 Synthesis of 2-deoxyhexose derivatives

2.1.1 Preparation of 2-deoxyhexopyranosides from glycals and glycal esters

The addition of various reagents across the double bond of enol-ether-type glycals (1,2-dideoxyhex-1-enopyranoses) (63–69) leads to 2-deoxyglycopyranose derivatives

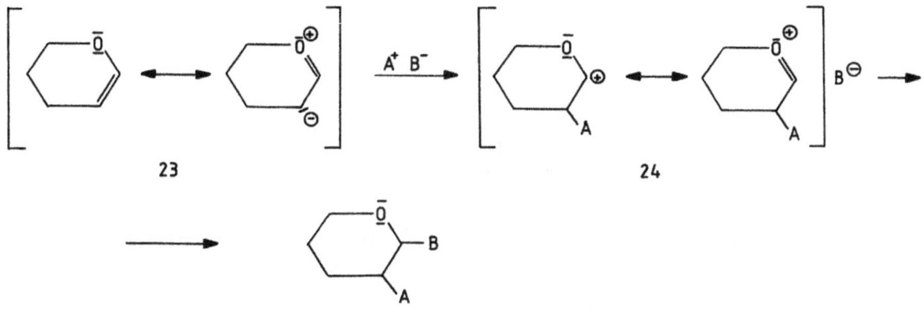

23 24 25

Scheme 1

either directly or following subsequent chemical transformations of the resulting addition products. With dissociable compounds (A^+B^-), the addition reaction takes place according to the mechanism (63, 64) depicted in Scheme 1.

In the presence of a mineral or Lewis acid catalyst, this "normal" addition process is generally accompanied or replaced by concurrent allylic rearrangement (63, 64) involving the elimination of the C-3 acyloxy substituent.

Addition of water and alcohols

The acid-catalyzed addition of water to glycals yields (70) a mixture of the α- and β-anomers of the respective 2-deoxyhexopyranoses and 2-deoxyhexofuranoses. However, glycals with different stereochemical structure yield highly variable proportions of such products because (particularly in the case of substrates with an axial C-3 hydroxyl group) an elimination reaction can proceed concurrently with the addition.

In the presence of a mineral or Lewis acid catalyst, the treatment of glycal esters carrying the C-3 and C-4 acyloxy groups *trans* to each other with nucleophiles, e.g., alcohols (63, 64, 73), yields hex-2-enopyranoses substituted at C-1. These reactions proceed (74–77) through an initial isomerization of the C-3 acyloxy group of the glycal ester molecule (allylic rearrangement; **23** → **26**) and the subsequent attack of the nucleophile on the intermediary allyloxocarbenium ion **27** (Scheme 2).

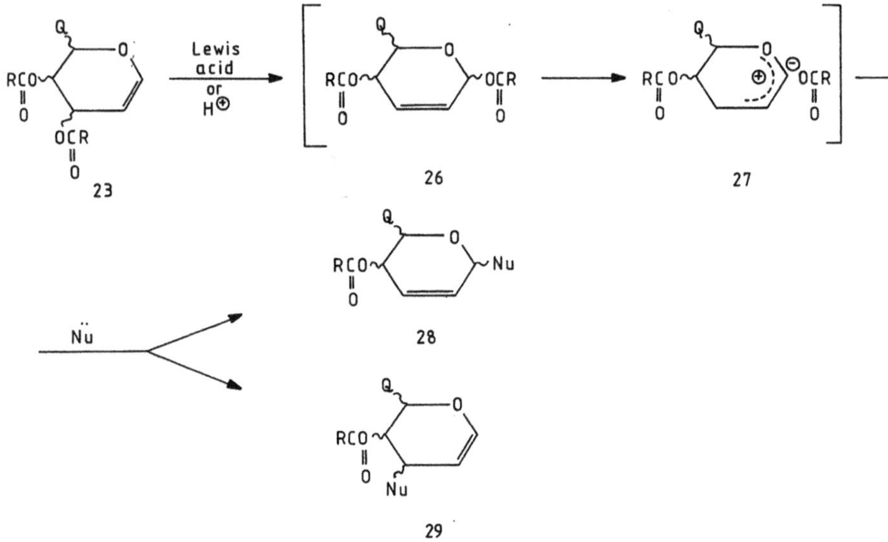

Scheme 2

This mechanism is most probable in strongly polar media where the addition reaction can result in 2,3-dideoxyhex-2-enopyranoside derivatives **(28)** or C-3-substituted glycals **(29)**. The alcoholate anion, considered (78) to be a "hard base", attacks ion **27** at C-1, which has the most pronounced "hard acid" character, producing the pseudoglycal-type hex-2-enopyranoside **28**. Carbon nucleophiles, being "soft bases", give products corresponding to the glycal **29**, whereas the attack of the azide ion, with a "borderline base" character (79), can take place concurrently at either C-1 or C-3 of ion **27** and lead to an equilibrium mixture of compounds **28** and **29**. The composition of this mixture is decisively determined by the experimental conditions used (77, 79, 80). A further explanation of the outcome of these reactions is given in Sect. 2.2.6.

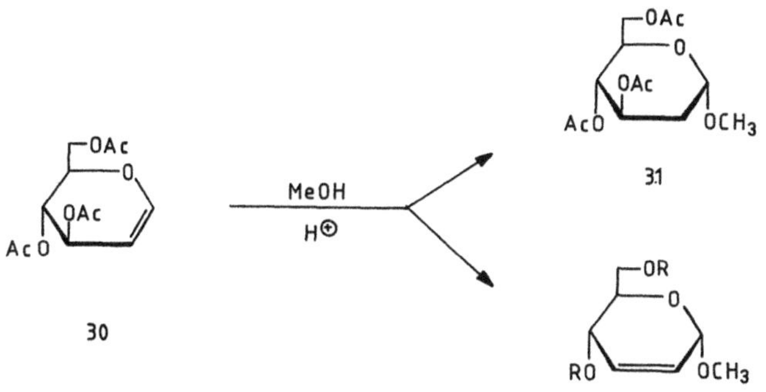

30

31

32 R = H

33 R = Ac

Scheme 3

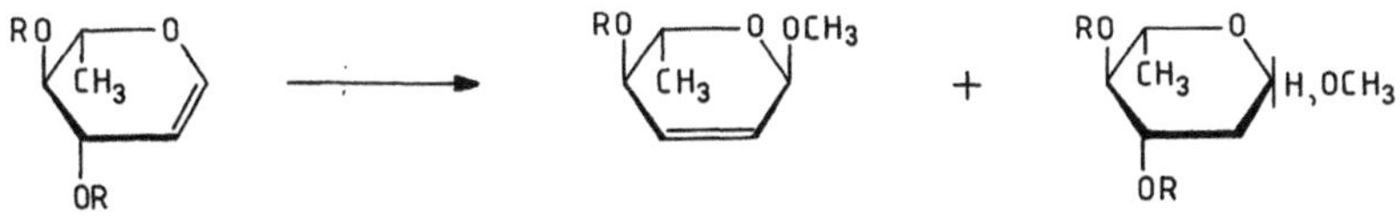

34 R = Ac

35 R = Bz

36 R = Ac

37 R = Bz

38 R = Ac or Bz

Scheme 4

On treatment with methanol in the presence of a mineral acid, 3,4,6-tri-O-acetyl-D-glucal **(30)** yields (81) 15–20% of the "normal" addition product (Scheme 3) methyl 3,4,6-tri-O-acetyl-2-deoxy-α-D-*arabino*-hexopyranoside **(31)** and 20–45% of the unsaturated **32** and its diacetate **(33)**. Further transformation of **32** and **33** leads (73) mostly to furane derivatives. The addition of methanol to **30**, catalyzed by boron trifluoride, also yields **31** along with the corresponding β-anomer (82, 83). From the same reaction of 3,4-di-O-acetyl-L-rhamnal **(34)**, 50% methyl 4-O-acetyl-2,3,6-trideoxy-α-L-*erythro*-hex-2-enopyranoside **(36)** along with its β-anomer and a small amount of the 2-deoxyglycoside **38** (R = Ac) were isolated (83) (Scheme 4). The reaction of the glycal **34** with methanol (84) in dichloromethane and in the presence of the Lewis acid catalyst tin (IV) chloride yielded as much as 90% of **36** and its β-glycoside in ca. 6:1 ratio. The use of a secondary alcohol (2-propanol) shifted the equilibrium more towards the α-anomer (α:β = 11:1), and only traces of the β-glycoside could be detected when *tert*-butanol was used. The addition of methanol to 3,4-di-O-benzoyl-L-rhamnal **(35)**, catalyzed with boron trifluoride yielded, (85) 20% methyl 3,4-di-O-benzoyl-2,6-dideoxy-α,β-L-*arabino*-hexopyranoside (**38**, R = Bz) and 73% of the unsaturated **37** (Scheme 4). The application of a cation exchange resin produced (86) mainly the "normal" addition products **38** (R = Bz).

For 3,4,6-tri-O-acetyl-D-galactal and 3,4-di-O-acetyl-L-fucal **(39)**, which contain the C-3 and C-4 acetoxy functions in the *cis* relationship that is less favored for allylic rearrangement, the 1,2-addition of alcohols predominated, using either mineral (63, 87) or Lewis acid (88) catalysts. Thus, methyl 2,6-dideoxy-α-L-*lyxo*-hexopyranoside **(40)** was readily obtained (88) from the fucal ester **39** in 51% yield (Scheme 5).

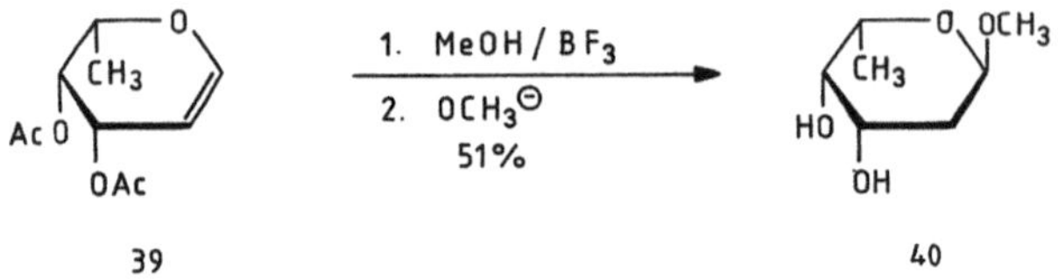

Scheme 5

With glycals carrying a "non-leaving" group (such as acylamino or alkoxycarbonylamino) at position C-3, allylic rearrangement does not take place and the corresponding 3-amino-2-deoxyglycosides are obtained in good yield. Such addition reactions are discussed in Sects. 2.2.6 and 2.6.1.

Addition of halogens. Alkoxy- and acyloxyhalogenation of glycal derivatives

Of the four isomeric 1,2-addition products expected to form upon addition of halogens at the double bond of glycal esters, the α-anomers (carrying the halogen atom at C-1 in the *axial* position) are generally predominant because they are more stable due to the anomeric effect. Glycosylation of such α-halogenoses using simple alcohols then provides the corresponding alkyl 2-deoxy-2-halogeno-β-glycosides as major products. These can be converted into the respective alkyl 2-deoxy-β-glycosides by reductive dehalogenation at C-2.

The addition of *chlorine* in nonpolar media generally proceeds through stabilized *syn* ion pair intermediates and leads, with high selectivity, to *cis* 1,2-addition products. The *bromination* reaction of glycal derivatives may – depending on the stereochemistry and electronic properties of the substituents – involve the participation of various ionic intermediates, such as *syn* ion pairs, cyclic bromonium ions, or open oxocations. The stereochemical outcome of both addition reactions is influenced by the polarity of the solvent, the structure of the glycal, and the nature of the halogen (89–92, 95).

Treatment of 3,4,6-tri-O-benzyl-D-glucal **(30a)** with chlorine in carbon tetrachloride furnished (92) exclusively the α-chloride **41** with D-gluco configuration (Table 2). The similar reaction of the corresponding diacetate **(30)** also proceeds with high stereoselectivity to yield **42**. Methyl-glycosylation of this product using the Koenigs-Knorr method gives methyl 2-deoxy-2-chloro-β-D-glucopyranoside **(43)** (90, 91).

Lemieux and Fraser-Reid reported (89, 90) that the bromination of the glucal ester **30** led to a 2:1 mixture of the α-D-gluco **(44)** and α-D-manno **(46)** dibromides in 90% yield. Later Boullanger and Descotes (92) observed the formation of the β-D-gluco isomer **45** as well. They found the ratio of the 1,2-dibromo compounds **44, 45,** and **46** to be 1:4:1. Methyl 2-deoxy-β-D-*arabino*-hexopyranoside **(47)** or its tri-O-

Table 2. Halogenation of D-glucal derivatives

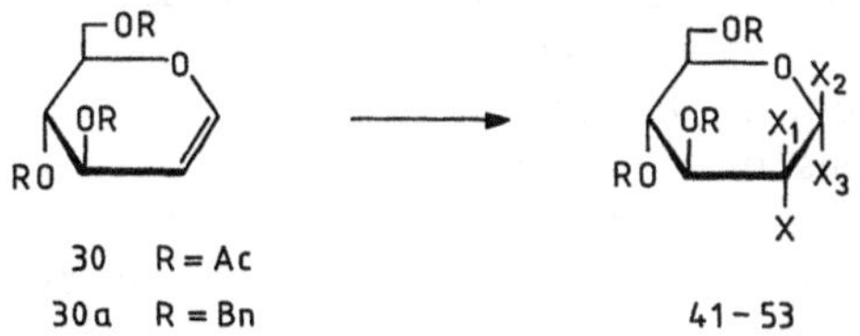

Compound	R	X	X_1	X_2	X_3
41	Bn	Cl	H	H	Cl (97%)
42	Ac	Cl	H	H	Cl (80%)
43	Ac	Cl	H	OMe	H (80%)
44	Ac	Br	H	H	Br
45	Ac	Br	H	Br	H
46	Ac	H	Br	H	Br
47	H	H	H	OMe	H
48	Ac	H	H	OMe	H
49	Ac	H	Br	H	OR_1
50	Ac	H	Br, Cl, I	H	OMe
51	Ac	Br, Cl, I	H	OMe	H
52	Ac	Cl	H	H	OMe
53	Ac	H	I	H	OBu^t

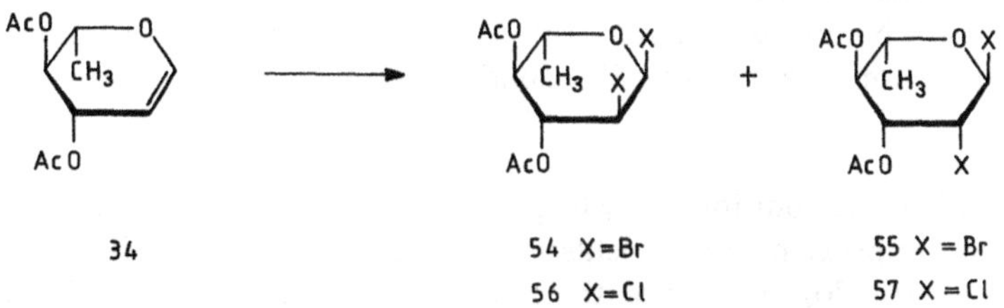

Scheme 6

acetate **(48)** were prepared in excellent yield by means of the Koenigs-Knorr methylglycosylation of the mixture of **44** and **46** followed by catalytic hydrogenation (89, 93) or reductive dehalogenation with tributyltin hydride (94).

Bromination of 3,4-di-O-acetyl-L-rhamnal **(34)** using bromine in carbon tetrachloride gave a 2:1 mixture of the 2-bromo-α-L-glucosyl- **(54)** and α-L-mannosyl- **(55)** bromides (Scheme 6). Chlorination yielded a higher proportion of the L-gluco isomer **(56)**. On the basis of ^{1}H-NMR spectral integrals, the ratio of L-gluco- **(56)** to L-manno- **(57)** dichloride was about 4:1; using nitromethane to increase the solvent polarity changed this distribution to 3:2. Bromination or chlorination of 3,4-di-O-acetyl-L-fucal **(39)** again yielded the 1,2-*cis* α-glycosyl halides, e.g., the α-galacto isomer, as the major products (95). Further details about the steric outcome and mechanism of the addition of halogens to various glycals have been most recently interpreted by Horton and coworkers (95).

Alkoxy- or *acyloxyhalogenation* of glycal esters under suitable conditions offer a highly stereoselective access to alkyl 2-deoxy-2-halogenopyranosides and their

1-O-acyl analogs, respectively, in a one-step procedure. Such compounds can be dehalogenated, by reduction or by photochemical transformation (96) under mild conditions, to yield the corresponding 2-deoxysugar derivatives. Contrary to halogen addition and subsequent glycosylation, the alkoxy- and acyloxyhalogenation of glycal esters result in high proportions of the α-anomers of 2-deoxyglycosides. Thus, methoxyhalogenation of 3,4,6-tri-O-acetyl-D-glucal **(30)** using bromine (89,90,92) or iodine (90) in a large excess of methanol and in the presence of silver acetate yields a 2:1 mixture of the *trans* addition products **50** and **51** (Table 2) in which the α-D-manno stereoisomer predomirates. When chlorine was used the ratio was found to be 5:1, and a small amount of a *cis* adduct **(52)** was also detected (90). In more recent studies (92) the proportion of methyl 2-chloro-2-deoxyglycosides **50**, **51**, and **52** formed by the similar methoxychlorination of glucal **30** was shown to be 4:7:2. The stereoselectivity of the alkoxy-iodination reaction is decisively determined by the steric bulk of the alkyl group of the alcohol used; under optimal conditions (97), *tert*-butyl 2-deoxy-2-iodo-3,4,6-tri-O-acetyl-α-D-mannopyranoside **(53)** was obtained in 85% yield.

The reaction of the glucal ester **30** with N-bromosuccinimide in alcohols (89–102) leads to *trans-diaxial* α-manno addition products (type **49**) in most cases. Similar results were obtained using protected 3-aminoglycal analogs (99). These reactions are discussed in Sects. 2.2.6 and 2.6.1.

By replacing the alcohol counterpart with a suitably protected sugar, carrying a single free hydroxyl function, numerous 2-deoxy-2-bromo- and 2-iodo-disaccharides with α-interglycosidic linkage have been synthesized (100, 103, 104) by means of the application of N-bromo- or N-iodosuccinimide. This procedure has now been extended (105, 106) for the preparation of antitumor 2′-iodoadriamycin derivatives as well.

Addition of hydrogen halides

Due to the anomeric effect, the addition of hydrogen halides across the double bond of glycal derivatives in nonpolar medium yields mainly the more stable 2-deoxy-α-glycosyl halides with *axial* halogen atom at the anomeric center. Similar products are also formed from 2-deoxyhexoses in the absence of bulky protecting groups at O-3 and O-4 upon treatment with hydrogen halides (107). When such reactive glycosyl halides, devoid of an O-acyl group at C-2, are treated with alcohols (108), the reaction proceeds via on "open glycosylium ion" intermediate. The ratio of the resulting anomeric 2-deoxyglycosides is strongly dependent upon the nature of the alcohol, the promoter, and the solvent. For example, benzoylated 2-deoxy-α-D-glycosyl bromides with *ribo* and *arabino* configuration can be successfully converted into β-linked 2-deoxyglycosides with high stereoselectivity by the application of silver zeolite as the promoter. The β-anomeric outcome of the glycosylation has been found to be higher in the 2-deoxy-*arabino*-hexopyranosyl series (109).

Studies made during the 1960s claimed (63, 110, 111) that in nonpolar solvents the addition of hydrochloric acid or hydrogen bromide to glycal esters bearing the C-3 and C-4 acyloxy groups *trans* to each other (xylal, glucal) results in both the "normal" addition products (i.e. 2-deoxyglycosyl halides) and in greater or lesser amounts of 3-halogeno-2,3-dideoxyglycosyl halides. However, in recent work (112–114) no formation of such C-3 halogeno compounds has been detected. This observation is similar to that with the corresponding *cis* isomers (arabinal, galactal) (115).

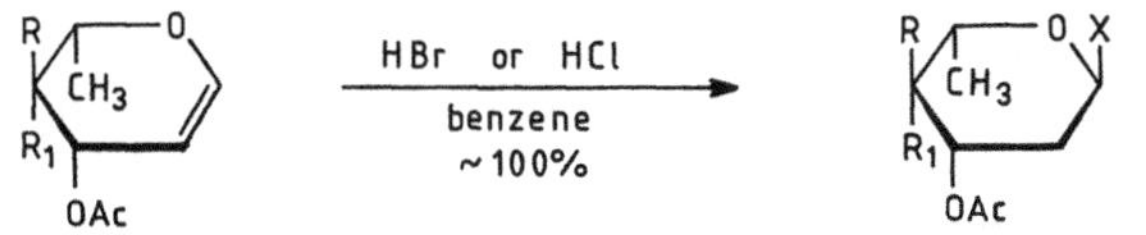

34 R = OAc; R_1 = H
39 R = H; R_1 = OAc

58 R = OAc; R_1 = H; X = Cl
59 R = OAc; R_1 = H; X = Br
60 R = H; R_1 = OAc; X = Cl
61 R = H; R_1 = OAc; X = Br

Scheme 7

For example, 3,4-di-O-acetyl-L-rhamnal **(34)** or 3,4-di-O-acetyl-L-fucal **(39)** yielded (112, 113) the corresponding 2-deoxy-α-glycosyl halides **(58–61)** nearly quantitatively upon the addition of hydrogen halides (Scheme 7). Koenigs-Knorr glycosylation of the chloride **58** using methanol resulted in a ca. 3:1 mixture of methyl 3,4-di-O-acetyl-2,6-dideoxy-β-L-*arabino*-hexopyranoside and the corresponding α-anomer (113).

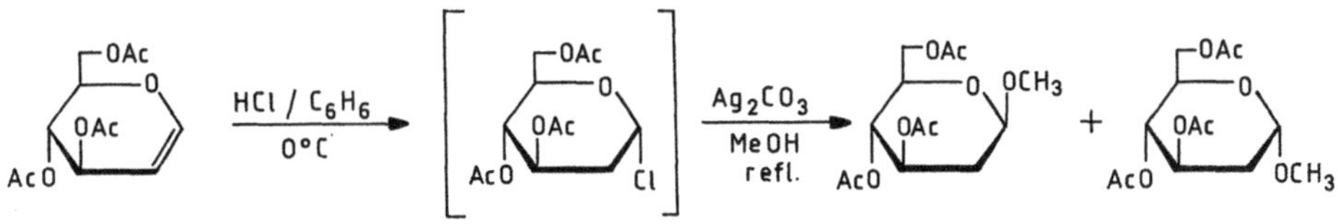

30 62 48 31

Scheme 8

Treatment of 3,4,6-tri-O-acetyl-D-glucal **(30)** with hydrochloric acid in dry benzene gave almost exclusively the corresponding α-chloride **(62)**. The glycosylation of **62** using methanol under Koenigs-Knorr conditions produced a 3:1 mixture of the β- **(48)** and α- **(31)** 2-deoxy methylglycosides (Scheme 8). From this mixture the β-anomer **48** could readily be isolated by crystallization in ca. 45% overall yield (based on **30**) (113, 114). In previous studies involving hydrogen bromide addition on **30** and subsequent methyl-glycosylation, the β-methyl glycoside **48** was prepared with only a 20% yield (111). When the aforementioned reaction was carried out in polar solvents (e.g., glacial acetic acid), the exclusive formation of "anomal" addition products was observed. The reactions of both 3,4,6-tri-O-acetyl-D-glucal **(30)** (115) and 3,4,6-tri-O-acetyl-D-allal (116) with hydrogen bromide followed by hydrolysis in the presence of silver carbonate gave the same product: 3-bromo-4,6-di-O-acetyl-2,3-dideoxy-α-D-*arabino*-hexopyranose.

Methoxy-, hydroxy- and acetoxymercuration and thallation of glycals

Methoxymercuration of glycals and glycal esters with mercury(II)acetate in a large excess of methanol yields a mixture of the 1,2-*diaxial* and 1,2-*diequatorial* methyl 2-acetoxymercurio-2-deoxyglycopyranosides (111, 117–120). After separation, these compounds can be readily demercurated into the corresponding methyl 2-deoxy-α- and -β-hexopyranoside derivatives. Takiura and Honda (118) have suggested that the mercuration reaction proceeds via cyclic "mercurinium ion" intermediates and that the proportion of the 1,2-*diaxial* and *diequatorial* addition products is strongly dependent on the steric disposition of the C-4 and C-5 substituents. Whereas in the presence of a *quasi-axial* C-4 acetoxy group, e.g., 3,4,6-tri-O-acetyl-D-galactal **(63)**, the 1,2-*diaxial* compound **(64)** is produced exclusively, from the glucal isomer **30** a 55:44 mixture of the methyl α-D-mannoside **(65)** and -β-D-glucoside **(66)** C-2 mercurio-derivatives is obtained (Scheme 9).

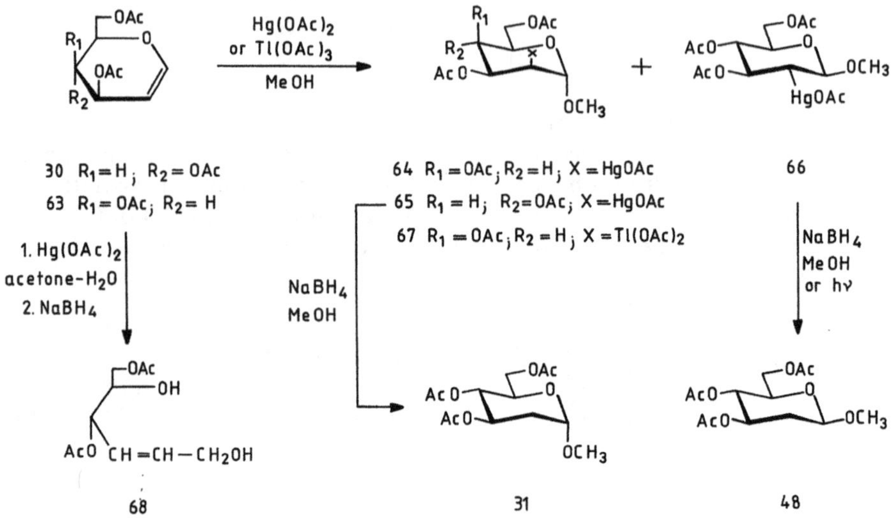

Scheme 9

Methoxythallation of 3,4,6-tri-O-acetyl-D-galactal **(63)** with thallium(III)acetate (121), produced exclusively the 1,2-*diaxial* addition product **67** with α-D-talo configuration. Methoxymercuration of glycal esters devoid of the *quasi-equatorial* C-5 substituent (e.g., 3,4-di-O-acetyl-D-xylal) forms preponderantly the corresponding 1,2-*diequatorial* acetoxymercurio compound (118). In the absence of ester protecting

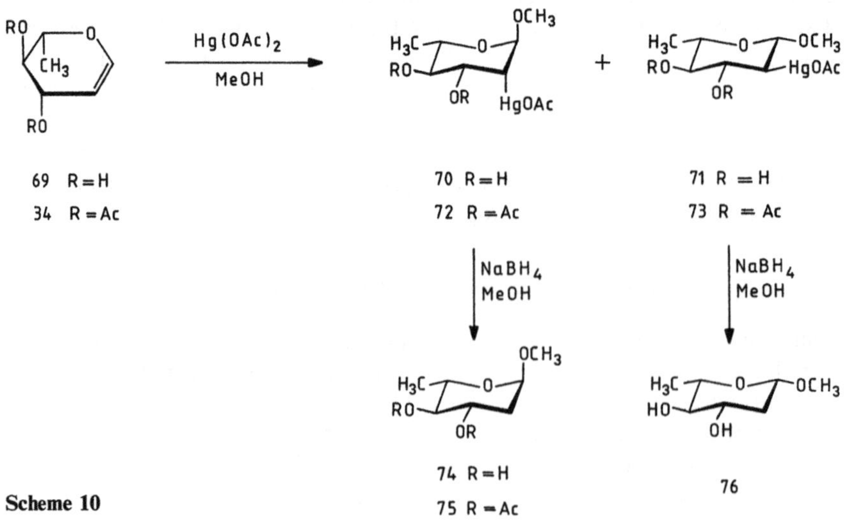

Scheme 10

groups on the glycal molecule, the proportion of the 1,2-*diaxial* addition product is enhanced: about 65–70% methyl 2-(acetoxymercurio)-2-deoxy-α-D-mannopyranoside is obtained from D-glucal (111, 113, 117). Subsequent reductive demercuration allows a preparatively useful synthesis of the deacylated analog (methyl 2-deoxy-α-D-*arabino*-hexopyranoside) of **31**.

Detailed ^{1}H-NMR studies of the product distribution of the methoxymercuration reaction (55, 113, 122, 123) of L-rhamnal **(69)** and its 3,4-di-O-acetate **(34)** have shown similar results (113) (Scheme 10). In both cases, the major product was the corresponding 1,2-*diaxial* 2-(acetoxymercurio)-methyl glycoside (**70** and **72** respectively). The ratios of **70:71** (derived from **69**) and of **72:73** (derived from **34**) were found to be 4:1 and 2:1, respectively.

Reductive demercuration of mercurials **65** and **66** (Scheme 9) and **70, 72,** and **73** (Scheme 10) using sodium borohydride provided easy access to the corresponding methyl 2-deoxyglycosides **31, 48** (118), **74** (113, 122, 123), **75** (55), and **76** (113), respectively. These methyl 2-deoxyglycosides are used as starting materials in numerous aminodeoxyhexose syntheses (vide infra). The photochemical conversion of **66** into **48** has also been reported (124). Methyl 2,6-dideoxy-α-L-*arabino*-hexopyranoside **(74)** and its β-anomer **(76)** were also prepared (125) by the acid-catalyzed glycosylation of the free sugar.

Hydroxymercuration (126) of 3,4,6-tri-O-acetyl-D-glucal **(30)** in aqueous acetone and subsequent reduction using sodium borohydride resulted in the unsaturated alditol derivative **68** (Scheme 9), whereas acetoxymercuration (126) in an indifferent solvent gave 1,3,4,6-tetra-O-acetyl-D-glucose in 50% yield.

Addition of O,O-dialkylphosphorodithioic acid on glycal esters

Polish chemists (127) have recently shown that the addition of O,O-dimethylphosphorodithioic acid **(77)** onto 3,4,6-tri-O-acetyl-D-glucal **(30)** proceeds regiospecifically and stereoselectively producing the 2-deoxy-1-O-α-(dimethyl-dithiophosphoroyl) derivative **78** exclusively. Other substituents on the phosphorous atom in **77** gave different α:β ratios of the edducts **78** (128) (Scheme 11). The nucleophilic displacement of the dimethyl-dithiophosphoroyl group in **78** by alcohols in the presence of bases takes place with a complete inversion of the configuration at the anomeric center. This offers the highest-yielding route to alkyl 2-deoxy-β-D-glucopyranosides **(79)** from D-glucal in small scale (mmole) experiments (127). However, according to the authors' own experience, this method furnished a ca. 3:1 mixture of methyl 2-deoxy-β- and α-D-*arabino*-hexopyranoside when the reaction was performed with 100 mmole quantities of the reactants.

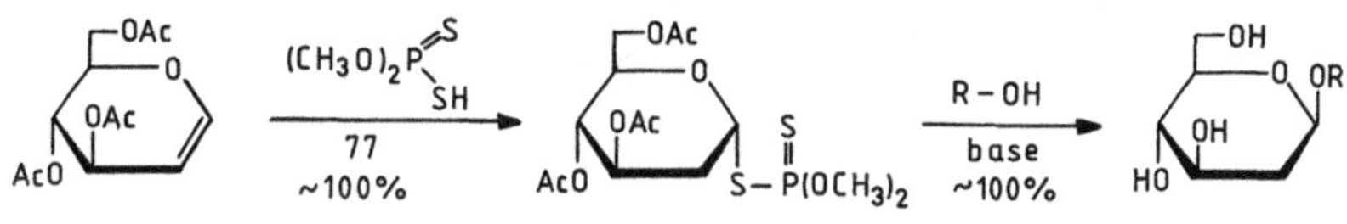

30 78 79

Scheme 11

Methyl 2-deoxy-α- **(31)** and -β-D-*arabino*- **(47)** hexopyranosides have been used for the production of various antibiotic 3-aminodeoxyhexoses via either the corresponding 4,6-O-benzylidene-3-O-sulfonate derivatives **148–150** (Sect. 2.2.1, Scheme 28), the C-3 *erythro*-uloside **100** (Sect. 2.5, Table 5), or the 4,6-O-benzylidene-3-oximinohexopyranosides **189–172** (Sect. 2.2.3, Table 3) derived from the C-3 α-hexopyranosid-ulose **100** and its β-anomer. Subsequent C-6 deoxygenation of the resulting aminodeoxyhexoses is discussed in Sect. 2.7.

Methyl 2,6-dideoxy-α-L-*arabino*-hexopyranoside **(74)**, obtained from L-rhamnal (**69**, Scheme 10), has found wide utility in syntheses of the derivatives of L-ristosamine (122, 225, 226), L-megosamine (55) (*via* the 3-O-tosylate **126**, Sect. 2.2.1, Scheme 26 and Fig. 4), as well as those of L-daunosamine (132, 155), L-acosamine (264, 266), and L-actinosamine (through the epoxide **174**, Sect. 2.2.2, Fig. 5, see also Sect. 2.8.1.). The β-anomer **(76)** of **74** was used to prepare methyl β-L-megosaminide (55) (**143**, Sect. 2.2.1, Fig. 4), and various derivatives (264) of the β-methyl glycoside of L-acosamine and its L-*xylo* stereoisomer (Sect. 2.2.2, Fig. 5).

2.1.2 Synthesis of 2-deoxyhexose analogs from oxirane and sulfonate ester derivatives

It is well known that the epoxide-ring cleavage of the 2,3-anhydrohexopyranosides with nucleophiles leads mainly to products having the C-2 and C-3 substituents in *trans-diaxial* orientation (129–131) (Fürst-Plattner rule) (132). Depending on the stereochemical properties the ring splitting of the oxirane function of the four stereoisomeric methyl 2,3-anhydro-4,6-O-benzylidene-α-D-hexopyranosides using lithium aluminum hydride provides useful yields of only the 3-deoxy analog having an *axial* hydroxyl group at C-2 or the 2-deoxy derivative with a C-3 *axial* hydroxyl substituent. Thus, methyl 4,6-O-benzylidene-2-deoxy-α-D-*ribo*- **(81)** (133) and α-D-*xylo*-hexopyranoside **(83)** (134) were prepared in excellent yield from the epoxides with D-allo- **(80)** and D-gulo **(82)** configurations, respectively (Scheme 12). Under similar conditions, the corresponding manno- and talo-epoxides gave 3-deoxy derivatives as major products. However, in a few cases different results have been reported (129).

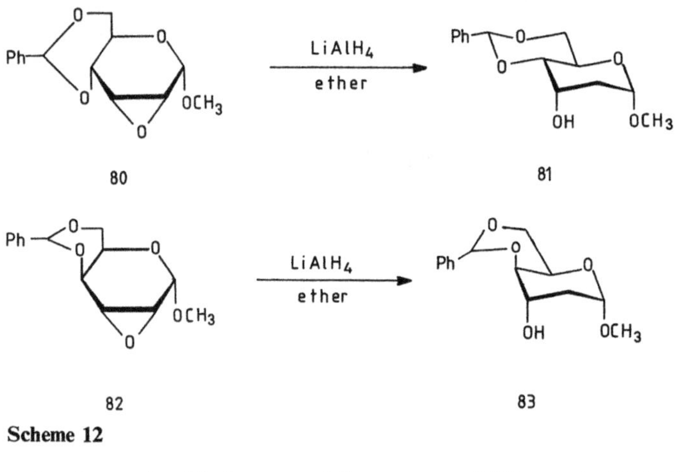

80 81 82 83

Scheme 12

As shown by Medgyes and Kuszmann (135), methyl 2,3:5,6-dianhydro-α-L-gulofuranoside **(84)** can be utilized (Scheme 13) for the preparation of 2,6-dideoxy sugar

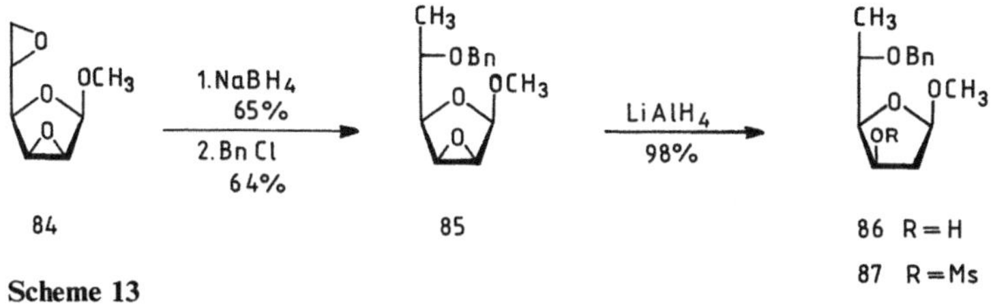

Scheme 13

derivatives, such as methyl 5-O-benzyl-2,6-dideoxy-3-O-methanesulfonyl-α-L-*xylo*-hexofuranoside **(87)**, which is a key intermediate in the synthesis of L-daunosamine (see Sect. 2.2.1, Scheme 30). Selective opening of the 5,6-anhydro ring of the diepoxide **84** using sodium borohydride and subsequent benzylation gave the 6-deoxy-2,3-monoepoxide **85**, which carries all of the substituents on one side of the tetrahydrofuran ring. As expected, this steric arrangement favored the almost exclusive formation of the 2-deoxyfuranoside derivative **86** upon reduction using lithium aluminum hydride. The 3-O-mesylate **87** was then prepared by conventional methods. An additional route to **86** and **87** is depicted in Scheme 23 (Sect. 2.1.5). It is to be noted that the oxirane ring of carbohydrates can be split using metal hydrides in the presence of other protecting groups also (129, 130, 136).

The ring-opening reaction of methyl 2,3-anhydro-hexo- and pentopyranosides with magnesium dibromide etherate or magnesium diiodide etherate produces the corresponding 2-halo-2-deoxy analogs as the major products through *diaxial* opening of the oxirane ring. Small quantities of the respective 3-halo-3-deoxy compounds are also produced (137, 138). Thus, methyl 4,6-O-benzylidene-2,3-anhydro-α-D-allopyranoside **(80)** was converted into methyl 4,6-O-benzylidene-2-iodo-2-deoxy-α-D-altropyranoside **(88)** in 97% yield by using magnesium diiodide etherate. Reduction of **88** using tributyltin hydride and sodium borohydride then gave the respective 2-deoxyglycoside **81** (Scheme 14).

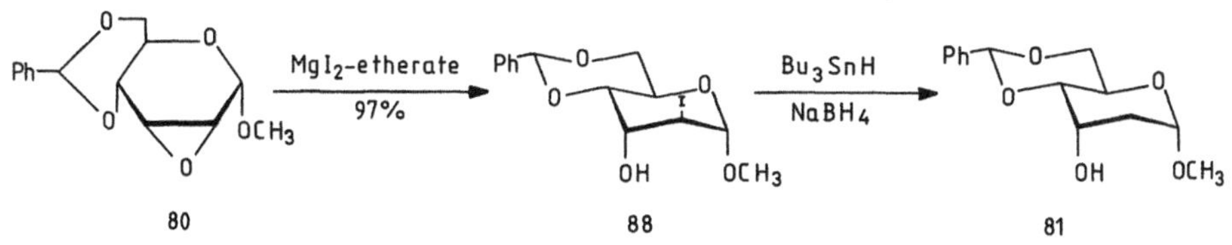

Scheme 14

An additional procedure (139) involving the ring cleavage of non-sugar epoxides using boron tribromide and subsequent oxidation of the resulting bromohydrin, produces an α-bromoketone that can also be utilized in the syntheses of functionalized 2-deoxyhexose derivatives.

It is well known that the primary sulfonyloxy groups of carbohydrates are readily reduced into terminal deoxy functions using lithium aluminum hydride, but secondary sulfonate esters usually suffer O-S bond cleavage (ester hydrolysis) under such conditions (140). In some cases, however, secondary sulfonic acid esters containing unsubstituted or acylated hydroxyl groups in the vicinity of the sulfonate function can be transformed by lithium aluminum hydride reduction into the corresponding deoxy derivatives (141–146). In these reactions the C-O bond scission may proceed by either intramolecular hydride displacement of the sulfonyloxy group involving an intermediary alkoxyaluminum hydride complex (R—O—AlH_3^-) (141, 142) or via an oxirane intermediate.

The regiospecifity of the reduction of the sulfonyloxy groups of the stereoisomeric methyl 4,6-O-benzylidene-2,3-di-O-tosyl-D-hexopyranosides using lithium aluminum hydride is decisively dependent on the steric position of the C-4 substituent and on the configuration of the anomeric centre; the major product is generally the corresponding 3-deoxy derivative. Only in the case of the α-idoside could a preparatively useful amount of the 2-deoxy analog **83** be isolated (ca. 40% yield). The reduction of phenyl 4,6-O-benzylidene-3-O-benzyl-2-O-*p*-toluenesulfonyl-β-D-glucopyranoside **(89)** using lithium aluminum hydride (Scheme 15) yielded (147) 55% phenyl 4,6-O-benzylidene-2-deoxy-β-D-*arabino*-hexopyranoside **(90)**. Investigations of the possible role of epoxide intermediates have revealed that no such compounds are involved in this reaction.

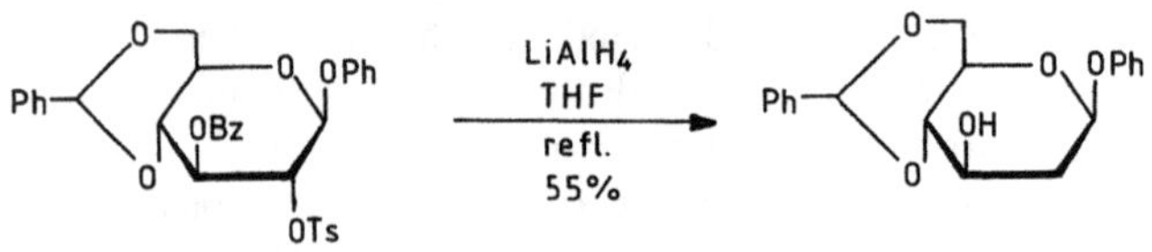

89 90

Scheme 15

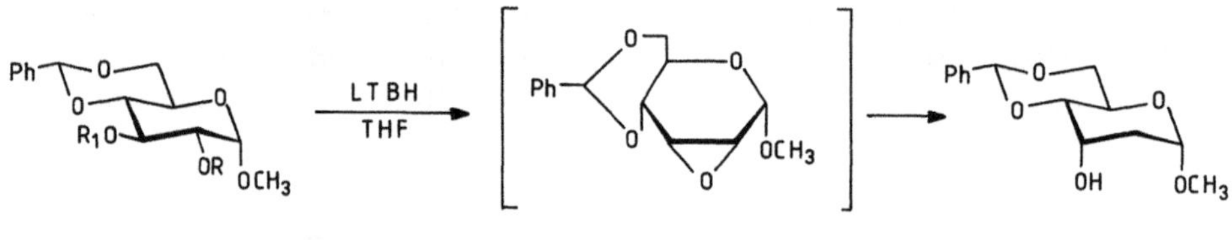

91 R = R1 = Ts
92 R = H; R1 = Ts
80
81

Scheme 16

In contrast, the reduction of the mono- and di-O-*p*-toulenesulfonates of methyl 4,6-O-benzylidene-α-D-glucopyranoside using lithium triethyl borohydride (LTBH) has been shown to proceed via 2,3-anhydro intermediates (148). The 2,3-di-O-tosylate **91** and the 3-O-monotosylate **92** both produced methyl 4,6-O-benzylidene-2-deoxy-α-D-*ribo*-hexopyranoside **(81)** in highly regiospecific and stereoselective reactions involving the intermediacy of the allo-epoxide **80** (Scheme 16). The corresponding 2-O-monotosylate gave the 3-deoxy-*arabino* isomer **81** via the respective α-D-manno epoxide, and for the corresponding β-anomers, the 3-O- and 2-O-monotosylates furnished (148) mixtures of regioisomeric 3- and 2-deoxyglycosides with β-D-*ribo*- and *arabino* configurations, respectively. The deoxygenation (148) of the above type of secondary sulfonate esters using lithium triethyl borohydride generally takes place with a much higher rate and yield than when lithium aluminum hydride is used, and involves less desulfonylation. In several cases LTBH reductions give rise to major products different from those available by using the aluminum hydride. As an example, reduction of the di-O-tosylate **91** with lithium aluminum hydride produced (149) 66% methyl 4,6-O-benzylidene-3-deoxy-α-D-*ribo*-hexopyranoside. The stereochemical dependence of the mechanism of such LTBH deoxygenation reactions has recently been discussed in detail by Baer et al. (150).

Methyl 4,6-O-benzylidene-2-deoxy-α-D-*ribo*- **(81)**, -*arabino*- **(111**, R=H) and α-D-*xylo*- **(83)** hexopyranosides were used in syntheses of various 3-aminodeoxyhexoses via either the 3-O-sulfonate esters **144**, **148**, **149**, or **153** (Sect. 2.2.1, Schemes 27–29) or the corresponding 3-uloses **100** and **385** Sect. 2.5, Table 5) and the α-D-*erythro*- **(189)** and *threo*- **(200)** 3-oximinohexopyranosides (Sect. 2.2.3, Table 3 and Scheme 36) followed by additional transformations that are described in Sects. 2.5, 2.7, and 2.8.

2.1.3 Synthesis of 2-deoxyhexose derivatives by means of the chain-elongation of pentoses

The well-known Fischer-Showden's one-carbon ascent (151–154) of L-arabinose has been applied (155) to the synthesis of methyl 2-deoxy-α-L-*arabino*-hexopyranoside, which was than transformed into methyl 2,6-dideoxy-α-L-*arabino*-hexopyranoside **(74)** by using standard procedures.

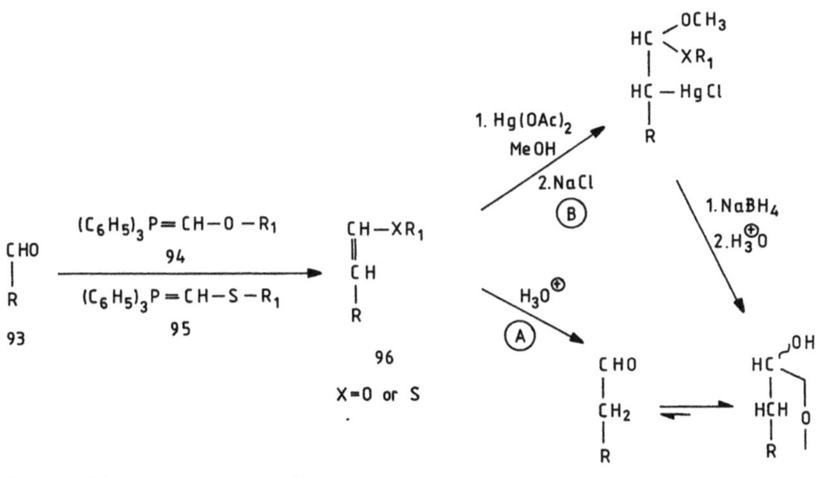

Scheme 17

The Wittig reaction (156) of protected *aldehydo* pentose derivatives is also useful in carbohydrate chemistry for the preparation of 2-deoxyhexose analogs (157). Thus, the reaction of such five-carbon sugars **(93)** with alkoxy- and aryloxymethylene triphenylphosphoranes **(94)** or with their mercapto analogs **(95)** gives rise (157) to the corresponding hex-1-enitol derivatives **96**. Further transformations of these products by means of method A or B result in the respective 2-deoxysugars **97** (Scheme 17). Bestman and Angerer (158) used this procedure to prepare 2-deoxy-D-

and -L-*arabino*-hexoses in ca. 70% yield, starting directly with unprotected D- and L-arabinose. Further applications of the Wittig reaction for the synthesis of functionalized 2-deoxysugars are discussed in Chap. 3.

2.1.4 Cleavage of the 2-phenyl-1,3-dioxolane-type acetal function of hexose derivatives

The Hanessian-Hullar method (159–162), introduced for the ring cleavage of carbohydrate 4,6-O-benzylidene acetals with N-bromosuccinimide and leading to 6-bromo-6-deoxy sugars (Sect. 2.7.2), has also been applied (163–165) to 2-phenyl-1,3-dioxolane-type monoacetals and to 2,3:4,6-di-O-benzylidene derivatives. Extensive studies have revealed (163, 164) that these diacetals are principally converted into 3,6-dibromo derivatives rather than into the corresponding 2,6-dibromo sugars.

Thus, treatment of methyl 2,3:4,6-di-O-benzylidene-α-D-mannopyranoside **(98)** with N-bromosuccinimide in refluxing carbon tetrachloride and in the presence of barium carbonate resulted exclusively in 3,6-dibromo- and 4,6-dibromo derivatives. This experimental finding has been well demonstrated by Monneret et al. (163) and then confirmed by Thiem and his group (164). Thus, the earlier report by Haga and coworkers (165) claiming the formation of an epimeric mixture of 2,6-dibromo compounds must be corrected because upon ring cleavage and subsequent reductive dehalogenation using lithium aluminum hydride, methyl 2,6-dideoxy-α-D-*arabino*-hexopyranoside **(99)** could not be isolated (163) (Scheme 18).

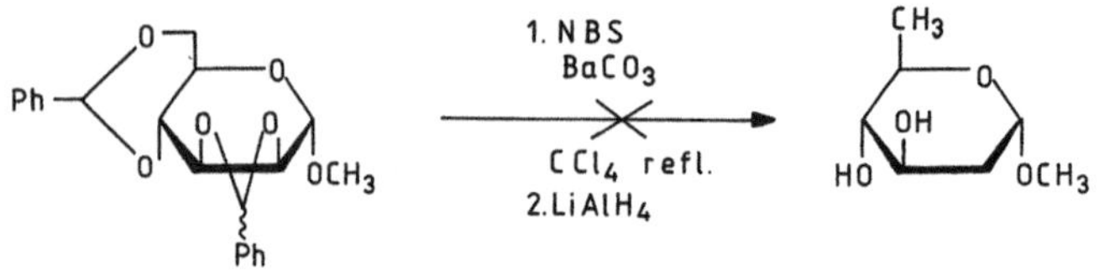

Scheme 18

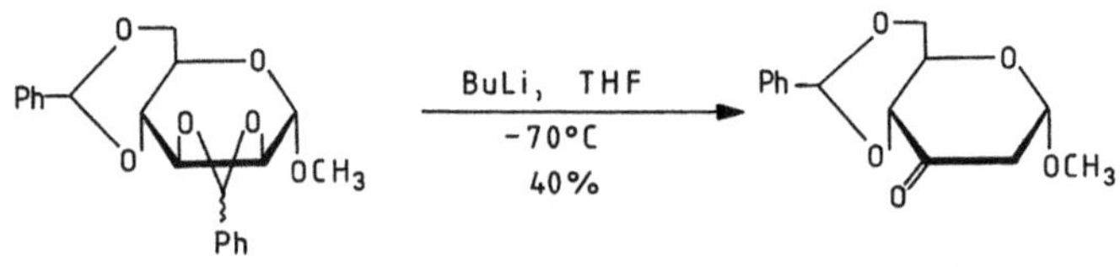

Scheme 19

When subjected to the same reaction conditions, methyl 2,3:4,6-di-O-benzylidene-α-D-allopyranoside produced (164) a mixture of methyl 3,4-di-O-benzoyl-2,6-dibromo-2,6-dideoxy-α-D-altropyranoside and methyl 2,4-di-O-benzoyl-3,6-dibromo-3,6-dideoxy-α-D-glucopyranoside in a ratio of 10:17 and in 60% overall yield. The corresponding di-O-benzylidene-glucopyranoside provided only the 6-bromo-6-deoxy compound (164), and similar ring cleavages of methyl 2,3:5,6-di-O-benzylidene-α-D-mannofuranoside, methyl 2,3-O-benzylidene-α-L-rhamnopyranoside, and 1,6-anhydro-2,3-O-benzylidene-D-mannose gave exclusively the appropriate C-3 bromo compounds (166).

Klemer and Rodemeyer reported (167) in 1974 that the 2-phenyl-1,3-dioxolane ring of methyl 2,3:4,6-di-O-benzylidene-α-D-mannopyranoside **(98)** can be chemoselectively split using butyllithium to produce methyl 4,6-O-benzylidene-2-deoxy-α-D-*erythro*-hexopyranosid-3-ulose **(100)** in 40% yield (Scheme 19). A similar fragmentation reaction (Scheme 20) of methyl 2,3-O-benzylidene-4-O-tetrahydropyran-2-yl-α-L-rhamnopyranoside **(101)** readily furnished (168) ca. 40% of the uloside **102**. The corresponding 4-O-methyl ether **(104)** was prepared by Horton et al. (169) from **103**. The 4-O-methoxymethyl- **(105)** and 4-O-(2-methoxyethoxymethyl) **(107)** ethers were converted (170, 171) into the corresponding uloses **106** and **108** (with 41 and 38% yields, respectively) by Brimacombe et al., but neither the O-4 unprotected 2,3-O-benzylidene acetal nor its 4-O-benzyl derivative underwent an analogous transformation (169). On the other hand, when the corresponding 4-O-*tert*-butyldimethysilyl ether derivative was subjected to the same reaction conditions, an inverse Klemer-Rodemeyer fragmentation occured (173) resulting in a C-2 ulose (methyl 4-O-*tert*-butyldimethylsilyl-3,6-dideoxy-α-L-*erythro*-hexopyranosid-2-ulose). This latter ketone was used for the preparation of the 4-O-*tert*-butyldimethylsilyl ethers of methyl 2-trifluoroacetamido-2,3,6-trideoxy-α-L-*ribo*- and *arabino*-hexopyranosides, the regioisomeric derivatives of L-ristosamine **(18)** and L-acosamine **(8)**, respectively.

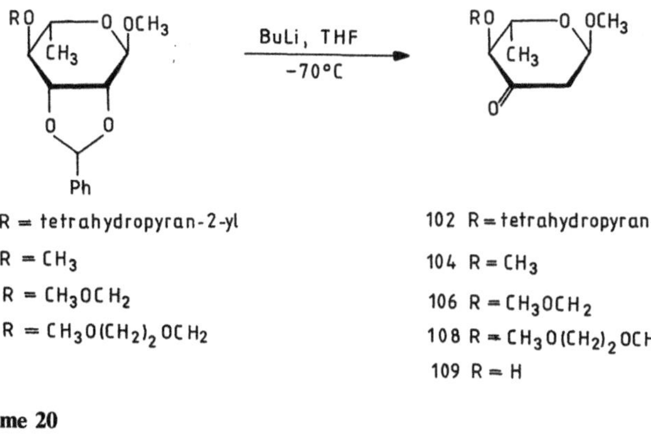

101 R = tetrahydropyran-2-yl
103 R = CH_3
105 R = CH_3OCH_2
107 R = $CH_3O(CH_2)_2OCH_2$

102 R = tetrahydropyran-2-yl
104 R = CH_3
106 R = CH_3OCH_2
108 R = $CH_3O(CH_2)_2OCH_2$
109 R = H

Scheme 20

The 4,6-O-benzylidene-ulose **100** was first used by the Horton group for the synthesis of L-daunosamine (172), D-acosamine (174), and D-ristosamine (175, 176) (see Sect. 2.2.3, Table 3 and Sects. 2.7.2 and 2.7.4, Tables 7 and 11). This keto sugar had also been utilized by Lukacs and coworkers in the first synthesis (177) of a vancosamine derivative (**411**, Sect. 2.5.2, Scheme 78 and Sect. 2.7.4, Table 13 entry 1) and for the preparation (55) of methyl α-D-rhodosaminide (**669**, Sect. 2.8.1, Table 14 entry 2). It was also used by Italian (178) and Japanese chemists (179) in two modified versions of the Horton's daunosamine synthesis. The Brimacombe and the Yoshimura groups also started with compound **100** in syntheses of D-evernitrose[1]

[1] For structures of the natural 3-nitro-3-C-methylhexoses L-evernitrose, L-decilonitrose, D-rubranitrose and D-kijanose see Sect. 2.2.4, Scheme 47.

(180, 181) (**414**, Sect. 2.5.4, Scheme 85), D-rubranitrose (182) (**241**, Sect. 2.5.3, Fig. 8), D-kijanose (182) (**437**, Sect. 2.5.3, Fig. 8 and Schemes 82, 83) and methyl N-acetyl-α-D-vancosaminide (183) (**682**, Sect. 8.1, Table 15 entry 5) derivatives. The L enantiomer of **100** has been used by Arcamone et al. (184) to prepare L-ristosamine analogs (Sects. 2.2.3 and 2.7.3, Schemes 37 and 112).

The 4-O-methoxymethyl- **(106)** and 4-O-(2-methoxyethoxymethyl)- **(108)** ethers served as key intermediates in the syntheses of L-ristosamine (170, 171) (Sect. 2.2.3, Scheme 41), methyl 3-acetamido-3-C-methyl-2,3,6-trideoxy-L-*arabino*- (183), -L-*ribo*- (185) and -L-*xylo*- (185) hexopyranosides (Sect. 2.5.2, Tables 5 and 6); the compound **(424)** is a derivative of the branched-chain aminodeoxy sugar component **(21)** of the vancomycin-type antibiotic A 35512 B. Derivatives **(417)** (186) of natural L-evernitrose **(240)**, L-acosamine **(138)** (187), and the free L-daunosamine **(12)** (187) have also been prepared from the C-3 uloses **104**, **106**, and **108**, respectively (cf. Sect. 2.5.2, Tables 5 and 6; 2.2.1, Fig. 3; and 2.8.1, Table 15 entry 2). Methyl 2,6-dideoxy-α-L-*erythro*-hexopyranosid-3-ulose **(109)** was also obtained in an independent way (Sect. 2.2.3, Scheme 38) and has been utilized in syntheses of substituted analogs of L-ristosamine **(209)** (188) and L-acosamine **(211)** (189–191) (Sect. 2.2.3, Scheme 39), L-daunosamine (190, 191) (**580**, Sect. 2.8.1, Table 15 entry 3). The C-4 epimer (methyl 2,6-dideoxy-α-L-*threo*-hexopyranosid-3-ulose) of **109** has been used to prepare methyl 3-benzamido-4-O-benzoyl-3-C-methyl-2,3,6-trideoxy-α-L-*xylo*-hexopyranoside (192) (**449**, Sect. 2.5.2, Table 6 entry 15). Methyl 4,6-O-benzylidene-2-deoxy-α-D-*erythro*-hexopyranosid-3-ulose **(100)** has also been obtained from methyl 4,6-O-benzylidene-2-O-*p*-toluenesulfonyl-α-D-*ribo*-hexopyranosid-3-ulose upon treatment with hydrazine (195) or trimethylphosphite followed by hydrolysis with sodium hydroxide (196) and by a photochemical transformation process (197). The corresponding benzyl glycoside has been synthesized (102) from benzyl 4,6-O-benzylidene-2-bromo-3-O-methanesulfonyl-α-D-mannopyranoside using hydrazinolysis.

As shown by Lipták and coworkers (198), the reductive ring cleavage of 2-phenyl-1,3-dioxolane-type carbohydrate benzylidene acetals using a lithium aluminum hydride-aluminum chloride reagent gives partially benzylated derivatives at O-2 or O-3. The site of the attack of the reagent is decisively determined by the steric position (*exo* or *endo*) of the phenyl substituent on the pyranose ring. Applications of this method of ring cleavage on 2-phenyl-dioxane-type 4,6-O-benzylidene acetals are discussed in Sect. 2.7.2.

An additional regioselective cleavage of the 2-phenyl-1,3-dioxolane acetal function has been reported (199). It involves the irradiation of the protected sugar and N-bromosuccinimide in the presence of water. In this case, regardless whether the phenyl group in the benzylidene acetal is *exo* or *endo*, only the thermodynamically less stable product is formed, i.e., the isomer with *axial* O-benzyl- and *equatorial* hydroxyl groups. These latter two methods have hitherto been widely used for the preparation of suitably protected starting materials for oligosaccharide syntheses, and they may find applications in the syntheses of deoxy- and aminodeoxy sugar derivatives as well.

2.1.5 Miscellaneous methods

Because 2-deoxyhexose derivatives have emerged as important starting materials for obtaining several types of organic compounds with potential biological significance, in recent years considerable effort has been devoted to the elaboration of additional novel strategies (200–212) for their syntheses. One of the most interesting of these — a procedure involving the reduction of dithiocarbonate ester derivatives — was reported by the Barton group (206). In it the reduction of compound **110** with tributyltin hydride (Scheme 21) furnished a 92% yield of the 2-deoxyglycoside **111** (R=Bz). On the other hand, when the cyclic 2,3-thiocarbonate **(112)** was sequentially treated with methyl iodide and then with tributyltin hydride, methyl 4,6-O-benzylidene-2-deoxy-α-D-*ribo*-hexopyranoside **(81)** was obtained in only 23% yield, and the major product (50%) was the 3-deoxy derivative with the D-*arabino* configuration (Scheme 22).

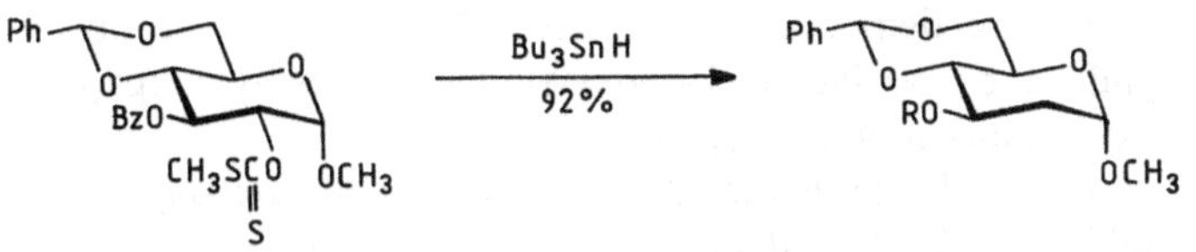

110 111

Scheme 21

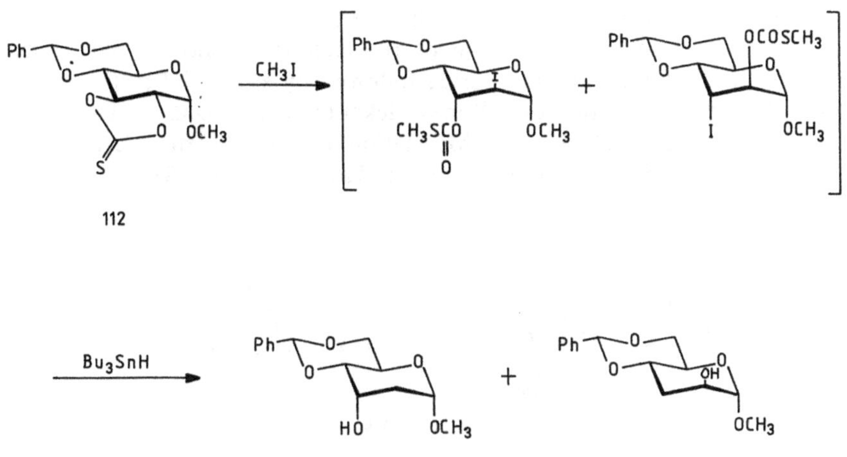

Scheme 22 81 (23%) (50%)

In an extension of the Barton method (206), methyl α-D-glucopyranoside was regioselectively converted (207) into the corresponding 2-O-thiocarbonate ester using phenoxythiocarbonyl chloride—dibutyltin oxide. From this methyl 3,4,6-tri-O-acetyl-2-deoxy-α-D-*arabino*-hexopyranoside **(31)** was prepared in good yield by acetylation and subsequent reduction using tributyltin hydride. Examples of the utilization of the Barton procedure for the generation of the 2-deoxy- and 4-deoxy functions of 3-aminodeoxyhexoses are given in Sects. 2.6.2 and 4.3.1, respectively.

Based on the reaction of aldonic acids with hydrogen bromide, the Pedersen group (208) developed convenient procedures for the production of various 2-bromo-2-deoxy- and 2,6-dibromo-2,6-dideoxyglycosyl bromides which are used in glycosylation reactions to obtain the 1,2-*trans* glycosides of 2-bromo-2-deoxy- and 2,6-dibromo-2,6-dideoxy-mono- and disaccharides. By using this method, each of the four stereoisomeric 2,6-dideoxy-D-hexoses and several derivatives thereof have been synthesized via the respective bromodeoxy aldonolactones.

An additional method (209) to be described is the conversion of diethyl mercaptals of protected *aldehydo*-hexoses into the corresponding 2-deoxysugars. This procedure has been used for the high-yielding syntheses of 2-deoxy-D-*arabino*- and D-*lyxo*-hexopyranose and their (2R)-2-deuterio analogs. For example, treatment (209) of 2,3:4,5-di-O-isopropylidene-D-glucose diethyl dithioacetal with potassium *tert*-butoxide in dimethylsulfoxide gave 2-deoxy-4,5-O-isopropylidene-D-*arabino*-hex-1-enose diethyl dithioacetal in 67% yield. Reduction of this latter ketene diethyl dithioacetal using lithium aluminum hydride proceeds via the intermediacy of an alkoxyaluminum hydride salt produced by the participation of the free C-3 hydroxyl group. It yields the 4,5-O-isopropylidene derivative of 2-deoxy-D-glucose diethylmercaptal, which gives the free 2-deoxy-D-glucose upon deprotection.

In an ingenious recent method (210) (Scheme 23) 1,2-O-isopropylidene-D-glucurono-6,3-lactone (**113**, R=OH; R_1=H) was deoxygenated at C-5 via the tributyltin hydride reduction of the 5-chloro-L-idurono lactone (**113**, R=H; R_1=Cl) obtained upon chlorination with sulfuryl chloride in pyridine. The overall yield was 87%. Acid-catalyzed methanolysis of the resulting 5-deoxylactone (**113**, R=R_1=H) followed by O-benzylation produced the methyl acetal **114**, which was converted into the diethyl mercaptal **115**. Reduction of the lactone function of this latter compound with diisobutyl aluminum hydride followed by sequential methanolysis and hydrogenolytic desulfurization over Raney nickel then gave methyl 5-O-benzyl-2,6-dideoxy-α-L-*xylo*-hexofuranoside **(86)**. Mesylation of **86** yielded the 3-O-mesylate **87**, a key intermediate (135) in the preparation of L-daunosamine (cf. Sects. 2.1.2 and 2.2.1, Schemes 13 and 30).

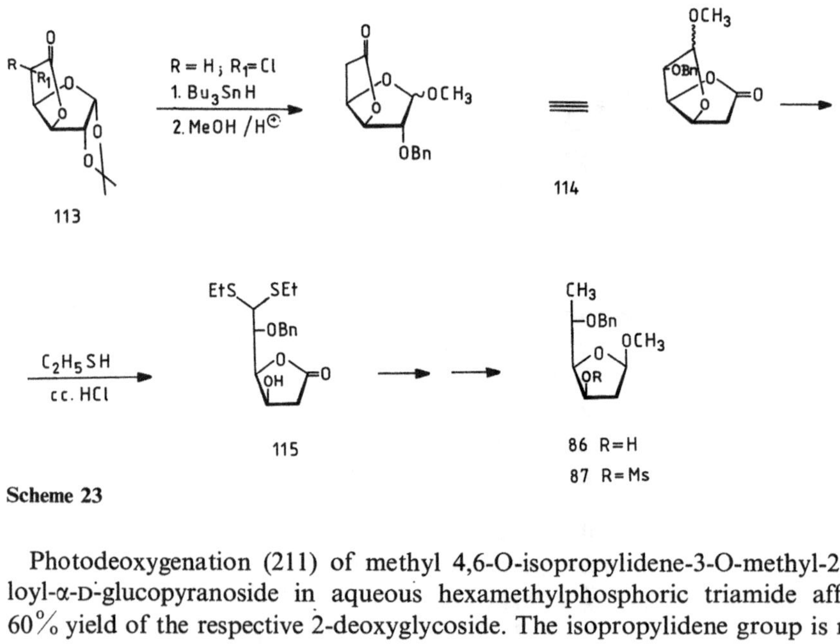

Scheme 23

Photodeoxygenation (211) of methyl 4,6-O-isopropylidene-3-O-methyl-2-O-pivaloyl-α-D-glucopyranoside in aqueous hexamethylphosphoric triamide afforded a 60% yield of the respective 2-deoxyglycoside. The isopropylidene group is resistant, but a benzylidene acetal function is easily cleaved under such conditions. Derivatives of methyl 2,6-dideoxy-α-D-*arabino*-hexopyranoside have been prepared in moderate yields from the corresponding 2,6-di-O-pivaloyl esters by the same method.

2,6-Dideoxyhexopyranosides containing the 3,4-*cis*-diol system (e.g., methyl α-L-digitoxoside and methyl 2-deoxy-α-L-fucoside) have been synthesized from the appropriate methyl hex-2-enopyranosides by the stereocontrolled conversion of the allylic alcohol function into a 2-iodo-3,4-cyclic carbonate moiety using iodonium dicollidine perchlorate and subsequent reductive dehalogenation (212). A similar strategy developed for the preparation of 3-amino-2,3,6-trideoxyhexoses carrying the C-3 amino- and C-4 hydroxyl functions in *cis*-relationship is discussed in Sect. 2.2.5.

2.2 Introduction of the C-3 amino function into mono- and dideoxyhexose derivatives

2.2.1 Displacement reactions of sulfonate esters and halogeno derivatives with nitrogen nucleophiles

One of the most frequently applied synthetic strategies for the introduction of a nitrogen function (azido- or amino group) into a carbohydrate molecule involves the bimolecular nucleophylic displacement of a sulfonate ester group or halogen atom by charged or neutral nitrogen nucleophiles. The rate and outcome of such a process are decisively affected by the conditions of the formation and transformation of the strongly polar complex in the transition state of the reaction. This in turn is influenced by favorable or unfavorable stereo-electronic (i.e., steric and dipolar) factors. Because there is less steric crowding in their transition states, the displacement reactions of primary sulfonate esters and halogen derivatives can be readily accomplished in the absence of an adjacent *cis-axial* substituent. However, the similar replacement of such leaving groups at secondary positions in cyclic carbohydrate derivatives is sometimes slow, and it may not take place at all (213–216).

The most important stereo-electronic factors affecting the energy-balance of the transition state in nucleophilic displacement reactions of carbohydrate sulfonates and halogeno derivatives have been thoroughly reviewed (from a qualitative point of view) by Richardson and his associates (217–220). They are therefore not described here. Examples are given to emphasize those factors that are particularly involved in bimolecular nucleophilic displacement processes that are useful for the development of a C-3 nitrogen function (azide and amine) in hexopyranosides and hexofuranosides.

Thus, because of the inhibitory effect of an adjacent *axial* electronegative functionality (218–222) at C-2 or C-4, and also because of the presence of a substituent (other than hydrogen) situated β-*trans-axial* to the leaving group (217–219), the nucleophilic displacement of C-3 sulfonate esters or halogeno compounds in the α-hexopyranoside series proceeds satisfactorily only for the allopyranoside-type stereoisomers. However, even in such cases a concurrent E_2C elimination (217), favored by the *axial* leaving group, may, to varying extent, accompany the displacement reaction, particularly under strong alkaline conditions.

However, the above steric factors are ultimately in operation only in conformationally rigid systems. In some cases these difficulties may be overcome by applying a more reactive leaving group (such as the trifluoromethanesulfonate) or an aprotic solvent of strong dipolar character or by performing the displacement reactions under more vigorous conditions. Also, many such processes that are unsuccessful with the negatively charged azide ion can be satisfactorily effected by using neutral nitrogen nucleophiles, e.g., ammonia, or preferably hydrazine (217).

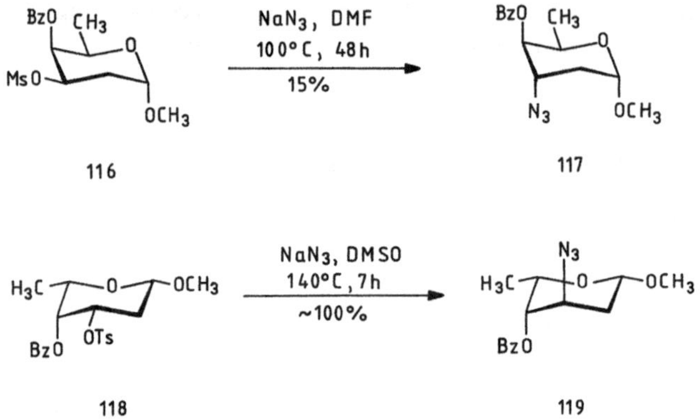

Scheme 24

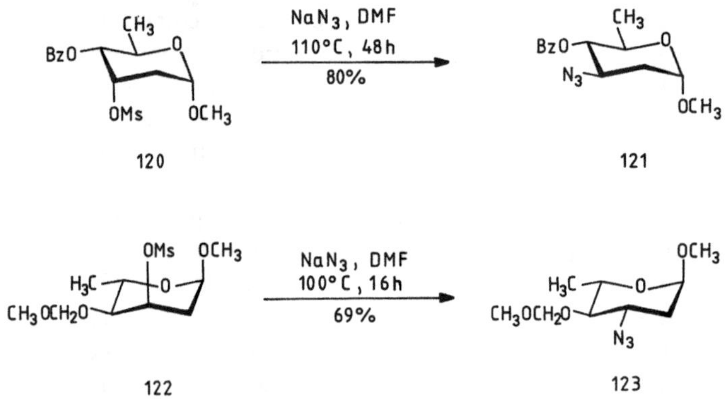

Scheme 25

For methyl 4-O-benzoyl-2,6-dideoxy-3-O-methanesulfonyl-α-D-*lyxo*-hexopyranoside (**116**), the azide displacement reaction (**116** → **117**) can be accomplished (Scheme 24) with only a 15% yield (55) because of the aforementioned two severe hindering factors, but the corresponding β-anomer in the L-series (**118**) provided an almost quantitative conversion (223) into the corresponding methyl 3-azido-β-L-*xylo* glycoside **119**. The similar reactions of methyl 4-O-benzoyl-2,6-dideoxy-3-O-methanesulfonyl-α-D-*ribo*-hexopyranoside (**120**) and methyl 2,6-dideoxy-3-O-methanesulfonyl-4-O-methoxymethyl-α-L-*ribo*-hexopyranoside (**122**) resulted in 80% and 69% yields, respectively, of the 3-azido-α-D- (**121**) (55) and α-L-*arabino*- (**123**) (187) methyl glycosides (Scheme 25). Methyl 3-azido-4-O-benzoyl-2,3,6-trideoxy-α-D-*arabino*-hexopyranoside **(121)** has also been obtained (85% yield) by the replacement of the 3-O-methanesulfonyl group of **120** with tetrabutylammonium azide in refluxing benzene (224). The azidolysis of the β-anomer of **120** produced (55) methyl 3-azido-4-O-benzoyl-2,3,6-trideoxy-β-D-*ribo*-hexopyranoside in only 33% yield.

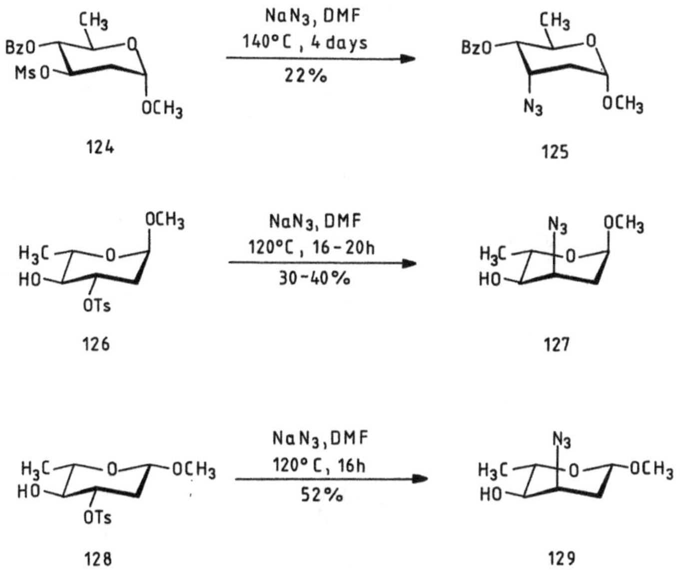

Scheme 26

Because of the β-*trans-axial* effect caused by the *axial* anomeric methoxy group, the displacement of the O-3 sulfonate ester function of the corresponding α-D-*arabino*- **(124)** (55) and α-L-*arabino*- **(126)** (55, 122, 225, 226) stereoisomers by azide ion leads to a moderate (22–36%) conversion into the α-*ribo*-D- and L-azidoglycosides **125** and **127** (Scheme 26). For the respective β-anomer **128** the inhibitory factor was not in operation, and the nucleophilic displacement reaction **128** → **129** proceeded with significantly higher (52%) yield (55).

A crucial problem in the above syntheses is associated with the selective sulfonylation of the respective 6-deoxy-carbohydrate 3,4-diol derivatives. Partial acylation, including tosylation of methyl 2,6-dideoxy-α- and β-L-*lyxo*-hexopyranoside with one equivalent of *p*-toluenesulfonyl chloride, resulted in regioselective substitution at the *equatorially* disposed C-3 hydroxyl group (223). In contrast, the corresponding α-L-*arabino*-analog **(74)**, carrying two *equatorially* attached hydroxyl groups, displays significantly more pure selectivity upon tosylation. The unimolar tosylation of **74** or of its enantiomer **99** has sometimes been reported without experimental details (269) (69% yield) or giving neither reaction conditions nor yield (123). Details for the mono-tosylation of **74** at O-3 (1.1 molar equivalent of tosyl chloride in pyridine for several days at 0 °C) have described (55, 122, 155) the production of the 3-O-*p*-toluenesulfonyl derivative **126** as the major product (50 to 55%) along with the 4-O-tosyl- and 3,4-di-O-tosyl compounds. The 3-O-tosyl-β-L-

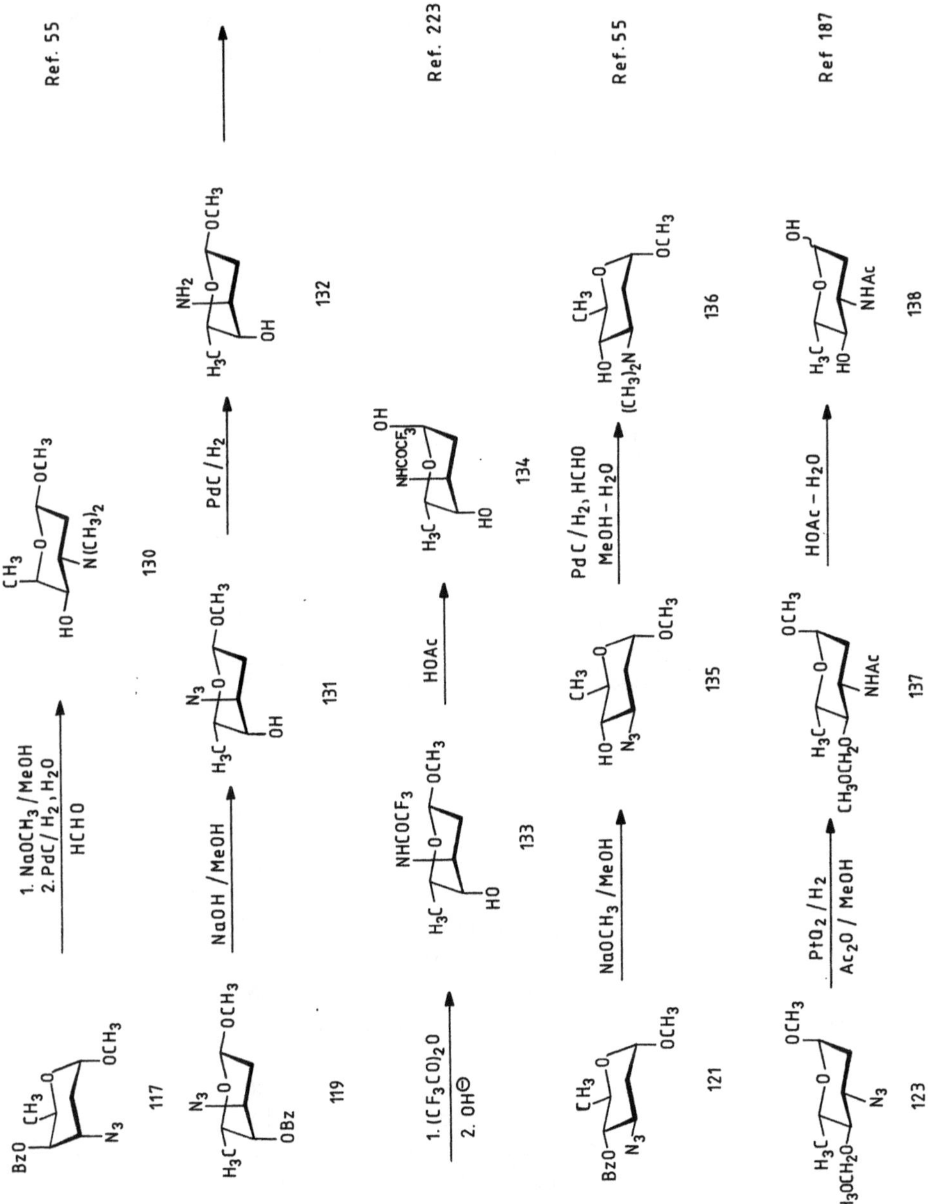

Fig. 3. Conversion of azidodeoxy hexopyranosides into antibiotic aminodeoxy hexoses (I)

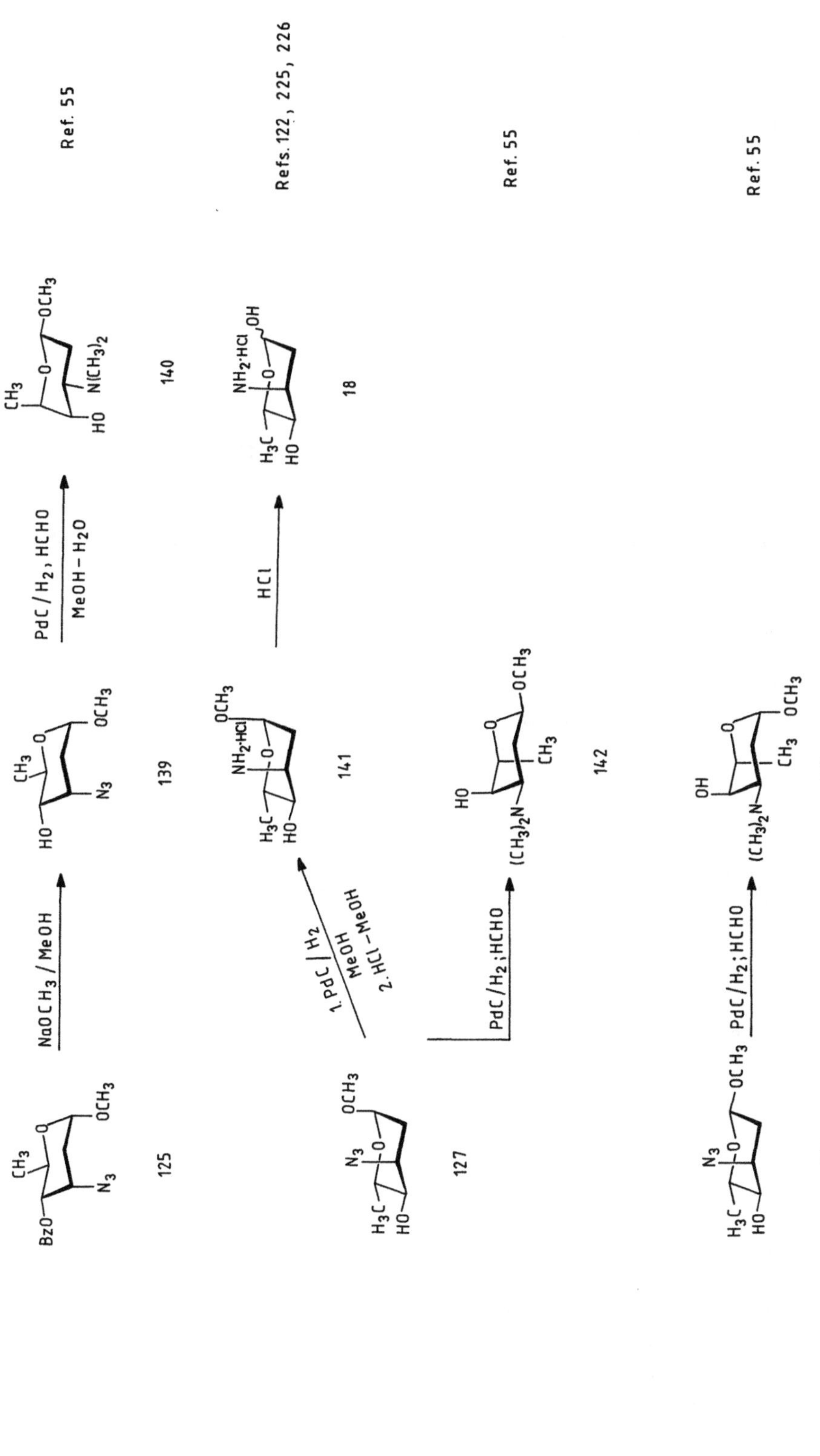

Fig. 4. Conversion of azidodeoxy hexopyranosides into antibiotic aminodeoxy hexoses (II)

arabino-hexopyranoside **128** has also been prepared (55), in significantly lower yield (39%), by means of the selective sulfonylation of **76**. An attractive improvement for the selective acylation, sulfonylation, or alkylation of such diol derivatives via their 3,4-O-stannylene acetals has recently been reported by Monneret et al. (226a). This strategy may also be of interest for the syntheses of C-3 or C-4 branched sugars from the appropriate ulose derivatives in cases where the distinction between the secondary hydroxyl groups at C-3 and C-4 is of supreme importance.

By means of the reduction and protection-deprotection methodology shown in Figs. 3 and 4, the azidoglycosides **117**, **119**, **121**, **123**, **125**, **127**, and **129** were converted into methyl 3-dimethylamino-2,3,6-trideoxy-α-D-*xylo*-hexopyranoside **(130)** (55), 3-trifluoroacetamino-2,3,6-trideoxy-α-L-*xylo*-hexopyranose **(134)** (223), methyl α-D-angolosaminide **(136)** (55), N-acetyl-L-acosamine **(138)** (187), methyl α-D-megosaminide **(140)** (55), L-ristosamine **(18)** (122, 225, 226), and methyl α-L- **(142)** and β-L-megosaminide **(143)** (55), respectively.

The azido compounds with D-*arabino*- **(121)**, L-*arabino*- **(123)**, and L-*ribo*- **(127)** configurations have also been used in syntheses of the corresponding C-4 epimeric aminodeoxy sugars: methyl α-D-rhodosaminide **(669)** (55), L-daunosamine **(12)** (187) (see Sect. 2.8.1, Table 14 entries 2 and 4), and methyl 3-trifluoroacetamido-2,3,6-trideoxy-α-L-*xylo*-hexopyranoside **(688)** (122) (Sect. 2.8.3, Table 17 entry 1). In addition, azide **135** (obtained from **121** by O-deacylation) has been utilized as a chiral template for the preparation (224, 269) of precursors of beta-lactam antibiotic analogs (Sect. 4.3.3).

^{1}H-NMR measurements clearly showed the reversed [$^1C_4$(D) and $^4C_1$(L)] conformation of each of the 3-dimethylamino derivatives **130**, **140**, **142**, and **143**, which would contain the C-3 substituent in the *axial* position in the usual [$^4C_1$(D) and $^1C_4$(L)] conformation.

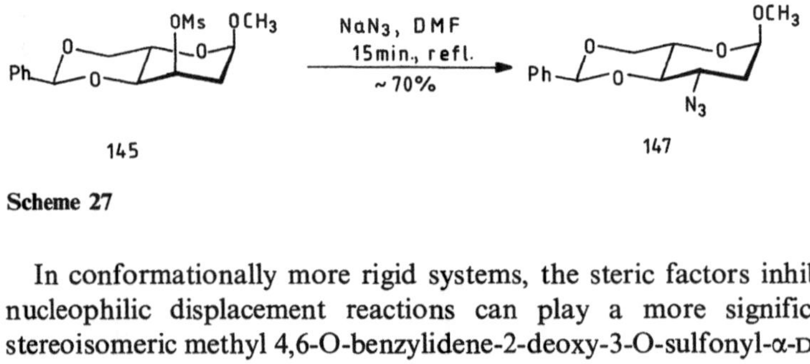

Scheme 27

In conformationally more rigid systems, the steric factors inhibiting bimolecular nucleophilic displacement reactions can play a more significant role. Of the stereoisomeric methyl 4,6-O-benzylidene-2-deoxy-3-O-sulfonyl-α-D-hexopyranosides, those exhibiting the *allo* steric arrangement are readily converted into the C-3 epimeric azido derivatives. Accordingly, methyl 4,6-O-benzylidene-2-deoxy-3-O-methanesulfonyl-α-D- **(144)** (174, 227–234) and -α-L-*ribo*- **(145)** hexopyranoside (235,

236) furnish the corresponding *arabino* azides **146** and **147** in over 70% yields. On the other hand, a similar azide substitution of the 3-O-mesylate **148** (237) or 3-O-*p*-bromobenzenesulfonate **149** (228) with the α-D-*arabino* configuration proceeded with only a 15–30% conversion[1] into the *ribo* azide **151**. However, in the case of the β-anomer **(150)** of **148**, where in the β-*trans-axial* effect is out of operation, an 84–92% yield of methyl 3-azido-4,6-O-benzylidene-2-deoxy-β-D-*ribo*-hexopyranoside **(152)** was prepared (114, 239, 240).

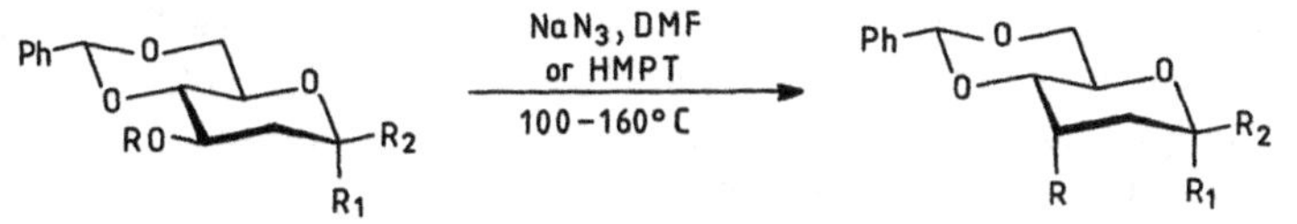

148 R = Ms; $R_1 = OCH_3$; $R_2 = H$
149 R = p-Br-$C_6H_4SO_2$; $R_1 = OCH_3$; $R_2 = H$
150 R = Ms; $R_1 = H$; $R_2 = OCH_3$

151 R = N_3; $R_1 = OCH_3$; $R_2 = H$
152 R = N_3; $R_1 = H$; $R_2 = OCH_3$
157 R = NH_2; $R_1 = OCH_3$; $R_2 = H$

Scheme 28

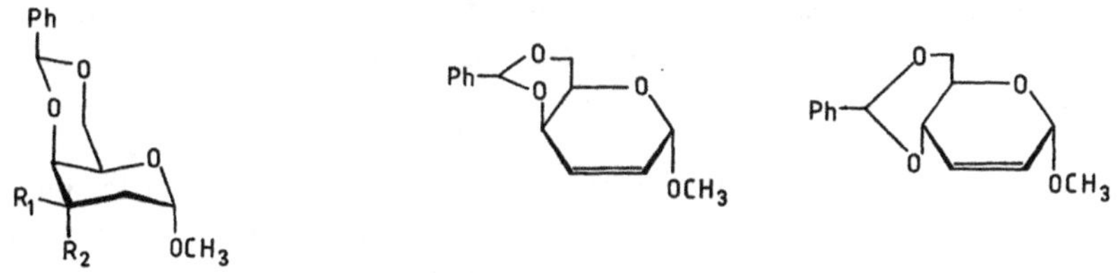

153 $R_1 = OTs$; $R_2 = H$
154 $R_1 = H$; $R_2 = N_3$
155 $R_1 = H$; $R_2 = OTs$

156

158

Scheme 29

Of the methyl 4,6-O-benzylidene-3-O-tosyl-α-D-hexopyranosides carrying an *qxially* bonded substituent at C-4, the *lyxo* isomer **(153)** afforded the C-3 *xylo* azide **154** in 29% yield (NaN_3, DMF, 140 °C, 6 d). Elimination (to give **156**), but no azide displacement reaction occured when the α-D-*xylo*-3-O-tosylate **155** was treated (228) under similar conditions (Scheme 29).

The reaction of mesylate **144** with ammonia in methanol produced only a 20% yield of the *ribo* amine **157**; the major product was methyl 4,6-O-benzylidene-2,3-dideoxy-α-D-hex-2-enopyranoside **(158)**. By performing the displacement on **144** using hydrazine followed by reduction, the aminosugar **157** (71%) was produced together with a 27% yield of the saturated analog of **158** (237).

Methyl 3-azido-4,6-O-benzylidene-2,3-dideoxy-α-D-*arabino*- **(146)**, α-L-*arabino*- **(147)**, α-L-*ribo*-, α-D-*xylo*- **(154)**, and β-D-*ribo*- **(152)** hexopyranosides have proved

[1] Under forced conditions (HMPT, 150 °C), this reaction has provided (238) a 79% yield. Also, Bargiotti et al. (236) described a 92% conversion of the L-enantiomer of **148** into the 3-azido-α-L-*ribo* methyl glycoside.

to be extremely versatile intermediates in the syntheses of numerous 3-amino-2,3-dideoxy- and 3-amino-2,3,6-trideoxyhexopyranose derivatives. The α-D-*arabino* azide **(146)** has also been utilized as a chiral intermediate for the preparation of 9-aza-prostaglandins (233) and thienamycin-type beta-lactam antibiotics (234) (see Sects. 4.3.2 and 4.3.3).

The Arcamone group (235, 236) converted methyl 3-azido-4,6-O-benzylidene-2,3-dideoxy-α-L-*arabino*-hexopyranoside **(147)** and its α-L-*ribo* analog (L-**157**) into 6-hydroxy-L-acosamine- (**672**, Table 15 entry 1) and 6-hydroxy-L-ristosamine- (**203**, Scheme 37) derivatives and also into their C-4 epimeric L-*lyxo*- (**679** and **683**, Sect. 2.8.1, Table 15 entry 1) and L-*xylo*- (**687** and **689**, Sect. 2.8.3, Table 16 entry 1) stereoisomers by catalytic hydrogenation of the azide function followed by inversion of the configuration at C-4, as shown in Sects. 2.8.1 and 2.8.3.

Methyl 3-azido-4,6-O-benzylidene-2,3-dideoxy-α-D-*arabino*-hexopyranoside **(146)** has been used to prepare D-acosamine **(10)** (174) and methyl N-acetyl-α-D-acosaminide **(544)** (231) with the same stereochemistry via the 6-bromo-6-deoxy analogs **561** and **562** (cf. Sect. 2.7.2, Table 7, entries 1 and 2). N-Benzoyl-D-daunosamine (229, 230) and methyl N-acetyl-α-D-daunosaminide (231) (**671** and **666**, respectively, Table 14 entry 6) and derivatives (**621**, **622**, and **625**, Table 11 entries 1 and 2) of methyl 3-amino-2,3,6-trideoxy-β-L-*xylo*-hexopyranoside **(132)** and N-benzoyl-L-ristosamine (**700**, Table 18 entry 1) have been synthesized (231, 232) by the inversion of the configuration at C-4 (Sect. 2.8.2) or C-5 (2.7.4), or at both asymmetric centres. Methyl 3-azido-4,6-O-benzylidene-2,3-dideoxy-β-D-*ribo*- **(152)** and α-D-*lyxo*-hexopyranoside **(154)** have been utilized in synthetic strategies (Sects. 2.7.2 and 2.7.1) leading to D-ristosamine (**20**, Table 7 entry 8) (114, 239, 240) and methyl 3-methylamino-2,3,6-trideoxy-α-D-*xylo*-hexopyranoside (**541**, Scheme 106) (228), respectively.

It has been already mentioned that for the sulfonate → azide displacement reactions the highly dipolar aprotic solvents (HMPT > DMSO > DMF) are most beneficially used (217, 241). The trifluoromethanesulfonyl (triflate) (242) > *p*-bromobenzenesulfonyl > benzenesulfonyl > tosyl > mesyl substituents are proposed (241) as leaving groups.

In recent years the trifluoromethanesulfonate group, being approximately 5.6×10^4 times as reactive as the mesylate group, has emerged frequently as the leaving group used in carbohydrate chemistry (242) for a variety of displacement processes for which other sulfonate esters have failed. Nucleophilic substitutions using triflates proceed quickly at low temperature, and thus there is generally little or no competition from undesired elimination reactions or rearrangements. An illustration of this in the field of 3-O-triflyl-2-deoxy sugars comes from our own experimental work (243), in which the azide displacement of both methyl 4,6-O-benzylidene-2-deoxy-3-O-trifluoromethanesulfonyl-α-D-*arabino*- and *ribo*-hexopyranoside proceeded quantitatively within 30 minutes in DMF as exothermic reactions. As has been discussed, the conversion of the corresponding mesylates **142** and **144** could be achieved with yields of 15–30% and 70%, respectively, at temperatures higher than 100 °C. Recently the Voelter group (244) has prepared several aminodeoxy sugar derivatives, involving 1,2:5,6-di-O-isopropylidene-3-amino-3-deoxy-α-D-gluco- and allofuranose, via the displacement of the respective triflates.

Steric hindrance due to the 1,2-O-isopropylidene acetal function in the *cis*-fused ring systems of 1,2:5,6-di-O-isopropylidene-3-O-methanesulfonyl-α-D-glucofuranose and 6-deoxy-1,2-O-isopropylidene-3-O-methanesulfonyl-β-L-idofuranose **(159)** has been demonstrated to considerably inhibit the nucleophilic displacement of the *exo*-oriented 3-O-methanesulfonyl function with azide ion by Whistler et al. (245) and by the Kuszmann group (135), respectively. Since no similar inhibitory effect is in operation in the molecule of a 2-deoxy derivative **(87)** of **159** (prepared as shown in Sect. 2.1.2, Scheme 30), the C-3 azide **160** with L-*lyxo* configuration could be obtained in excellent yield. Catalytic hydrogenation of this latter product followed by acid hydrolysis then readily furnished L-daunosamine hydrochloride **(12)** (135). Methyl 3-acetamido-5-O-benzyl-2,3,6-trideoxy-α-L-*lyxo*-hexofuranoside **(161)** was prepared by the selective reduction of the azido group in **160** (Scheme 30). Derivatives of **160** and **161** were used (264) for the synthesis of furanoside representatives of the anthracycline antibiotics. In 1982 the El Khadem group (247) reported the first method for the synthesis of similar furanose derivatives of L-daunosamine based on the production and cyclization of the acyclic diethyl dithioacetal **162** prepared from N-trifluoroacetyl-L-daunosamine (Sect. 4.4, Scheme 189). Prior to this work, Japanese investigators (248) had synthesized methyl 3-amino-2,3,6-trideoxy-α,β-L-*lyxo*-hexofuranoside **(533)** by means of an independent procedure discussed in Sects. 4.6.2 and 4.7.4 (Schemes 99 and 116).

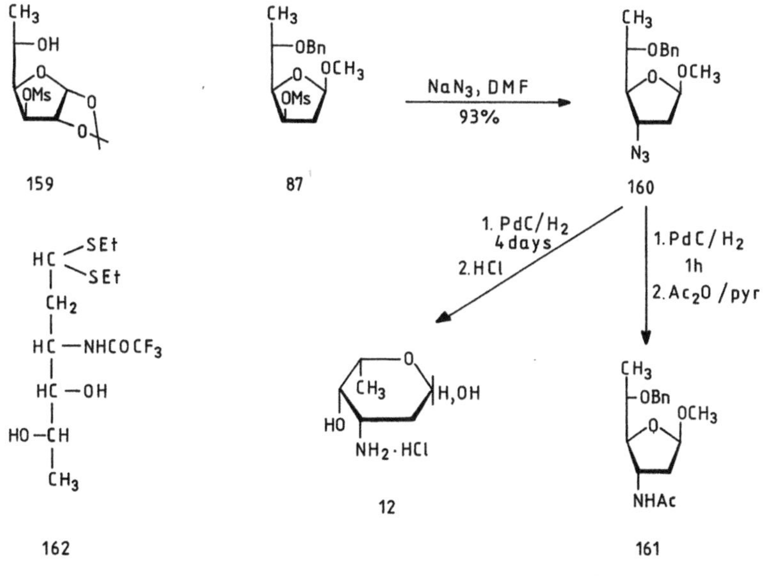

Scheme 30

Apart from catalytic hydrogenation (over palladium on carbon or barium sulfate, Raney nickel, platinum, Lindhlar catalyst, etc.) several additional methods for the conversion of azido sugars into aminodeoxy sugar derivatives have been described. Metal hydride reagents ($LiAlH_4$ and $NaBH_4$), hydrazine—Raney nickel (217), and zinc in warm DMF (249) have commonly been used for this purpose. Additional recent methods (250), elaborated for the reduction of non-carbohydrate azides, can also play a role, particularly when the preparation of aminohexose with temporary protecting groups or functionalities that are sensitive to the above conventional reducing agents is required.

By using a recent synthetic "trick", 2-, 3- and 6-phthalimidodeoxy hexose derivatives have been prepared in good yields from the corresponding azido sugars in a one-pot operation by heating them with triphenylphosphine and phthalic anhydride in benzene or toluene in the presence of a catalytic amount of tetrabutylammonium cyanide (251).

2.2.2 Ring opening of anhydro- and epiminohexopyranosides with nitrogen nucleophiles

The ring opening of pyranoside anhydro sugar derivatives (129, 252) using nitrogen nucleophiles is also a widely used strategy in the syntheses of aminodeoxy sugars. The products carrying the two new functional groups in a *trans-diaxial* relationship (Fürst-Plattner rule) (132) usually predominate.

Due to the flexibility of the half-chair conformers (129, 253) [$^{0}H_{5}$(D), $^{5}H_{0}$(L) and $^{0}H_{1}$(D), $^{1}H_{0}$(L)] of the "monocyclic" 2,3- or 3,4-anhydrohexopyranosides, they may easily adopt the alternative conformational status. Thus, the proportion of the "normal" *diaxial* and the "abnormal" *diequatorial* ring-opened products, produced upon attack by nitrogen nucleophiles, is much more dependent on the behavior of the nucleophile and on the reaction conditions than is that observed for the products formed conformationally more rigid "bicyclic" carbohydrate epoxides (254).

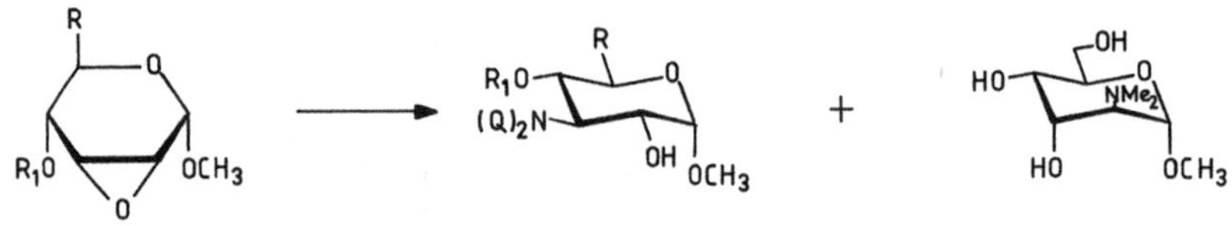

163 R = CH_2OH; R_1 = H — 164 R = CH_2OH; R_1 = H; Q = CH_3 — 165

166 R = CH_3; R_1 = Bz — 167 R = CH_3; R_1 = Q = H

Scheme 31

In representatives of the "monocyclic" methyl 2,3-anhydro-hexopyranosides, "abnormal" *diequatorial* ring cleavage occurs, particularly with the allo-isomers, and furnishes considerable quantities of the respective 3-amino-3-deoxy-glucose derivatives (255–260). Thus, Foster et al. (256) reported that treatment of methyl 2,3-anhydro-α-D-allopyranoside **163** with dimethyl amine (Scheme 31) led to a 1:1 mixture of methyl 3-deoxy-3-dimethylamino-α-D-glucopyranoside **(164)** and methyl 2-deoxy-2-dimethylamino-α-D-altropyranoside **(165)**. The gluco isomer **164** was then used (256) for a transformation into micaminose (**226**, 3,6-dideoxy-3-dimethylamino-D-glucopyranose, cf. Sect. 2.2.4 Scheme 43), a common building block of non-polyene macrolide-type antibiotics. Cleavage of the anhydro function of the 6-deoxy analoge **(166)** of **163** gave (257) a 52% yield of the "abnormal" product **167**. Similarly, the reaction of methyl 2,3-anhydro-6-deoxy-α-L-allopyranoside **(168)** with dimethyl amine resulted in (259, 260) methyl α-L-micaminoside **(169)** in near 70% yield (Scheme 32). On the other hand, analogous ring opening reactions of the corresponding α- and β-L-manno epoxides are highly stereospecific and yield 2,3-*diaxial* products (261–263). Thus, upon ammonolysis and subsequent N-acetylation, epoxide **170** gave the amine **171** in 81% yield (Scheme 33).

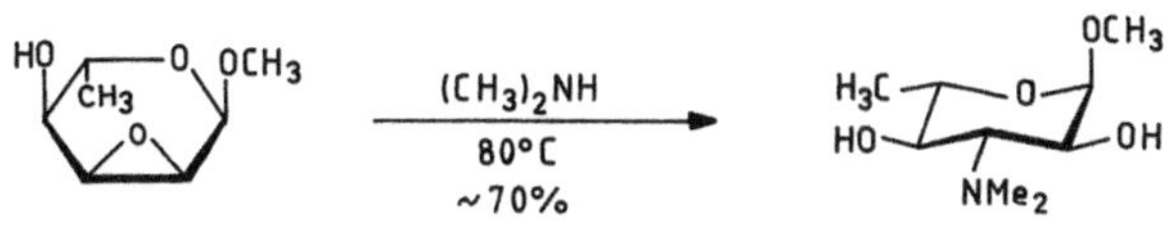

168 169

Scheme 32

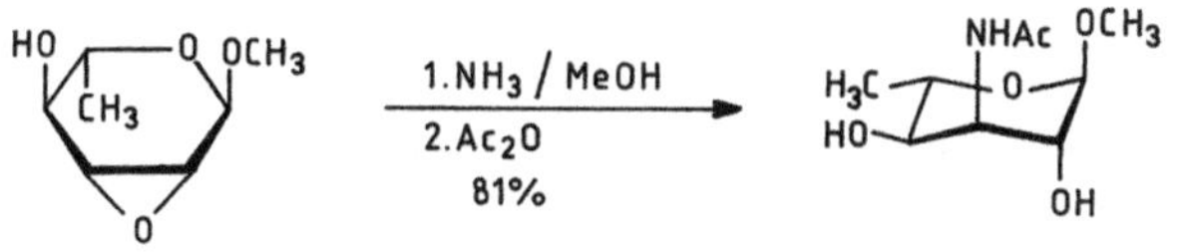

170 171

Scheme 33

Detailed studies have demonstrated (264) that the ring cleavage of "monocyclic" methyl 3,4-anhydro-2,6-dideoxyhexopyranosides (123, 155, 265, 266) with α- and β-L-*lyxo* (**172** and **173**), and α- and β-L-*ribo*- (**174** and **175**) configurations using nitrogen nucleophiles (azide ion, ammonia, dimethyl amine) is also highly regio-specific and stereoselective. Apart from the reaction of the β-L-*ribo* isomer **175** with ammonia and azide, the reagent attacks predominantly at C-3. Additionally, exclusive or predominant *diaxial* ring opening takes place with the *lyxo* isomers **172** and **173** to yield compounds **131**, **132**, **176–179**. In the L-*ribo* series the reaction of the α-anomer (**174**) with the aforementioned nucleophiles and the reaction of the β-anomer **175** with dimethyl amine result mainly (60–70% yields) in the "abnormal" products **180–182** and **185**. Similar 3,4-*diequatorial* products were isolated earlier from the reactions of methyl 3,4-anhydro-6-deoxy-α-D-galactopyranoside (267), methyl 3,4-anhydro-2-O-benzoyl-6-deoxy-α-D-talopyranoside (268), and methyl 3,4-anhydro-2,6-dideoxy-α-D-*ribo*-hexopyranoside (269, 270). Methyl 3-azido-2,3,6-trideoxy-α-D-*arabino*-hexopyranoside (**135**, the enantiomer of **180**) has been used (224, 269, 270) as a chiral intermediate in syntheses of precursors to thienamycin (Sect. 4.3.3).

The regio- and stereo-selectivity of the ring cleavage reactions of the above 3,4-epoxides have been explained in terms of steric and conformational factors (129, 264, 271). The more frequent formation of *diequatorial* products in the cases of such 3,4-anhydro derivatives can be attributed to the small energy difference (129) between the two alternative conformers ($^{0}H_{1}$ and $^{1}H_{0}$), which allows the reaction to proceed through either of the corresponding transition states.

In conformationally more rigid systems, such as in the 1,6:3,4-dianhydro-D-galactose derivative **186**, the opening of the 3,4-epoxide function follows the Fürst-Plattner rule to yield (272) the C-4 amine **187** in 94% yield (Scheme 34). From the latter, methyl 4,6-dideoxy-3-C-methyl-4-(N-acetyl-N-methyl)-amino-α-D-altropyranoside, the C-3 epimer of methyl N-acetylsibirosaminide, was prepared.

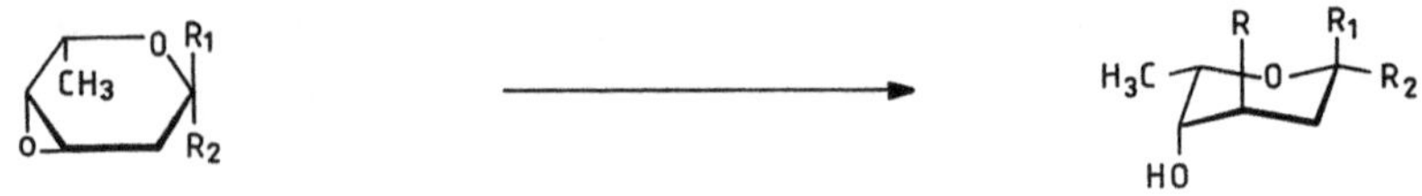

172 $R_1 = OCH_3$; $R_2 = H$
173 $R_1 = H$; $R_2 = OCH_3$

176 $R_1 = OCH_3$; $R_2 = H$; $R = N_3$ (90%)
177 $R_1 = OCH_3$; $R_2 = H$; $R = NH_2$ (100%)
178 $R_1 = OCH_3$; $R_2 = H$; $R = NMe_2$ (100%)
131 $R_1 = H$; $R_2 = OCH_3$; $R = N_3$ (100%)
132 $R_1 = H$; $R_2 = OCH_3$; $R = NH_2$ (100%)
179 $R_1 = H$; $R_2 = OCH_3$; $R = NMe_2$ (100%)

174 $R_1 = OCH_3$; $R_2 = H$
175 $R_1 = H$; $R_2 = OCH_3$

180 $R_1 = OCH_3$; $R_2 = H$; $R = N_3$ (100%)
181 $R_1 = OCH_3$; $R_2 = H$; $R = NH_2$ (~60%)
182 $R_1 = OCH_3$; $R_2 = H$; $R = NMe_2$ (100%)
183 $R_1 = H$; $R_2 = OCH_3$; $R = N_3$ (20–25%)
184 $R_1 = H$; $R_2 = OCH_3$; $R = NH_2$ (20–25%)
185 $R_1 = H$; $R_2 = OCH_3$; $R = NMe_2$ (75%)

Fig. 5. Epoxide-ring opening of 3,4-anhydro-2,6-dideoxyhexopyranosides

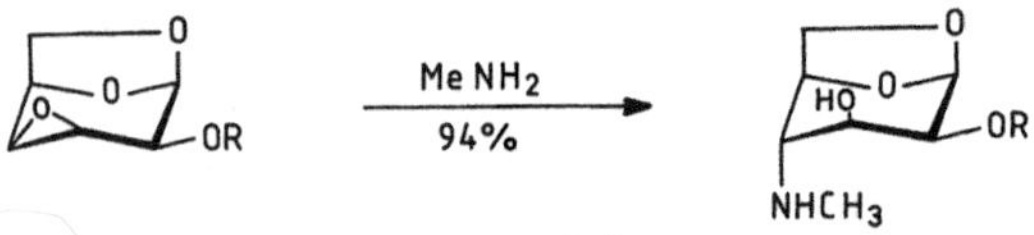

$R = (CH_2)_2OC_2H_5$

186 187

Scheme 34

As has been shown previously, the cleavage of the epoxide ring of the 3,4-anhydro-hexopyranose derivatives **172–175** using nitrogen nucleophiles allows the preparation (Fig. 5) of methyl α- and β-L-acosaminide **(181** and **184)** and their N-methylated derivatives **(182** and **185)**, the same compounds in the α- and β-L-*xylo*-hexopyranoside series **(177, 132, 178,** and **179)**, and the corresponding azido analogs with both the L-*arabino*- **(180, 183)** and the L-*xylo* configurations **(176** and **131)**.

Methyl 3-azido-2,3,6-trideoxy-α-L-*arabino*-hexopyranoside **(180)** has been used by Goodman et al. (265) and Gupta (266) for the synthesis of L-acosamine **(8)**, methyl N-acetylactinosaminide (**188**, Scheme 35) and other acosamine derivatives. The first synthetic method (123) for preparing L-daunosamine **(12)** by means of the inversion of the configuration of compound **180** at C-4 (see Sect. 2.8.1, Table 14 entry 1) was also based on this azido sugar. In 1983 Grethe et al. (155) applied essentially similar procedures for the preparation of L-acosamine and L-daunosamine and their

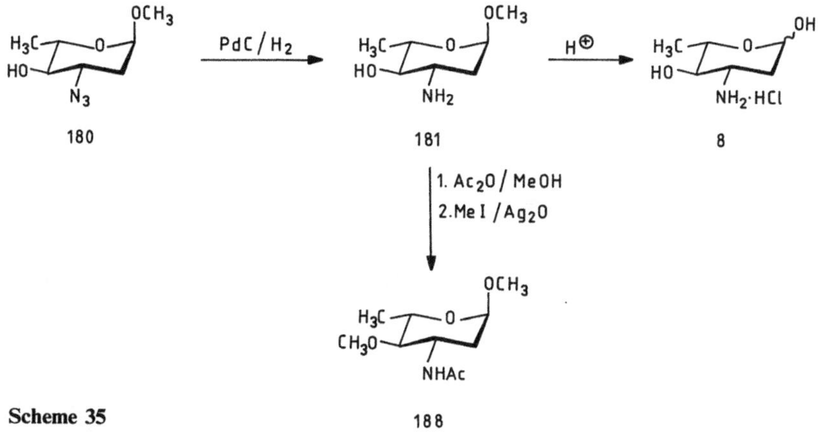

Scheme 35

derivatives. The only difference was that the key intermediate epoxide **(174)** was synthesized from cyclopentadiene by means of an asymmetric seven-step process (see Sect. 3.2.1, Scheme 132). The 4-O-*p*-nitrobenzoyl ester of azide **180** has been used by the Bognár group (273) for the synthesis of phenyl β-acobioside [phenyl 2-O-(3-amino-2,3,6-trideoxy-α-L-*arabino*-hexopyranosyl)-β-D-glucopyranoside], a derivative of the disaccharide component of the actinoidin antibiotic variants.

Recent studies concerning the preparation of polycyclic β-amino alcohols have shown (274) that halomagnesium alkylamides react with epoxides under mild conditions to give, after hydrolysis, products with vicinal amino and hydroxy functions in good yields. This method of epoxide-ring cleavage might also be adapted to the production of aminodeoxy sugar derivatives. Although the generation of an aziridine functionality in hexopyranoside derivatives (275) requires the prior introduction of a nitrogen substituent (amine or azide) into the molecule, the ring opening of such epimino sugars using nucleophiles sometimes provides convenient access to aminodeoxyhexoses (276–278) that are otherwise only elaborately available.

The reaction of alkyl 2,3-epimino-hexopyranosides with the azide ion is a generally useful method (220, 278–280) for the synthesis of 2,3-diamino-2,3-dideoxyhexoses. The reductive ring cleavage of the latter type of aziridine derivatives, which leads to 3-amino-2,3-dideoxy sugar analogs via the simultaneous development of the C-3 amino- and C-2 deoxy functions, is discussed in Sect. 2.3.

2.2.3 Reduction of 3-oximino-2,3-dideoxyhexopyranosides

Nowadays, numerous methods (281–290) are available for the efficient oxidation of the secondary hydroxyl groups of suitably protected carbohydrates; they provide a popular alternative route to form aminodeoxy sugars *via* the reduction of the oxime-, hydrazone-, or phenylhydrazone derivatives of the uloses produced. Such a strategy, involving the utilization of 2-deoxyhexopyranosid-3-uloses, which are readily available either by the aforementioned oxidation methods or (with more moderate yield) by means of the Klemer-Rodemeyer fragmentation reaction (167) of 2-phenyl-1,3-dioxolane acetals of hexopyranosides (cf. Sect. 2.1.4, Schemes 19 and 20), has been often used for the synthesis of 3-amino-2,3-dideoxy- and 3-amino-2,3,6-trideoxyhexose derivatives.

Table 3. Steric outcome of the reduction of the oxime derivatives **189–192**

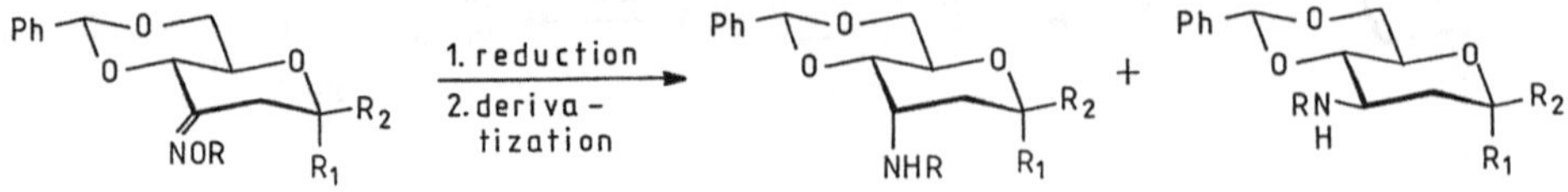

189 $R=R_2=H; R_1=OCH_3$
190 $R=R_1=H; R_2=OCH_3$
191 $R=Ac; R_1=OCH_3; R_2=H$
192 $R=Ac; R_1=H; R_2=OCH_3$

157 $R=R_2=H; R_1=OCH_3$
194 $R=R_1=H; R_2=OCH_3$
196 $R=Ac; R_1=OCH_3; R_2=H$
198 $R=COCF_3; R_1=OCH_3; R_2=H$

193 $R=R_2=H; R_1=OCH_3$
195 $R=R_1=H; R_2=OCH_3$
197 $R=Ac; R_1=OCH_3; R_2=H$
199 $COCF_3; R_1=OCH_3; R_2=H$

Entry	Oxime	Reducing agent	Reduction product		*Ribo*:*Arabino* ratio	Refs.
			ribo	*arabino*		
1	189	$LiAlH_4$[a]	196	197	70:10	174
2	189	$LiAlH_4$[a]	196	197	87:12	172
3	189	$LiAlH_4$	157	193	86:14	178
4	189	$LiAlH_4$[b]	198	199	88:12	179
5	189	$LiAlH_4$	157	193	100:0	228
6	L-189	$LiAlH_4$	L-157	L-193	85:10	184, 236
7	189	Vitride*[b] (Red-Al)	198	199	99:1	179
8	190	$LiAlH_4$	194	195	47:53	178
9	192	$LiAlH_4$	194	195	ca. 1:1	291
10	189	Al/Hg	157	193	100:0	178
11	190	Al/Hg	194	195	74:26	178
12	189	PtO_2/H_2	157	193	100:0	228
13	189	Ni/H_2	157	193	40:24	228
14	191	$BH_3 \cdot THF$[a]	196	197	10:30	292

* Sodium dihydrobis-(2-methoxyethanolato)-aluminate-1
[a] Followed by acetylation
[b] Followed by trifluoroacetylation

Although the reduction of 3-deoxy-3-oximinohexopyranosides usually leads to a mixture of the corresponding C-3 epimeric amino compounds, in the majority of cases remarkable diastereoselectivity has been observed upon the application of different reducing agents. The steric outcome of the reduction of the oximes and acetoximes **189–192**, derived from methyl 4,6-O-benzylidene-2-deoxy-α- **(100)** and β-D-*erythro*-hexopyranosid-3-ulose (extremely versatile starting materials for the synthesis of numerous 3-amino- and 3-C-methylhexoses), accomplished by catalytic hydrogenation or chemical reduction methods is demonstrated by the data in Table 3.

Thus, hydrogenation of the α-D-*erythro* oxime **189** over Raney nickel (entry 13) led (228) to a 40:24 mixture of the *ribo*- **(157)** and *arabino*- **(193)** amines. When the reduction was performed in the presence of Adams catalyst (228) (entry 12) or aluminum amalgam (178) (entry 10) the exclusive formation of the *ribo* compound **157** was observed, but when lithium aluminum hydride was used (entries 1–4 and 6) the *arabino* amine **193** was also produced (172, 174, 178, 179, 184, 236) in at least 10% yield. An almost quantitative conversion of oxime **189** into the *ribo* compound **198** has recenty been carried out (179) by applying sodium dihydrobis(2-methoxyethanolato)-aluminate-1 (Vitride; Red-Al[R]) and subsequent trifluoroacetylation (entry 7). The reduction of the β-anomers **(190** and **192)** of oximes **189** and **191** with lithium aluminum hydride (entries 8 and 9) or with aluminum amalgam (entry 11) is not nearly so stereoselective (178, 191) and results in a considerable amount (25–50%) of methyl 3-amino-4,6-O-benzylidene-2-deoxy-β-D-*arabino*-hexopyranoside **(195)**. On the other hand, the *arabino* amino compound was observed to form as the major product when oxime **189** or its acetyl derivative **191** was reduced (292) with the borane–tetrahydrofuran complex; after acetylation of the reaction mixture, the ratio of the *ribo-* **(196)** and *arabino-* **(197)** C-3 acetamido sugars was found to be 10:30 (entry 14).

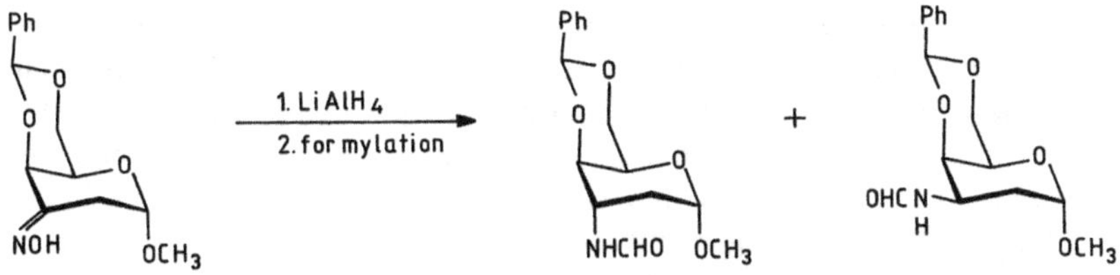

200 201 202

Scheme 36

Reduction of methyl 4,6-O-benzylidene-2,3-dideoxy-3-oximino-α-D-*threo*-hexopyranoside **(200)** using lithium aluminum hydride produced (228) a 55:36 mixture of the *xylo-* **(201)** and *lyxo-* **(202)** isomeric C-3 formamido compounds after formylation (Scheme 36). By removal of the dioxane acetal function of methyl 3-amino-4,6-O-benzylidene-2,3-dideoxy-α-L-*ribo*-hexopyranoside **(L-157)** and subsequent N-trifluoroacetylation (Scheme 37), the Arcamone group (184, 236) prepared methyl 2,3-dideoxy-3-trifluoroacetamido-α-L-*ribo*-hexopyranoside **(203)**. Inversion of the configuration at C-4 of **203** (Sect. 2.8.3), or deoxygenation at C-6 (Sect. 2.7.3) produced methyl 2,3-dideoxy-3-trifluoroacetamido-α-L-*xylo*-hexopyranoside (**689**, Table 16 entry 1) (236) or methyl N-trifluoroacetyl-α-L-ristosaminide (**597**, Scheme 112) (184), respectively.

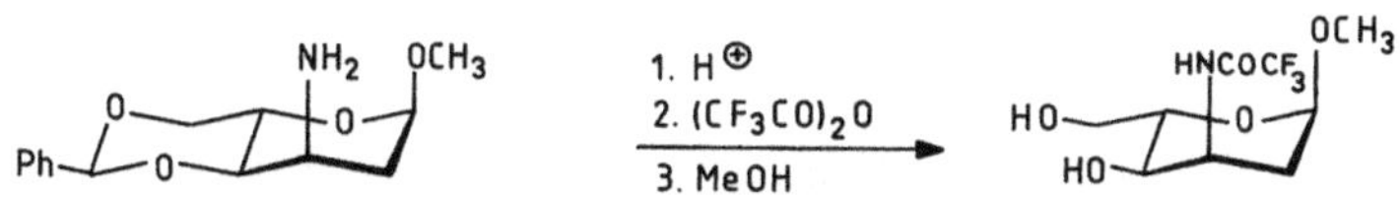

L-157 203

Scheme 37

Horton and associates converted methyl 3-acetamido-4,6-O-benzylidene-2,3-dideoxy-α-D-*ribo-* **(196)** and *arabino* **(197)** hexopyranosides into D-ristosamine **(20)** (175) and D-acosamine **(10)** (174) and their partially protected derivatives, respectively, by means of deoxygenation at position C-6 (Sect. 2.7.2, Table 7 entries 5 and 2). The 6-bromo-6-deoxy sugar **565**, derived form the *ribo* amine **196** (Sect. 2.7.4, Table 11 entry 3) served as the key intermediate in the first preparative synthesis of L-daunosamine **(12)** having a good yield. This synthesis was also done by Horton et al. (172) via the inversion of the configuration at C-5. In modified versions of this method (Table 11 entry 4) an analogous strategy was applied (178, 179) for the production of N-trifluoroacetyl-L-daunosamine **(580)**.

Upon conversion of the terminal hydroxyl group of amine **201** into a deoxy moiety, with concomitant reduction of the formylamino group into an N-methyl function (Sect. 2.7.1, Scheme 106), the Overend group (228) prepared methyl 3-methylamino-2,3,6-trideoxy-α-D-*xylo*-hexopyranoside **(541)**.

As mentioned in Sect. 2.1.4, the O-4 protected derivatives **(102, 104, 106,** and **108)** of methyl 2,6-dideoxy-α-L-*erythro*-hexopyranosid-3-ulose **(109)** can be prepared in near 40% yield by the Klemer-Rodemeyer fragmentation (167) of methyl 2,3-O-benzylidene-α-L-rhamnopyranosides **101, 103, 105**, and **107** (Scheme 20). The O-4 unprotected C-3 ulopyranoside **109**, is available in better yield (60–70%) by means of the chemoselective oxidation (293) of the allylic hydroxyl group of L-rhamnal **(69)** using manganase dioxide (294) or the Fètizon-reagent (188–191) (silver carbonate on Celite) or (even more conveniently) (190, 191) using barium manganate (295) or pyridinium dichromate (296) and the subsequent Michael-type addition of methanol (297) onto the resulting enone **204**. It has become a versatile starting material for the preparation of each of the four stereoisomeric 3-amino-2,3,6-trideoxyhexoses in the L-series (122, 188–191).

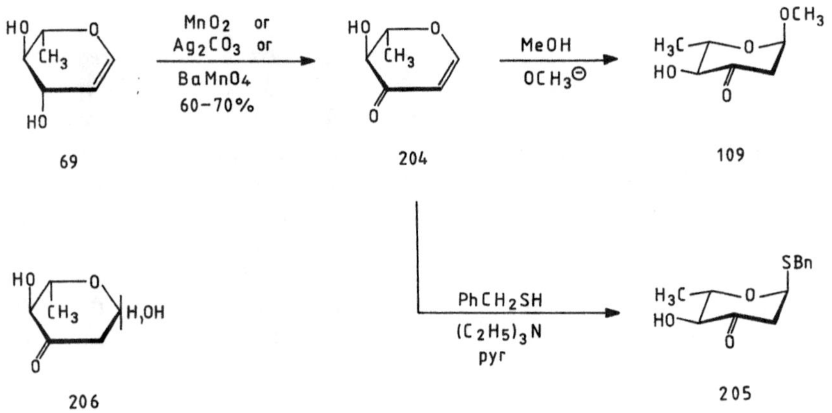

Scheme 38

Although Thiem and Elvers (297) reported the exclusive formation of the α-methyl glycoside **109** upon Michael methanol addition to **204**, our own ^{1}H-NMR studies (188) clearly indicate that ca. 6% of the corresponding β-anomer of **109** is also produced. The Michael addition of phenylmethanethiol onto the enone **204** proceeded with high stereoselectivity to yield benzyl 2,6-dideoxy-1-thio-α-L-*erythro*-hexopyranosid-3-ulose **(205)**, and no trace of the corresponding β-anomer could be detected (189, 190) in the reaction mixture (Scheme 38). In 1983 2,6-dideoxy-L-*erythro*-hexopyranos-3-ulose **(206)** was first found (298) in nature as a component of the antifungal polyene macrolide-type antibiotic candidoin.

Catalytic hydrogenation of oxime **207**, derived from **109**, over Raney nickel produced (188) a 79:14 mixture of methyl N,O-diacetyl-α-L-ristosaminide **(209)** and methyl N,O-diacetyl-α-L-acosaminide **(210)** after acetylation. The products are

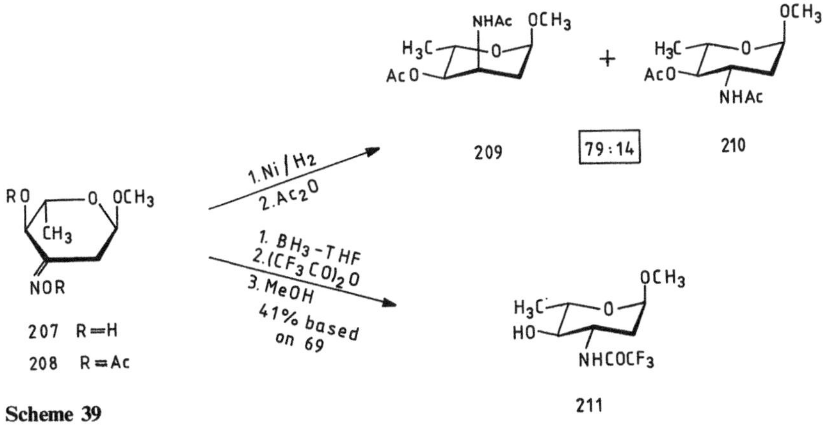

Scheme 39

readily separable by means of column chromatography (Scheme 39). On the other hand, reduction (189–191) of the di-O-acetyl derivative **208** of oxime **207** using the borane-tetrahydrofuran complex followed by trifluoroacetylation of the crude reduction mixture, produced methyl 2,3,6-trideoxy-3-trifluoroacetamido-α-L-*arabino*-hexopyranoside (**211**, methyl N-trifluoroacetyl-α-L-acosaminide) as the exclusively isolable product in 41% overall yield based on L-rhamnal (**69**), and only traces of the corresponding *ribo* isomer could be detected.

Inversion of the configuration of the C-4 hydroxyl group of **211** and that of the N-trifluoroacetyl derivative **597** of **209** (Sects. 2.8.1 and 2.8.3) by the reduction of the intermediary hexopyranosid-4-uloses **676** and **691** then led to the additional two stereoisomers: L-*lyxo* (**580**, Table 15 entry 3) and L-*xylo* (**688**, Table 17 entry 1), respectively (190, 191, 122). Consequently, this method allows an easy access to each of the four stereoisomeric 3-amino-2,3,6-trideoxy-L-hexoses starting with the common "synthon" **109** and with relatively simple operations.

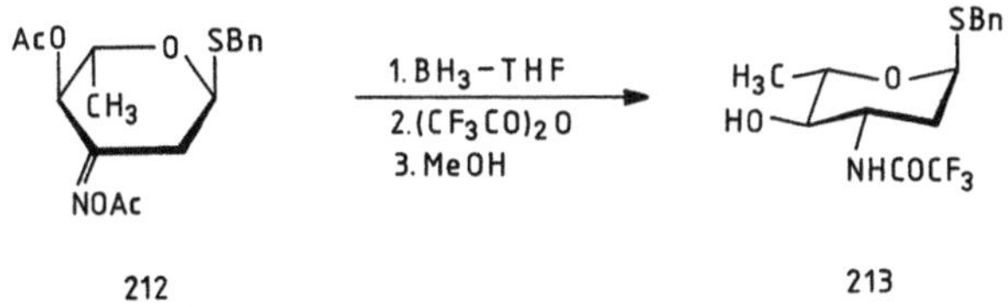

Scheme 40

An essentially similar strategy, involving the reduction of the oxime **212** derived from the benzyl thioglycoside **205** by using the borane–tetrahydrofuran complex, produced (189–191) benzyl 1-thio-2,3,6-trideoxy-3-trifluoroacetamido-α-L-*arabino*-hexopyranoside (**213**), a unique 1-thio analog of L-acosamine, as the sole product (Scheme 40).

In an alternative method, Brimacombe and coworkers (170, 171) applied uloses **106** and **108**, obtained by means of the Klemer-Rodemeyer procedure (167). Catalytic hydrogenation of the oximes **214** and **215**, prepared from **106** and **108**, respectively, over Adams catalyst, followed by acetylation resulted in 78% and 100% of the L-*ribo* aminoglycosides **216** and **217**, respectively (Scheme 41). Acid hydrolysis of

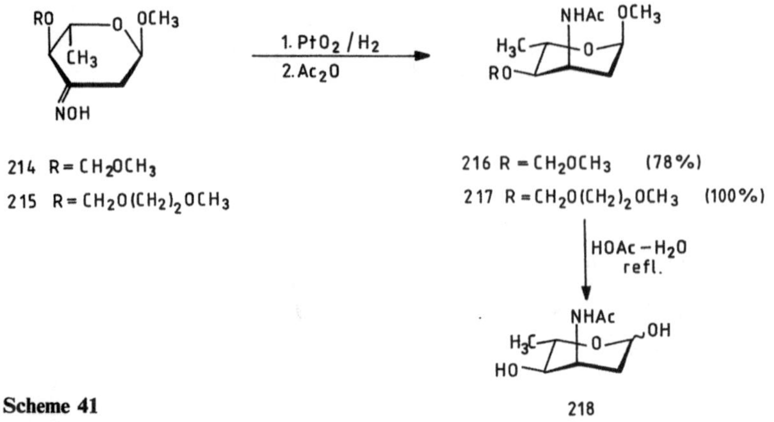

Scheme 41

the protecting groups of either **216** or **217** then gave N-acetyl-L-ristosamine (**218**, 3-acetamido-2,3,6-trideoxy-L-*ribo*-hexopyranose).

2.2.4 Synthesis and transformation of 3-deoxy-3-nitrohexopyranosides into 3-amino-3-deoxyhexose derivatives

During the past twenty-five years, several representatives of the 3-amino-3-deoxyhexoses in both the D- and L-series have been prepared by means of the Baer-Fischer procedure (299) involving the aldol condensation of "sugar-dialdehydes" with nitromethane to produce 3-deoxy-3-nitrohexopyranosides.

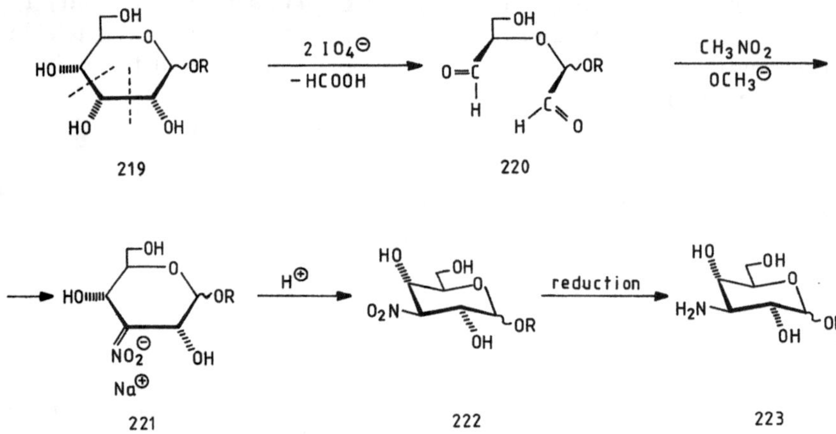

Scheme 42

Upon base-catalyzed cyclization of dialdehydes **220**, readily available by the periodate oxidation of aldohexopyranosides (**219**), with nitromethane, the methyl group of the reagent is incorporated into the pyranose ring. Acidification of the resulting mixture of the 3-*aci*-nitro-3-deoxyhexopyranoside stereoisomers **221** yields a mixture of 3-deoxy-3-nitrohexopyranosides (**222**) in which the thermodynamically more stable products predominate (300, 301). Catalytic hydrogenation (Pt, Pd/C, Ni) of these products then provides the corresponding 3-amino compounds (Scheme 42).

Because the cyclization of dialdehyde **220** produces two new chiral centers and an additional one develops upon acidification, the whole process theoretically involves (299–303) the formation of four stereoisomeric nitronates **(221)** and eight isomeric nitrosugars **(222)**. However, in practice, the Baer-Fischer condensation reaction always exhibits a remarkable stereoselectivity and usually results in only four stereoisomeric nitro compounds. Although a *trans* relationship is observed between the C-1 glycosidic group and the C-2 hydroxyl function of the nitronates initially formed, epimerization of these primary products in the alkaline medium leads to an equilibrium mixture of the C-3 nitronate salts **(221)**. This is clearly demonstrated by the experimental fact that the composition of the reaction mixture generally changes with the reaction time (300).

Extensive studies of the protonation of 3-*aci*-nitrohexopyranosides **221** show a pronounced tendency of the 3-nitro group to occupy an *equatorial* orientation under thermodynamically controlled conditions (300, 304). In the 3-deoxy-3-nitrosugar series, only a single example has been reported (305) of the formation of a small amount of a nitro sugar (methyl 3,6-dideoxy-3-nitro-β-D-idopyranoside) bearing an *axially* disposed nitro group in the Baer-Fischer nitromethane condensation reaction.

The exclusive development of an *equatorial* nitro function in the C-3 nitronate series is explained (300) by the more favored approach of the proton from the *axial* side of the ring because on the *equatorial* side this protonation process is significantly inhibited by the vicinal *axial* C-2 and C-4 bondings. Such steric and electronic factors are unambigously responsible for the development of four nitro compounds **(222)** instead of the theoretically expected eight stereoisomers. On the other hand, recent investigations (306) have demonstrated that the 2-deoxy-2-nitro-hexopyranoside isomers carrying *axial* nitro function are thermodynamically more stable than the respective C-2 epimers having *equatorially* oriented nitro groups. Starting with methyl 6-deoxy-α-D-glucopyranoside **(224)**, Richardson (307) applied this method to obtain methyl 3,6-dideoxy-3-nitro-α-D-glucopyranoside **(225)** as the major product in the nitroglycoside mixture. Reduction of this product gave the C-3 aminoglucoside **167** in ca. 25% yield, based on **224**. Compound **167** was then transformed into micaminose (**226**, 3,6-dideoxy-3-dimethylamino-D-glucose), the

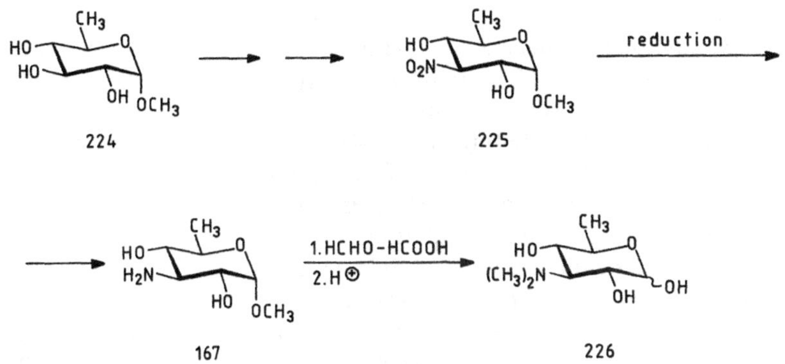

Scheme 43

aminodeoxy sugar component of the antibiotic magnamycin (308), according to standard procedures (Scheme 43). Alternative routes for obtaining derivatives of this latter sugar are shown in Sect. 2.2.2 (Schemes 31 and 32).

Analogous cyclization (309) of L′-methoxy-L-methyldiglycolic aldehyde, arising from methyl α-L-rhamnopyranoside, using nitromethane produced a mixture of four isomeric methyl 3,6-dideoxy-3-nitro-α-L-hexopyranosides with gluco, manno, talo, and galacto configurations in a ratio of 18:8:3:1. Shortening the reaction time changed this ratio to 80:80:23:17. This indicates an epimerization of the nitroglycosides initially formed, as the reaction proceeds under the alkaline conditions, as explained above. Using nitroethane instead of nitromethane makes the Baer-Fischer method a valuable strategy for the synthesis of C-3 branched-chain nitro-, and aminodeoxy hexoses carrying the nitrogen functionality in the *equatorial* orientation (Sect. 2.5.1).

In an additional aspect of the synthetic utilization of nitro sugar derivatives, Baer and his associates (310–317) adopted the Schmidt-Rutz reaction of 1-nitro-1-deoxyalditol acetates (154) for the elimination of the C-2 or C-4 acetoxy functions of acetylated alkyl 3-deoxy-3-nitrohexopyranosides or both. Selective catalytic hydrogenation of the resulting unsaturated 3-nitro-2- or 3-enopyranosides produced alkyl 2,3-dideoxy- or 3,4-dideoxy-3-nitro-hexopyranosides which were found to be versatile starting materials for the production of highly functionalized aminodeoxy sugar analogs.

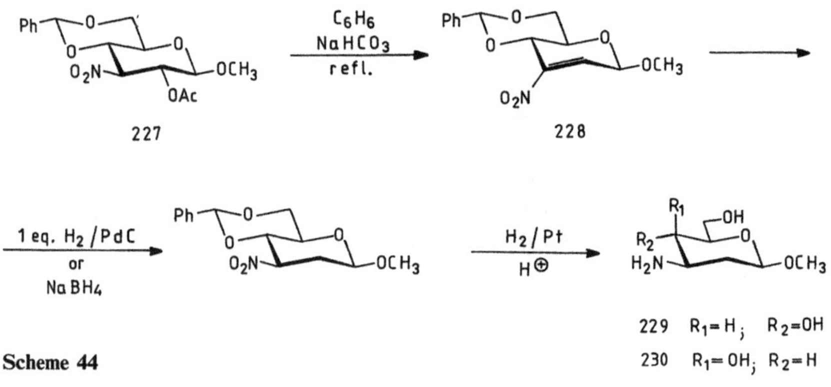

Scheme 44

Methyl 2-O-acetyl-4,6-O-benzylidene-3-deoxy-3-nitro-β-D-glucopyranoside **(227)** was converted in this way into methyl 3-amino-2,3-dideoxy-β-D-*arabino*-hexopyranoside **(229)** (310, 317), and the galacto isomer of **227** afforded (312, 317) the corresponding β-L-*lyxo* steroisomer **(230)** (Scheme 44). Deoxygenation of these two last named sugars (**229** and **230**) at C-6 (Sect. 2.7.1) produced (318) N-acetyl-D-acosamine **(546)** and N-acetyl-D-daunosamine **(545)**, respectively (Scheme 108).

The unsaturated nitro sugar **228** and its additional structural and stereoisomers also readily provide the corresponding methyl 2,3-dideoxy-3-nitro-D-glycopyranosides upon sodium borohydride reduction (319). However, when the reduction is performed with zinc in acetic acid 2,3-dideoxy-3-oximino-hexopyranosides are isolated (319). By the extension of the method used for the preparation of amino glycosides **229** and **230**, Baer and coworkers developed the first synthesis (Scheme 45) of D-ango-

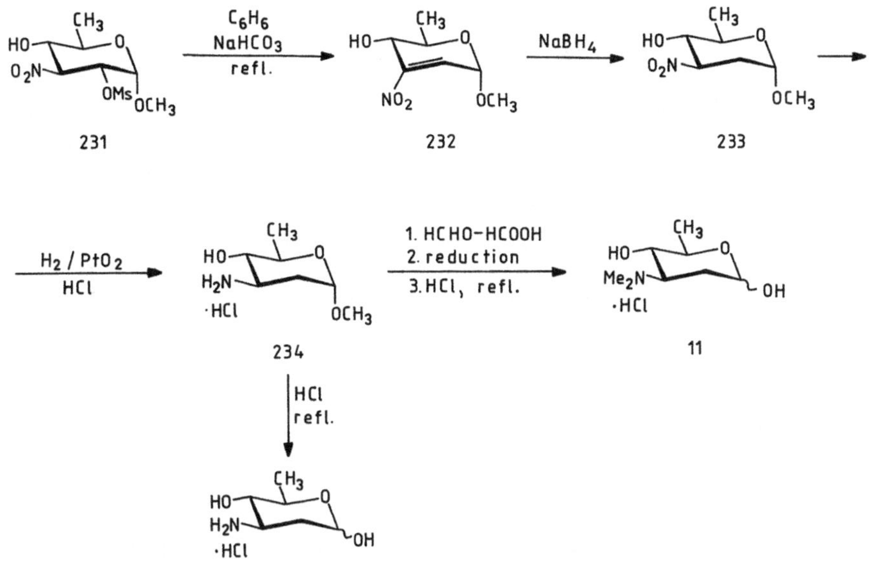

Scheme 45

losamine (**11**, 3-dimethylamino-2,3,6-trideoxy-D-*arabino*-hexopyranose) starting from methyl 3,6-dideoxy-2-O-methanesulfonyl-3-nitro-α-D-glucopyranoside **(231)**. Treatment of this nitro sugar under Schmidt-Rutz conditions resulted in the unsaturated nitro glycoside **232**, which was reduced to **233** using sodium borohydride. This saturated nitro compound **(233)** was alternatively obtained by the direct sodium borohydride reduction of **231** without isolation of the 3-nitro-hex-2-enopyranoside **232** (Scheme 45). Catalytic hydrogenation followed by N-methylation of the resulting

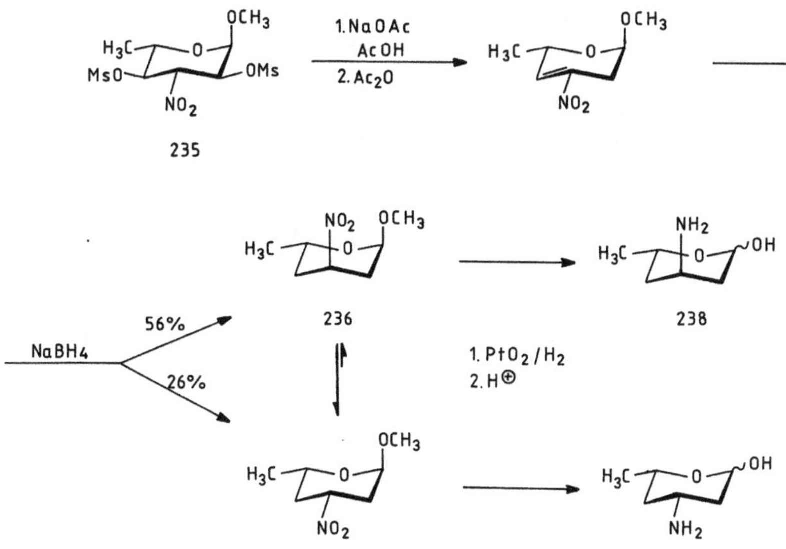

Scheme 46

amino glycoside **234** (methyl α-D-acosaminide) then furnished D-angolosamine hydrochloride **(11)** after removal of the glycosidic function. D-Acosamine hydrochloride **(10)** was also prepared upon mild acid hydrolysis of the glycoside **234**. Based on this experience, these Canadian chemists (315, 316) then performed sequential deoxygenation of methyl 3,6-dideoxy-2,4-di-O-methanesulfonyl-3-nitro-α-D-glucoyranoside **(235)** at C-4 and C-2 for the synthesis of 4-deoxy-L-ristosamine **(238)** and 4-deoxy-L-daunosamine **(239)** (Scheme 46).

It is to be noted that the thermodynamically unstable α-L-*erythro* glycoside **236** could be readily epimerized into the *threo*-nitro compound **237** in neutral to slightly basic medium, and by making preparative use of this process a 71% yield of **237** was obtained along with traces of **236**.

A similar methodology had previously been applied by the Baer group (320) for the synthesis of L-desosamine (3-dimethylamino-3,4,6-trideoxy-L-*xylo*-hexopyranose), the enantiomer of a common aminodeoxy sugar component (D-desosamine, **553**, Scheme 109) of several non-polyene macrolide antibiotics.

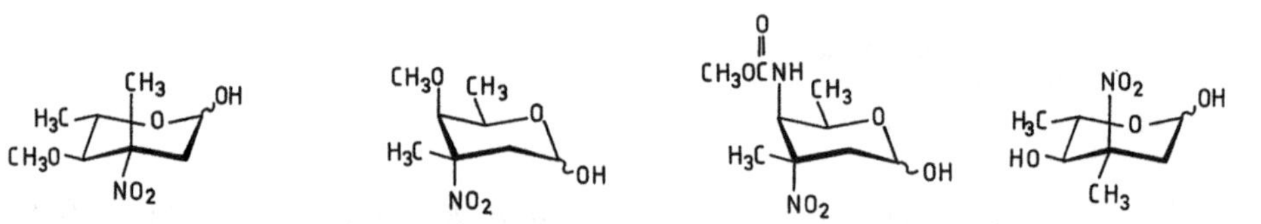

240 241 242 243

Scheme 47

Until now four branched-chain 3-nitro-hexose derivatives have been found in nature as building units of antibiotic substances (Scheme 47). Thus, evernitrose (**240**, 3-C,4-O-dimethyl-3-nitro-2,3,6-trideoxy-L-*arabino*-hexopyranose) (321, 322), the first naturally occurring nitro sugar to be isolated, is a component of the oligosaccharide antibiotics everninomycin B, C, and D. Rubranitrose (**241**, 3-C,4-O-dimethyl-3-nitro-2,3,6-trideoxy-D-*xylo*-hexopyranose) (323, 324) was obtained from rubradirin. Kijanose (tetronitrose) [**242**, 4-(methoxycarbonyl)-amino-3-C-methyl-3-nitro-2,3,4,6-tetradeoxy-D-*xylo*-hexopyranose] (325) is the branched hexose element of the antitumor antibiotics kijanimycin and tetrocarcins A and B, and decilonitrose (**243**, 3-C-methyl-3-nitro-2,3,6-trideoxy-L-*ribo*-hexopyranose) (326, 327). It was isolated from the novel anthracycline antibiotics decilorubicin, arugomycin, and viriplanin[2]. Synthetic aspects of these and related C-3 branched nitro- and amino-sugars are discussed in Sect. 2.5.

2.2.5 Preparation of 3-amino-3-deoxyhexoses from hex-2-enopyranoside (pseudoglycal) derivatives

The Sharpless reaction (328), introduced for the *cis*-oxyamination of unsaturated compounds, has been adopted by Dyong et al. (329–333), Heyns et al. (334), and

[2] Recent studies by the Zeeck group have revealed that viriplanin contains 3-C-methyl-3-nitroso-2,3,6-trideoxy-D-*ribo*-hexopyranose, isolated in form of D-decilonitrose [Zeeck A, Hütter H, Kind R (1987) 4th European Carbohydrate Symposium, 12–17 July, Darmstadt Abstr. C-31].

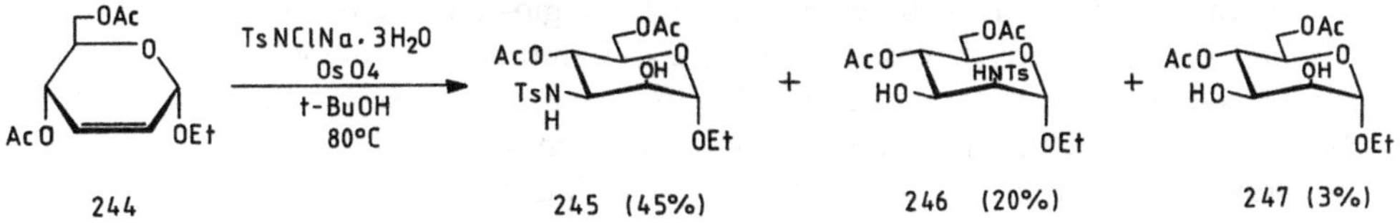

Scheme 48

Zamojski et al. (335) for the synthesis of 3-amino-3-deoxyhexose derivatives from the readily available alkyl hex-2-enopyranosides (84, 336–342). Treatment of ethyl 4,6-di-O-acetyl-2,3-dideoxy-α-D-*erythro*-hex-2-enopyranoside **(244)** with chloramin T in the presence of osmium tetroxide (329, 330) resulted in a 45% yield of the 3-amino-mannopyranoside **245** as the major product accompanied by the glycosides **246** and **247** (Scheme 48). Detailed studies by Zamojski et al. (335) have established that this process is highly regiospecific and stereoselective, always furnishing a high quantity of the *cis*-addition product carrying the amino group at C-3 *trans* to the glycosidic moiety. By making practical use of this reaction in the field of antibiotic aminodeoxy sugars, Dyong et al. converted the tosylamide **245** into 3-amino-3-deoxy-D-mannose (2-epi-kanosamine) (329) by reduction and subsequent acid hydrolysis. In an essentially similar way, methyl 4-O-acetyl-2,3,6-trideoxy-α-D-*erythro*-hex-2-enopyranoside was used (330) for a synthesis of N-acetylmycosamine (3-acetamido-3,6-dideoxy-D-mannose), a common sugar component of macrolide antibiotics. The extension of this method to 3,4-unsaturated componds, such as *trans*-2-benzyloxy-3,6-dihydro-6-methyl-2H-pyran **(248)**, led to (343) a mixture of the *cis*-edducts **249** and **250**, both carrying the nitrogen function at position C-4 (Scheme 49).

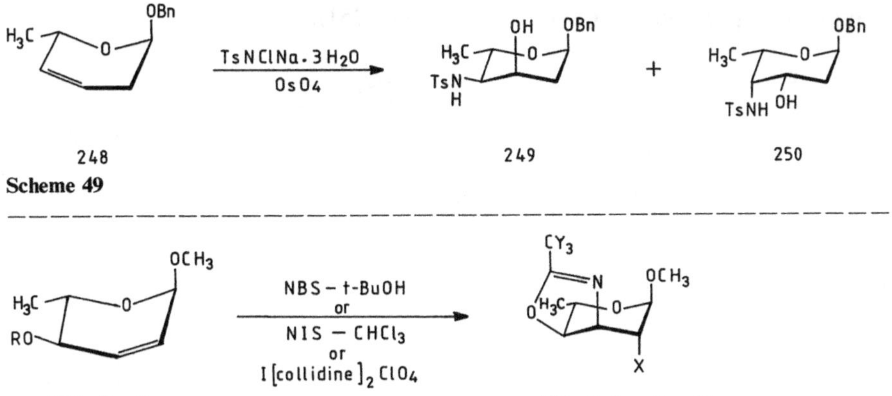

Scheme 49

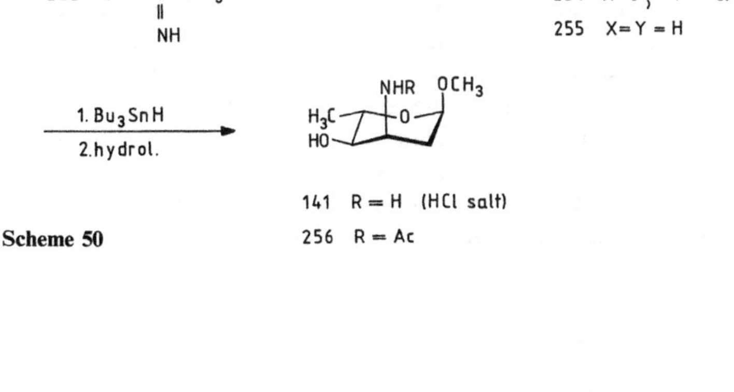

Scheme 50

A simple and efficient method, useful for the regio- and stereoselective synthesis of 3-amino-2,3,6-trideoxyhexoses bearing the C-3 amino- and C-4 hydroxyl functions in a *cis*-relationship (ristosamine and daunosamine) has recently been worked out (344–349, 351) by utilizing the allylic hydroxyl group of methyl 2,3,6-trideoxy-hex-2-enopyranosides in a novel *cis*-oxyamination reaction. Bromonium or iodonium ion-induced cyclization of the trichloroacetimidate ester **252** of methyl 2,3,6-trideoxy-α-L-*erythro*-hex-2-enopyranoside **(251)** either with iodonium dicollidine perchlorate (344, 349), N-bromosuccinimide (345), or N-iodosuccinimide (345, 347) resulted in an almost quantitative yield of the corresponding 2-bromo- **(253)** or 2-iodo-oxazoline **(254)** derivatives, respectively (Scheme 50). Reductive dehalogenation of these last named compounds using 1.5 equivalents (345, 349) of tributyltin hydride followed by hydrolysis, produced methyl α-L-ristosaminide **(141)** (345). Racemic **141** has been synthesized (347) according to an essentially similar procedure. Exhaustive dehalogenation of the iodo-oxazoline **254** with a large excess (4.5–5 equivalents) of the reducing agent smoothly furnished (344–349) methyl N-acetyl-α-L-ristosaminide **(256)** via the acid hydrolysis of an intermediary dehalogenated dihydro-oxazole **255**.

A crucial question in the adoption of this procedure for the synthesis of daunosamine (i.e., L-*lyxo*) derivatives is the efficient preparation of the C-4 epimeric α-L-*threo* isomer **(257)** of the *erythro*-hex-2-enopyranoside **251**. Because no selective reduction of the corresponding C-4 pyranulose into the *threo* 4-hydroxy compound **257** has been developed, this problem was solved by the epimerization at C-4 of the readily available *erythro* isomer **251** either by (346, 347, 349) the Mitsunobu reaction (350) or by the nucleophilic displacement of the respective 4-O-methanesulfonate ester of **251** using acetate ion on a polymeric support (65% yield) (348), an excess of anion-exchange resin in the CO_3^{2-} form (80% yield), (348) or (even more effectively) cesium propionate in DMF (99% yield) (351).

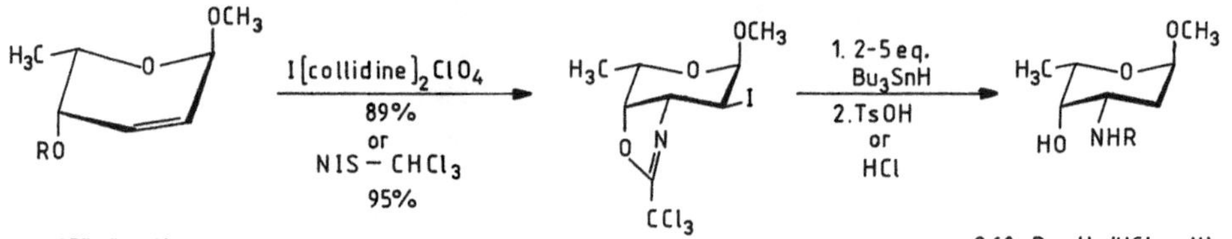

257 R=H

258 R = –C(=NH)–CCl3

259

260 R = H (HCl salt)

261 R = Ac

Scheme 51

Cyclization of the imidate **258** derived from **257** with iodonium dicollidine perchlorate (346, 349) or with N-iodosuccinimide (348) and a subsequent reductive dehalogenation – hydrolysis strategy led to methyl α-L-daunosaminide **(260)** (348) and methyl N-acetyl-α-L-daunosaminide **(261)** (346, 349) (Scheme 51). The outcome of the reduction step is dependent upon the quantity of the reducing agent (Bu_3SnH), as shown. Cyclization of the imidate **258** with iodonium dicollidine perchlorate proceeded much more slowly than that of the C-4 epimeric **252**, presumably because the *syn* approach of the iodonium ion to the double bond is sterically hindered by the anomeric OCH_3 group (349). Nevertheless, this procedure appears to be one of the most effective routes to L-daunosamine derivatives with an overall yield of 55%, based on the *erythro*-hex-2-enopyranoside **251**. Essentially the same procedure, performed (351) using the benzyl glycoside analog of **257**, produced benzyl 3-acetamido-2,3,6-trideoxy-α-L-*lyxo*-hexopyranoside (benzyl N-acetyl-α-L-daunosaminide). In an alternative to this process (Scheme 52) the generation of an imidate with the α-D,L-*threo* configuration (±**262**) was carried out directly on the racemic *erythro* glycoside ±**251** (cf. Sect. 3.1, Scheme 129). Treatment of this imidate with triphenylphosphine, diethyl azodicarboxylate, and dibenzoylimine (modified Mitsunobu conditions) produced a 93% yield of the imidate ±**262**. The reaction of this latter imidate with N-bromosuccinimide in the presence of ethanol and in situ hydrolysis of the intermediary orthoamide ±**263** produced under acidic conditions, gave methyl 2-bromo-N,O-dibenzoyl-α-D,L-daunosaminide (±**264**). Removal of the bromine substituent upon tributyltin hydride reduction followed by standard alkaline and acid hydrolysis (for complete deprotection) afforded racemic daunosamine (±**12**) with an overall yield higher than 30%.

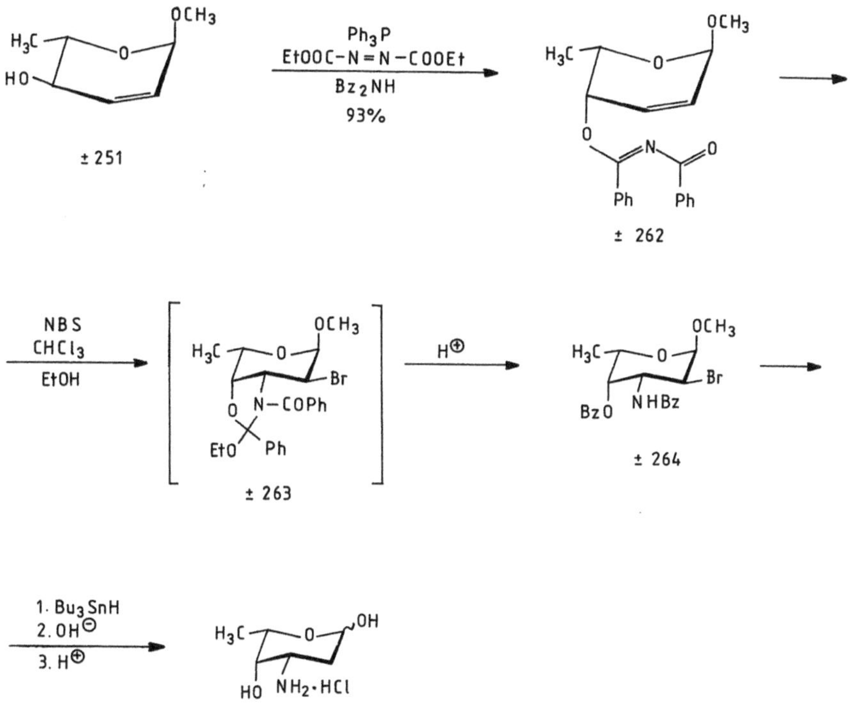

Scheme 52

This simple and high-yielding route to 3-amino-2,3,6-trideoxy-L-*ribo*- and L-*lyxo*-hexopyranoses and their racemic forms, carrying *cis*-related amino- and hydroxyl functionalities, has recently been extended (352) to the preparation of the 3-C-methyl-branched 3-amino- and 3-nitro analogs of the corresponding stereoisomers in the D-*ribo* series, thus providing an easy access to rubranitrose and its derivatives. The details of this method, as well as of that utilizing a thermal rearrangement

(353, 354) of other carbohydrate trichloroacetimidates for the synthesis of structurally related branched amino sugar derivatives, are discussed in Sect. 2.5.5.

Palladium-catalyzed allylic amination of 2,3-unsaturated hexopyranoside acetates, proceeding with high regio- and stereoselectivity and affording 4-amino-4-deoxy-hex-2-enopyranosides has been utilized by the Baer group (355–357) for the preparation of rare 4-amino-polydexyhexoses, such as forosamine [2,3,4,6-tetradeoxy-4-(dimethylamino)-D-*erythro*-hexopyranose], present in the macrolide-type antibiotic spiramycin (357).

The addition of toluene-*p*-sulfonamide to olefins in the presence of anhydrous mercury(II)nitrate, followed by sodium borohydride reduction (sulfonamidomercuration and demercuration) has been reported (358) to produce the corresponding N-alkylsulfonamides in high yields. Adaption of this procedure to prepare hex-2-enopyranoside derivatives seems to offer an additional simple route to 2- or 3-amino-2,3-dideoxyhexopyranosides.

2.2.6 Synthesis of 3-amino-3-deoxyhexose derivatives by means of the allylic azide rearrangement reaction

As discussed in Sect. 2.1.1 (Scheme 2), the Lewis acid-catalyzed addition of the azide ion of "borderline base" character (79) to the double bond of glycal esters

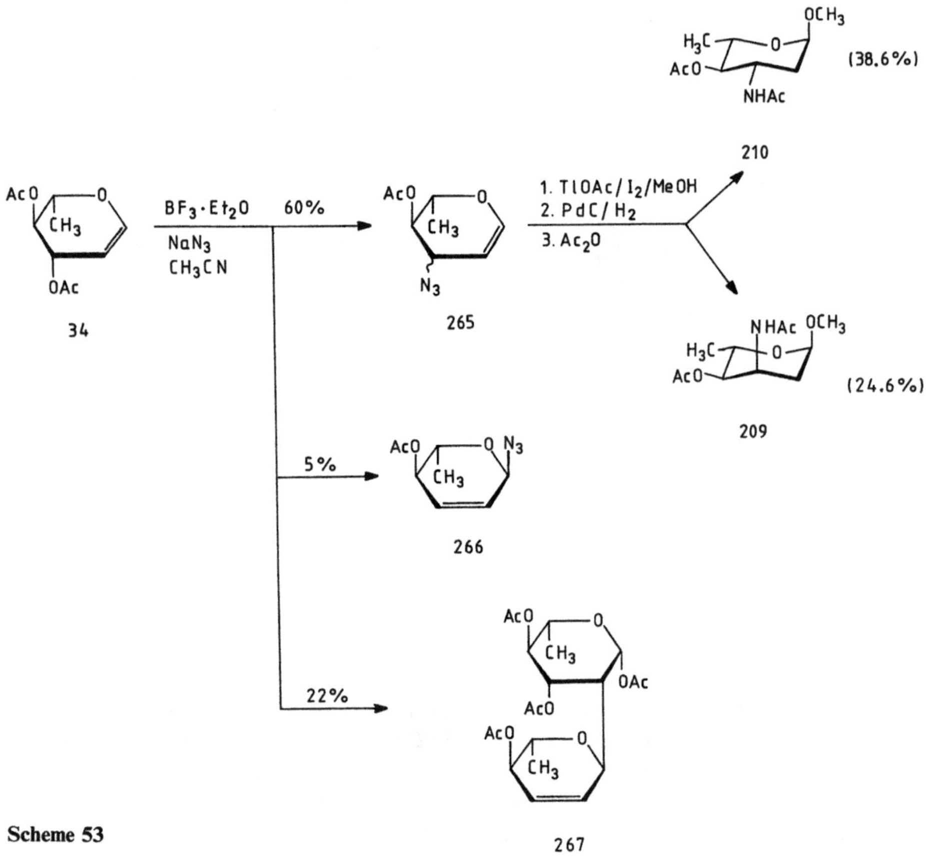

Scheme 53

(23) may take place concurrently on either carbon C-1 or C-3 of an intermediary allyoxocarbenium ion **(27)** and results in an equilibrium mixture of 2,3-unsaturated glycosyl azides (**28**; Nu=N_3) and 3-azidoglycals (**29**; Nu=N_3). This process was extensively investigated by Heyns (77, 359–761), Monneret (362, 363), Guthrie (79, 80, 364), and Thiem (351, 365) and their associates during the last decade.

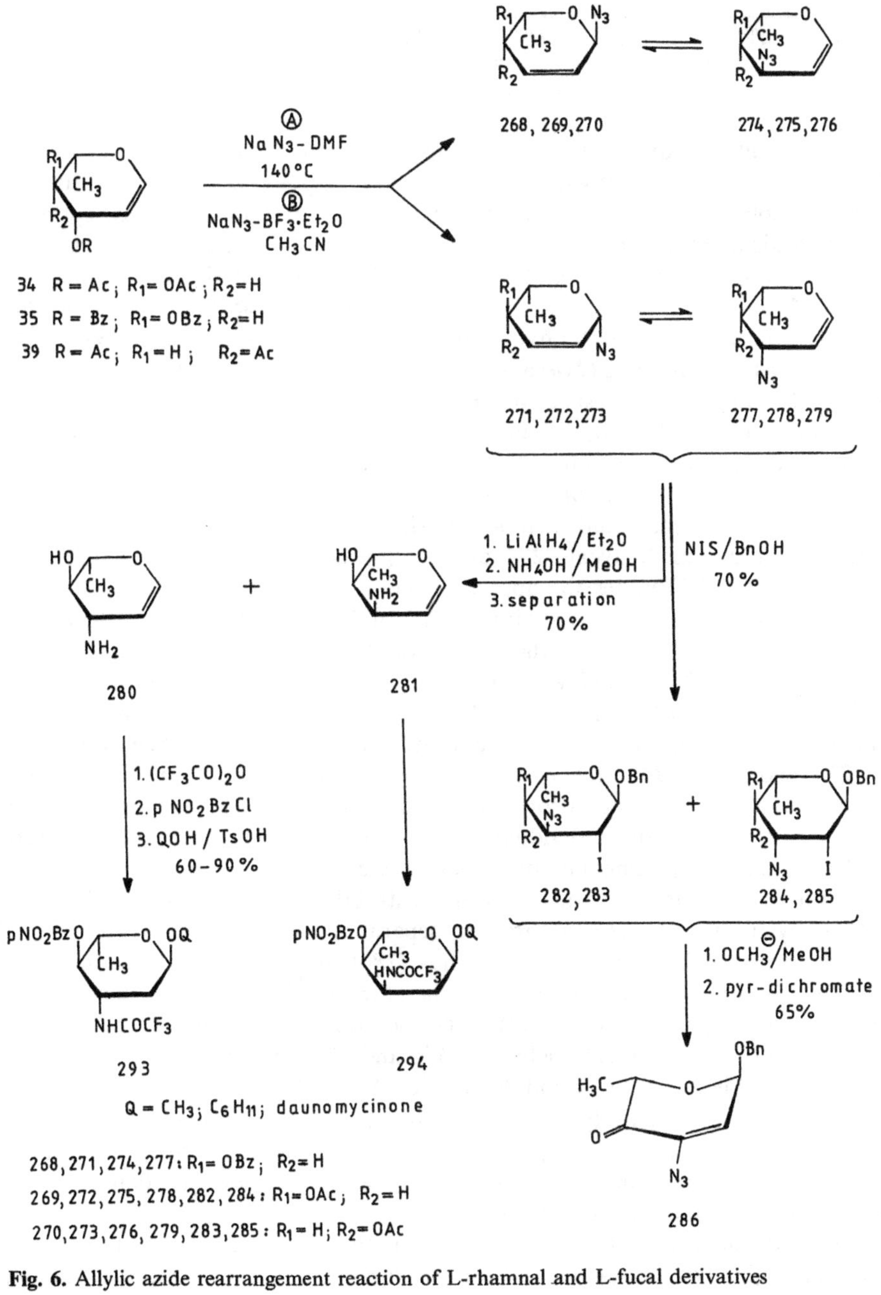

Fig. 6. Allylic azide rearrangement reaction of L-rhamnal and L-fucal derivatives

Table 4. Composition of the equilibrium mixtures obtained from the reaction of the glycal **35** with sodium azide

Product	**271**	**277**	**268**	**274**
Method A	6%	38%	16%	40%
	44%		56%	
Method B	4%	25%	20%	51%
	30%		70%	

The first results in this field were reported by Heyns et al. (359), who treated 3,4-di-O-acetyl-L-rhamnal **(34)** with sodium azide in the presence of boron trifluoride etherate and isolated, as the major product, a 60% yield of a mixture of the epimeric C-3 azidoglycals **(265)** together with a 5% yield of 1-azido-4-O-acetyl-2,3,6-trideoxy-α-L-*erythro* hexopyranose **(266)** and a 22% yield of the dimerization product C-glycoside **267**. Iodomethoxylation of the azidoglycal mixture **265** with thallous acetate and iodine, followed by catalytic hydrogenation and acetylation gave (359) methyl N,O-diacetyl-α-L-acosaminide **(210)** and methyl N,O-diacetyl-α-L-ristosaminide **(209)** in yields 38.6% and 24.6%, respectively (Scheme 53).

The composition of the mixture **265** of azido compounds produced in the [3.3]-sigmatropic rearrangement reaction has been thoroughly investigated by Monneret et al. (362, 363). By using 3,4-di-O-benzoyl-L-rhamnal **(35)**, the formation of each of the four possible regio- and stereo-isomeric addition products (glycosyl azides **268** and **271** and 3-azidoglycals **274** and **277**) was detected in the equilibrium mixture (cf. Sect. 2.1.1) under the applied reaction conditions (Methods A and B; Fig. 6). Thus, upon addition of sodium azide on **35** in DMF (Method A) and in the presence of boron trifluoride etherate (Method B), the ratios of the equilibrated products **271** + **277**:**268** + **274** were found to be 9:11 and 3:7, respectively. The azidoglycals **274** and **277** can be readily separated by HPLC, but they quickly rearrange (in the absence of sodium azide) into the respective equilibrium mixtures **268** ⇌ **274** and **271** ⇌ **277** (for equilibrium values see Table 4) indicating that isomerization of allylic azides proceeds via an intramolecular rearrangement. In contrast, conversion of the individual products in the original reaction mixture called for the initial experimental conditions of Methods A or B.

Reduction of the four-component mixtures with lithium aluminum hydride led — after O-deacylation and chromatographic separation — to a mixture (70%) of 3-amino-1,5-anhydro-2,3,6-trideoxy-L-*arabino*- **(280)** and -L-*ribo*-hex-1-enitol **(281)** with an equilibrium ratio corresponding to the aforementioned product-distribution values. Thus, aminoglycals **280** and **281**, corresponding to L-acosamine and L-ristosamine, were obtained in overall yields of 34% and 39%, respectively, by means of Method A and yields of 18% and 52% with Method B. No formation of the corresponding unsaturated glycosylamines (reportedly very unstable) (366) was observed. This was apparently explained (362) by the mechanism of the lithium aluminum hydride reduction of 2,3-unsaturated glycosyl azides, as demonstrated by the example of compound **271** (Scheme 54).

According to this mechanism the carbon-substituted nitrogen atom of the azido function reacts (367) as the specific site for electrophilic attack by lithium aluminum

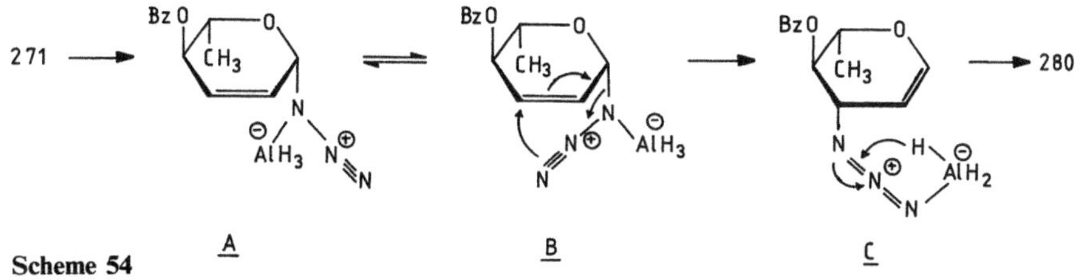

Scheme 54

hydride. This gives a highly ordered transition state *A* that is closely related to those previously described by Fraser-Reid et al. (368) to explain the formation of 3-deoxyglycals via a reductive rearrangement of 2,3-unsaturated methyl pyranosides. In the present case, however, instead of the delivery of the hydride ion in *A* to C-3 (which would lead to a 3-deoxyglycal upon the simultaneous breaking of the extremely unstable C_1—N bond), rotation around this bond can take place to yield intermediate *B*. Thus, intramolecular nitrogen attack at C-3 of *B* combined with the migration of the double bond, provides *C*, which is then reduced into a 3-aminoglycal, such as **280**.

Following the same azide-addition sequence (Method B), it was recognized (363) that methyl 4-O-acetyl-2,3,6-trideoxy-α-D-*erythro*-hex-2-enopyranoside provides exactly the same addition products as 3,4-di-O-acetyl-6-deoxy-D-glucal. The former compound and its *threo* analog (obtained by C-4 epimerization) were used to synthesize 1,5-anhydro-3-amino-2,3,6-trideoxy-hex-1-enitols in the D-*arabino*-, D-*ribo*-, and D-*xylo* series. Note, however, that the D-*lyxo*-compound, which corresponds to the glycal of D-daunosamine, was not detected in the reaction mixture arising from the α-D-*threo* isomeric hex-2-enopyranoside (vide infra). The allylic azide rearrangement reactions of 3,4-di-O-acetyl-L-rhamnal **(34)** and 3,4-di-O-acetyl-L-fucal **(39)** were studied in detail by Thiem and Springer (351, 365) along with the N-iodosuccinimide-benzyl alcohol glycosylation procedure (103). Treatment of **34** with sodium azide in the presence of boron trifluoride etherate (Method B) gave the α- **(269)** and β-L-*erythro*- **(272)** glycosyl azides, which equilibrated with the 3-azido-L-*ribo*- **(275)** and L-arabino-glycal **(278)**, respectively (Fig. 6). Treatment of this equilibrium mixture with N-iodosuccinimide and benzyl alcohol produced a 70% yield of a mixture of the 2-iodo-3-azido-α-L-altro- **(282)** and manno- **(284)** benzyl glycosides in a ratio of 4:1.

A similar procedure, performed with 3,4-di-O-acetyl-L-fucal **(39)**, resulted (351, 365) exclusively in benzyl 4-O-acetyl-3-azido-2,3,6-trideoxy-2-iodo-α-L-idopyranoside **(283)**; the corresponding α-L-talo-C-3 epimeric compound **285** was not detected. This stereoselective formation of **283** suggested the exclusive preliminary production of glycosyl azide **270** and thus the absence of **273** or **279** or both.

Both electronic control involving the participation of the vicinal 4-O-acetyl group to give an acetoxonium ion and preferential *axial* attack on the allylic system have been evoked to explain (369) the different behaviors of **34** and **39** and the corresponding hex-2-enopyranosides. In **34** an *axial* attack by N_3-BF_3*(a)* produces a situation *(D)* where the C_1^+—N_3—BF_3^- and C_3—O dipoles are parallel and are aligned in the same direction, whereas an attack from the β-face — although unfavourable — results in the formation of dipoles which are aligned in opposite directions and decrease the energy of the reaction. On the other hand, in **39**, a

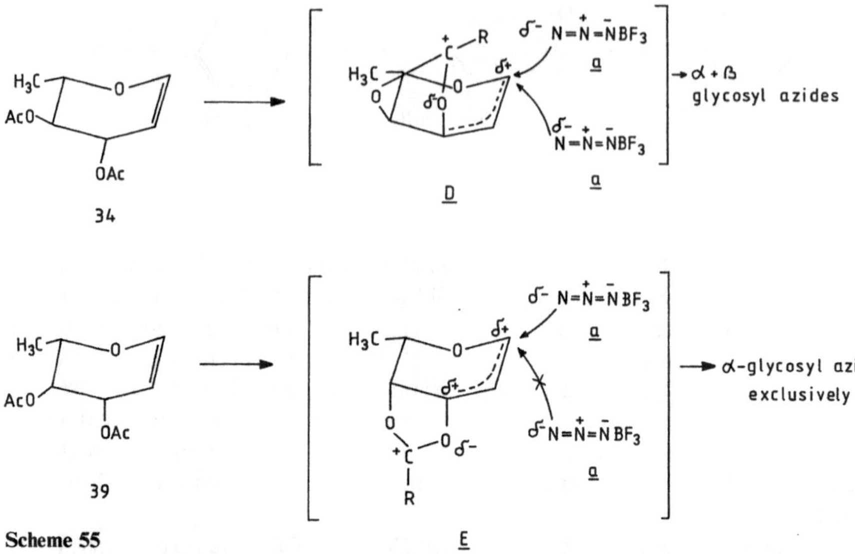

Scheme 55

pseudo-axial attack by $N_3—BF_3$ leads to an intermediate (*E*) in which the dipoles are parallel and aligned in opposite directions, and the reaction affords exclusively α-glycosyl azides (369) (Scheme 55).

With a goal of obtaining L-daunosamine from **284**, a sequence involving O-deacylation, inversion of the configuration at C-4 through an intermediary ulose, and subsequent reduction was attempted (351, 365). Oxidation of the C-4 hydroxyl group of either the O-deacylated derivative of **284**, or the mixture of this latter with the deacylated analog of **282** with pyridinium dichromate produced, however, an unexpected product, benzyl 3-azido-2,3,6-trideoxy-α-L-*glycero*-hex-2-enopyranosid-4-ulose **(286)** in 65% yield (Fig. 6). Sodium borohydride reduction of the carbonyl group of **286** at ambient temperature gave a separable 7:2 mixture of the *erythro*- **(287)** and *threo*- **(288)** azides (or stereospecifically **287** at —20 °C) (Scheme 56). The azidoglycoside **257** with the *erythro* configuration were completely converted into the *threo* compound **288** via a sequence involving mesylation at O-4 followed

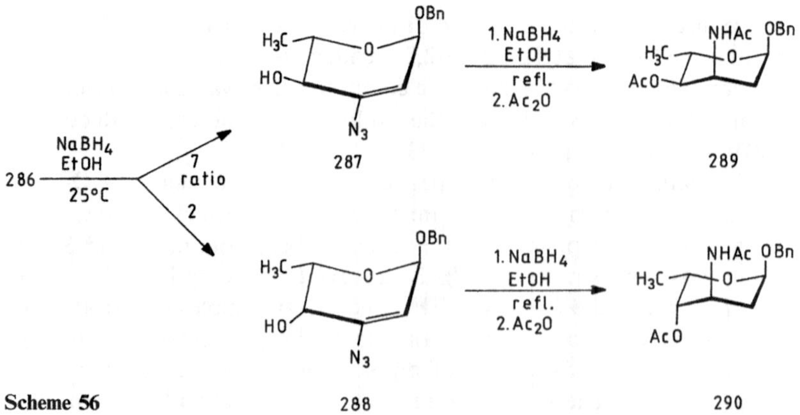

Scheme 56

by treatment with cesium propionate and subsequent deacylation. Compounds **287** and **288** were then subjected to further reduction and acetylation to obtain benzyl N,O-diacetyl-α-L-ristosaminide **(289)** and its α-L-*xylo* isomer **(290)**, respectively; both carry the C-3 amino function in the *axial* position. The failure to obtain the desired amino glycosides with an *equatorial* amino group (the *lyxo* and *arabino* isomers) has been explained by the addition of hydrogen from the less-hindered lower (β) side of the ring of the respective C-3 imine derivatives that are presumably involved as intermediates in the reduction process. Allylic substitution in tri-O-acetyl- and tri-O-benzyl-D-glycals and related compounds with sodium azide in acetonitrile under boron trifluoride catalysis have brought results (79, 80, 360) similar to those described above.

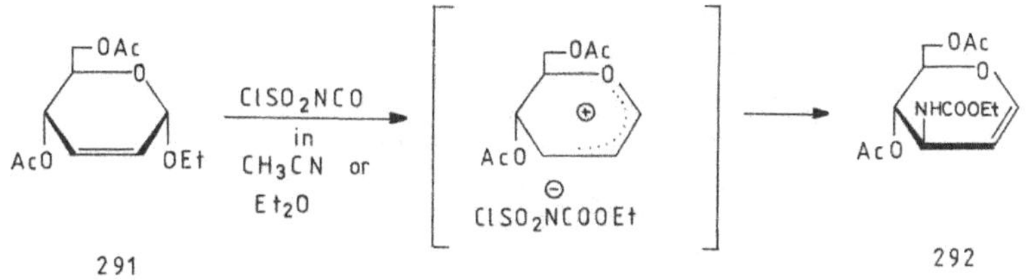

291 292

Scheme 57

In a closely related allylic displacement reaction, Hall et al. (370, 371) obtained the C-3 ethoxycarbonylamino glycal **292** as the major product upon treatment of the hex-2-enopyranoside **291** with chlorosulfonyl isocyanate (Scheme 57).

The aforementioned C-3 aminoglycals have recently emerged with great synthetic value in strategies to prepare semisynthetic antibiotic analogs, primarily in the field of aminocyclitol and anthracycline glycoside antibiotics. This is apparently because of the advantage that glycosylation with such protected 3-aminoglycals proceeds with high stereoselectivity to yield only or nearly only α-glycosides. In contrast, glycosylation reactions using 3-amino-2,3-dideoxy- and 3-amino-2,3,6-trideoxy-hexopyranosyl halides, either under Koenigs-Knorr conditions or in the presence of silver salts that are soluble in organic solvents, generally yield mixtures of the corresponding α- and β-glycosides, particularly in the case of bulky aglycones. In addition, being the acylated (N-acetyl-, N-trifluoroacetyl-, N-ethoxycarbonyl-, etc.) amino function is a rather sluggish leaving group, allylic rearrangement — frequently observed with glycal esters, and resulting in "pseudoglycal" derivatives (cf. Sect. 2.2.1) — does not take place in glycosylations with protected 3-aminoglycals (see also Sect. 2.6.1).

To illustrate these advantages through an example taken from our results (362, 363), the aminoglycals **280** and **281** were readily converted, with a 60–90% yield, into protected derivatives of methyl and cyclohexyl α-L-acosaminide (**293**; Q=CH_3 or C_6H_{11}) and α-L-ristosaminide (**294**; Q=CH_3 or C_6H_{11}), respectively, by means of the acid-catalyzed addition of methanol or cyclohexanol (Fig. 6). The glycosylation reactions between 3-aminoglycals and simple alcohols had been studied earlier by Japanese (99–101, 372, 373) and English (374) researchers. These investigations were intended primarily to obtain methyl glycosides of 3-amino-2,3-dideoxy- and 3-amino-2,3,6-trideoxy-hexoses via deoxygenation at C-2 of protected 3-amino-3-deoxy- and

3-amino-3,6-dideoxy sugars by means of the classical Fischer's glycal synthesis methodology and the subsequent addition of methanol to the resulting 3-amino-glycal. This alternative method for the generation of the 2-deoxy function of hexoses already bearing the C-3 amino moiety is discussed in Sect. 2.6.1.

The first example of the glycosylation of a complicated and bulky alcohol-type aglycone using a C-3 amino-glycal was reported by Daniels et al. (375) in 1973, although Lourens and Jordaan (376) had used glycal **292** in the synthesis of 3′-amino-2′-deoxyglucopyranosyl-purine nucleoside analogs a year earlier. Daniels et al. (375) treated tetra-N-benzyloxycarbonylgentamine-C_1 **(295)** with 3-acetamido-4,6-di-O-acetyl-1,2,3-trideoxy-D-*arabino*-hex-1-enopyranose **(296)** in the presence of *p*-toluenesulfonic acid catalyst to prepare 3-amino-2,3-dideoxy-α-D-*arabino*-hexopyranosyl-gentamine-C_1 **(297)** after removal of the blocking groups (Scheme 58).

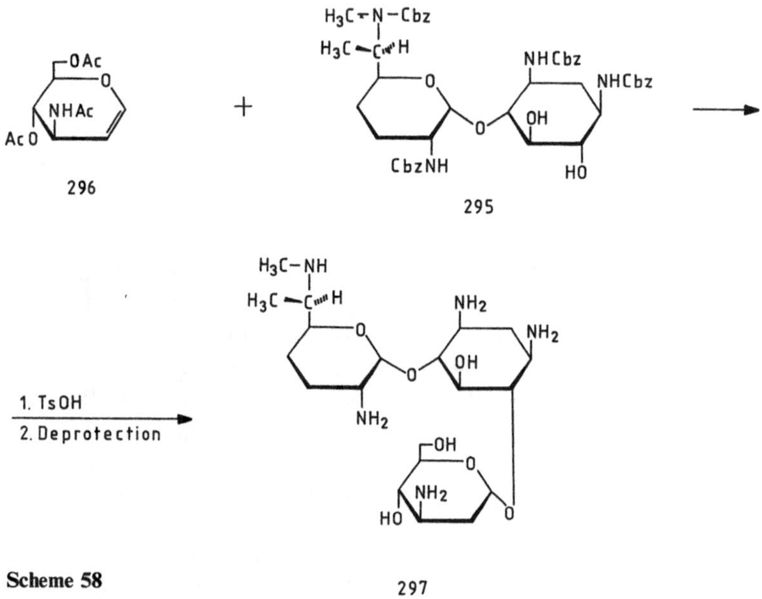

Scheme 58

A similar glycosylation reaction performed by Reid et al. (377) with 3′,4′-di-O-*p*-nitrobenzoyl-1,2′,3,6′-tetra-N-trifluoroacetylneamine and the glycal **296** resulted in an additional novel semisynthetic aminocyclitol antibiotic analog: 2″-deoxykanamycin B. In both cases, the glycosylation reaction using the aminoglucal **296** proceeded with high stereoselectivity and gave exclusively α-glycosides.

To provide similar stereoselective α-glycosylation reactions in the anthracycline antibiotic series, several methods for the production of 3-aminoglycal derivatives directly from 3-amino-2,3-dideoxy- and 3-amino-2,3,6-trideoxyhexoses **(298–301)** have been worked out by the groups of Arcamone (235, 378), Horton (176, 379), and Lukacs (380) (Fig. 7). Acid-catalyzed glycosylations (176, 235, 362, 363, 378–380) of daunomycinone with the protected C-3 aminoglycals **302–305** led smoothly to the corresponding daunomycinone-glycosides, exclusively or almost exclusively with an α-glycosidic linkage. The only exception observed so far (363) was the reaction of 1,5-anhydro-4-O-*p*-nitrobenzoyl-2,3,6-trideoxy-3-trifluoroacet-

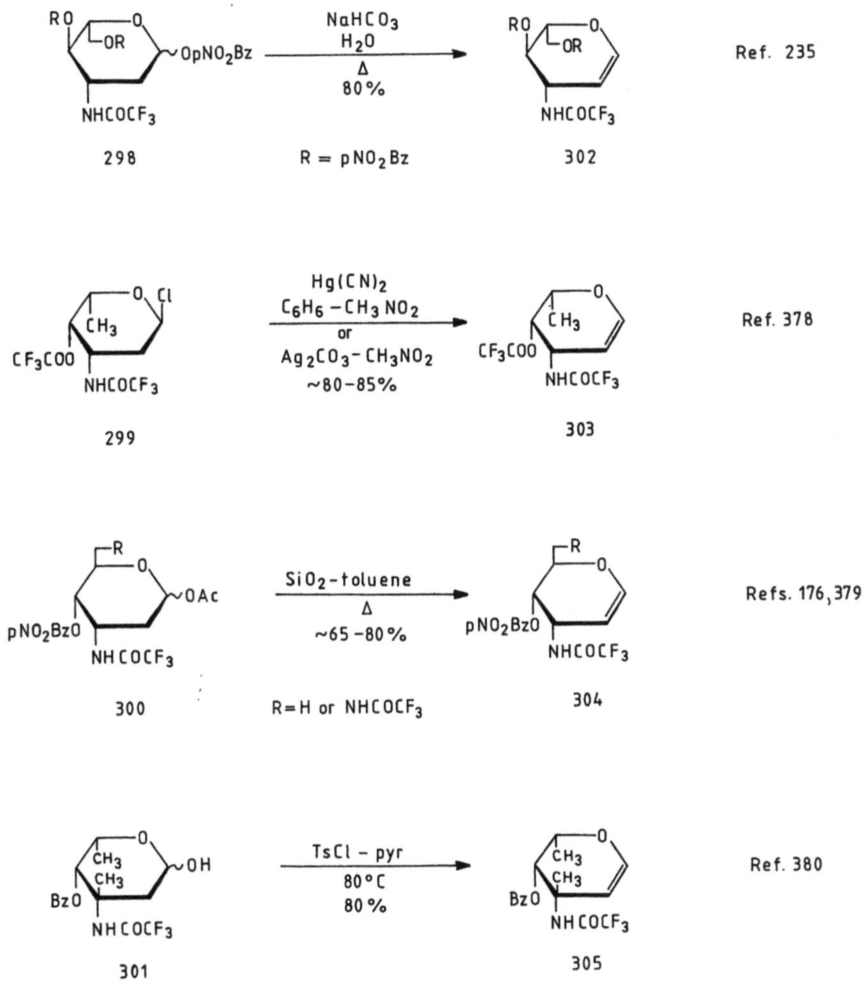

Fig. 7. Synthesis of 3-aminoglycal derivatives from 2-deoxyhexose analogs

amido-D-*arabino*-hex-1-enitol with daunomycinone, which gave a 59:41 mixture of the respective α- and β-anthracyclinone glycoside analogs. A fascinating extension of the allylic azide rearrangement reaction, involving a 2-methyl-glycal derivative and used for the total synthesis of N-acetylactinobolamine, has recently been reported by the Danishefsky group (381). The substituted 3-aminoglycal derivatives, hitherto prepared by the allylic azide rearrangement reaction or other procedures, are included in Table 21 (see Appendix).

2.2.7 Miscellaneous methods

"Sugar dialdehydes", which are readily accessible by means of the periodate oxidation of hexopyranosides (cf. Sect. 2.2.4), are suitable targets for cyclization

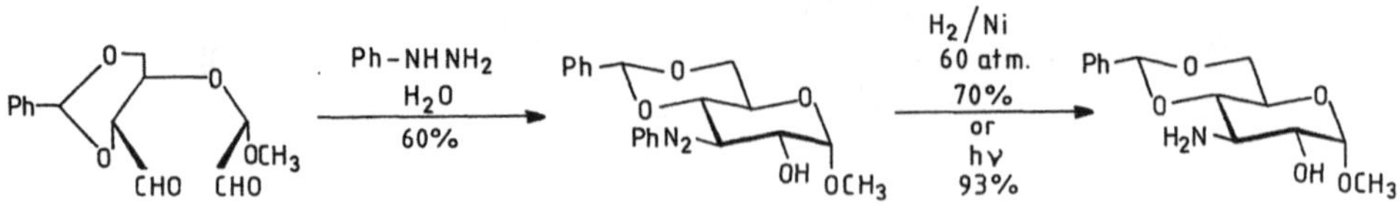

306 307 308

Scheme 59

using phenylhydrazine to obtain phenylazo-hexopyranoside derivatives. Guthrie and Johnson (382) used this procedure to transfom dialdehyde **306**, made from methyl 4,6-O-benzylidene-α-D-glucopyranoside, into methyl 4,6-O-benzylidene-3-deoxy-3-phenylazo-α-D-glucopyranoside **(307)** in 60% yield (Scheme 59). Hydrogenation of this product over Raney-nickel catalyst and under pressure then produced methyl 3-amino-4,6-O-benzylidene-3-deoxy-α-D-glucopyranoside **(308)**. Alternatively, the photolysis of the phenylazo-glucoside **307** with benzophenone as sensitizer also produced (383) a 93% yield of the aminoglucoside **308** and did so in a more convenient and effective way. The amine **308** was used by Richardson (384) as the key intermediate in the first synthesis of D-desosamine **(553**, 3-dimethylamino-3,4,6-trideoxy-D-*xylo*-hexose, see Sect. 2.7.1, Scheme 109), the naturally occurring amino-deoxy hexose component of several macrolide antibiotics.

Patroni and Stick (383) used the Barton deoxygenation procedure (206) (cf. Sect. 2.1.5) to convert the N-acetyl derivative **(529)** of **308** into the corresponding 2-deoxy analog **(197)**, which is an important and popular material for the syntheses of various highly functionalized aminodeoxy sugar derivatives. The details of this method are given in Sect. 2.6.2 (Scheme 97).

The nucleophilic *trans* opening of the 1,6-anhydrogalactobenzoxonium ion **310** (derived from the 3,4-O-benzylidene acetal **309**) by the vicinal *trans* benzimidate group as an intramolecular nucleophile leads to the formation of the N-methyloxa-zolium ion **311** (Scheme 60). The subsequent hydrolysis of this latter substance using

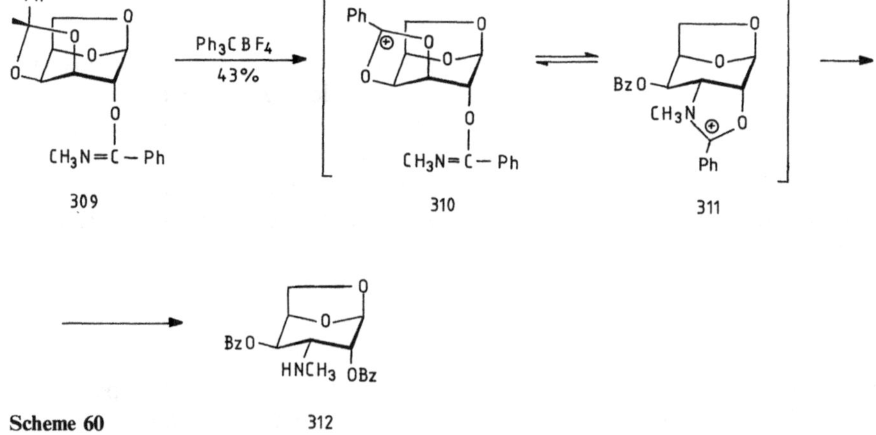

309 310 311

312

Scheme 60

hydrogen fluoroborate readily produced (385) a 43% overall yield of 1,6-anhydro-4-O-benzoyl-3-deoxy-3-methylamino-β-D-gulopyranoside **(312)**, a potential candidate for further transformations into rare 3-amino-3-deoxy sugars.

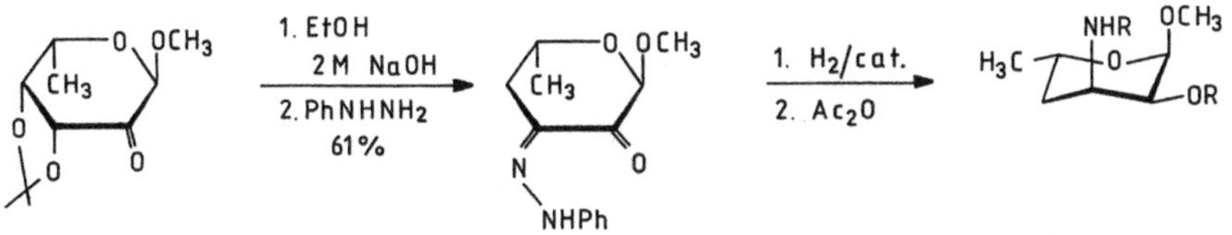

313 314 315 R = Ac 316 R = H (HCl salt)

Scheme 61

Another method for the production (386) of a C-3 aminohexose from a 3,4-O-acetal is based on the reaction of methyl 6-deoxy-3,4-O-isopropylidene-α-L-*lyxo*-hexopyranosid-2-ulose **(313)** with phenylhydrazine under alkaline conditions. Reduction and subsequent acetylation of the resulting phenylhydrazone derivative **(314)** results in methyl 3-acetamido-2-O-acetyl-3,4,6-trideoxy-α-L-*ribo*-hexopyranoside **(315)**. Removal of the protecting groups of **315** gives rise to aminosugar **316**, a structural isomer of methyl α-L-ristosaminide **(141)**. This reaction sequence (Scheme 61) is said to be useful for the preparation of aminodeoxy sugars containing a $-CH_2-CH(NH_2)-$ structural system. However, methyl 6-deoxy-2,3-O-isopropylidene-α-L-*lyxo*-hexopyranosid-4-ulose **(317)** failed to provide an aminodeoxy hexose derivative. Instead, the action of a base on ketone **317** — even in dilute solution — produced (386) maltol (**318**, 3-hydroxy-2-methyl-4H-pyran-4-one) (Scheme 62), in agreement with earlier results (387) that had shown the transformation of **317** into maltol upon treatment with either a cation or an anion exchange resin, via splitting off of the isopropylidene acetal function, followed by the rapid β-elimination of water and methanol.

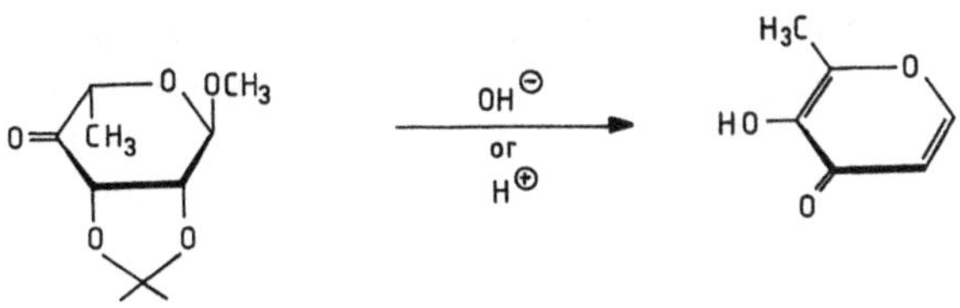

317 318

Scheme 62

2.3 Simultaneous generation of the C-3 amino and C-2 deoxy functions

Catalytic hydrogenation or chemical reduction of 3-azido-3-deoxyhexopyranosides carrying a reducible C-2 substituent readily furnish the corresponding 3-amino-2,3-dideoxyhexopyranoside derivatives. Christensen and Goodman (388) converted methyl 3-azido-2-S-benzyl-4,6-O-benzylidene-3-deoxy-2-thio-α-D-altropyranoside **(321)**, obtained from **319** *via* the epi-sulfonium intermediate **320**, into methyl 3-amino-4,6-O-benzylidene-2,3-dideoxy-α-D-*ribo*-hexopyranoside **(157)** by applying such a methodology (Scheme 63).

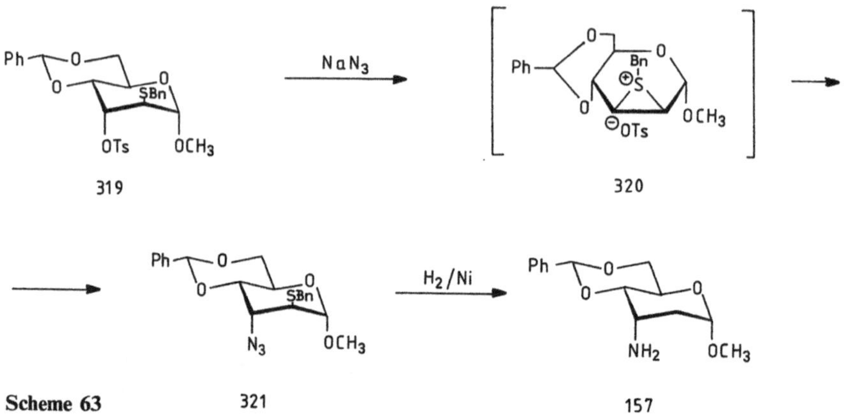

Scheme 63

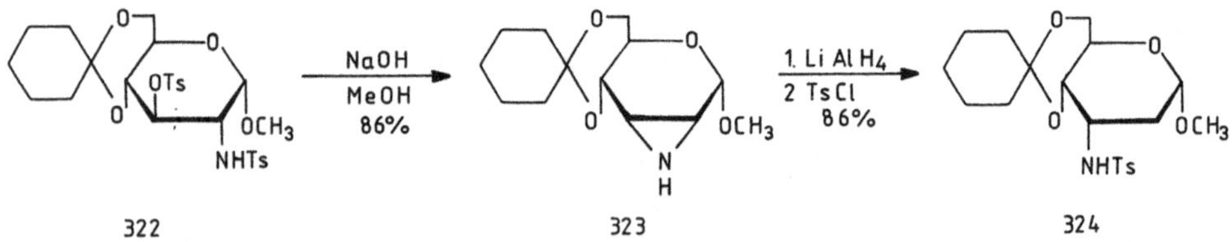

Scheme 64

The simultaneous generation of the 3-amino- and 2-deoxy functions of a hexopyranoside can also be accomplished by means of the reductive ring opening of 2,3-epiminohexopyranosides. As an example, the Umezawa group (277) isolated an 86% yield of methyl 4,6-O-cyclohexylidene-3-*p*-toluenesulfonamido-2,3-dideoxy-α-D-*ribo*-hexopyranoside **(324)** upon lithium aluminum hydride reduction of the allo-epimine **323** and subsequent tosylation (Scheme 64). The amino compound **324** was also obtained in good yield (85%) by the direct treatment of the 2,3-N,O-ditosyl-glucosaminide **322** with lithium aluminum hydride. The similar aziridine ring opening (276) of the 4,6-O-benzylidene analog of **323** with Raney nickel likewise occurs, but in the case of the corresponding manno epimine the reaction produced methyl 2-amino-4,6-O-benzylidene-2,3-dideoxy-α-D-*arabino*-hexopyranoside as the major product.

2.4 2-Halogeno analogs of 3-amino-3-deoxy-hexopyranosides

Early studies of the synthetic availability of 3-amino-2-halogeno-2,3-dideoxyhexose derivatives were stimulated by their possible utilization (276) in cancer chemotherapy. The preparation of 2,3-epiminohexopyranosides (278, 389–393) and their ring opening reactions (276, 278, 394) with halogeno nucleophiles were examined in detail.

In recent years interest in the C-2′-halogenated analogs of the anthracycline glycoside antibiotics which contain 3-amino-2-halogeno-2,3,6-trideoxy-hexopyranoses as their sugar moiety, was awakened by the expectations that the strong electron-withdrawing effect of the C-2′ halogen atom (particularly fluorine) might enhance the stability of the glycosidic linkage of the antibiotic molecule toward in vivo hydrolysis. In an effort to synthesize the C-2′-halogeno-anthracyclines and especially to introduce fluorine (with a steric bulk comparable to that of hydrogen) into amino-sugar molecules, investigations (395–399) were carried out according, in many cases, to the pioneering results derived from the studies of the 2,3-epiminohexopyranosides. Several additional methods (345–349, 351, 400–402) have been developed for the preparation of 3-amino-2,3-dideoxy-2-halogeno-hexopyranoside derivatives, some of which have also served as candidates for reductive dehalogenation at C-2 thereby providing an alternative route to 3-amino-2,3,6-trideoxyhexoses (see Sects. 2.2.5 and 2.6.3).

2.4.1 Syntheses via 2,3-epiminopyranoside intermediates

Richardson et al. (278) showed that methyl 4,6-O-benzylidene-2,3-dideoxy-2,3-epimino-α-D-allopyranoside **(325)** undergoes hydrolysis of the 4,6-O-bezylidene acetal function without rupture of the epimine ring when treated with hydrochloric acid in acetone. Upon the reaction of epimine **325** with ammonium chloride, bromide, or iodide under neutral conditions in DMF (278), an anomalous *diequatorial* cleavage of the aziridine ring occurs and yields the C-3-halogeno-2-aminoglucopyranosides **326–328** (Scheme 65). When the N-acetyl- **(329)** or N-benzoyl- **(330)** derivative of epimine **325** was subjected to the above reaction conditions (Scheme 66), completely different results were obtained. Thus, treatment of **329** or **330** with an equimolar amount of acid (HCl or HI) in acetone resulted predominantly in a *diaxial* aziridine ring opening prior to the hydrolysis of the acetal function to yield (276, 278) the 3-acylamino-2-halogenoaltropyranosides **331–333**. By applying an excess hydrochloric acid, the benzylidene moiety was also split off to give **(334)** and **(335)**. The reaction of the N-acylepimines **329** and **330** with ammonium salts in hot DMF formamide under essentially neutral conditions (278, 395) resulted, likewise, in a preponderantly *diaxial* cleavage of the epimine ring to produce the corresponding N-substituted 2-halogeno-2,3-dideoxy-α-D-altropyranoside analogs **(331–333)** in near 40% yield along with a 3–15% yield of the 2-amino-3-halogenoglucosaminides **326–328**.

Opening of the aziridine ring of benzyl 4,6-O-benzylidene-2,3-benzoylepimino-2,3-

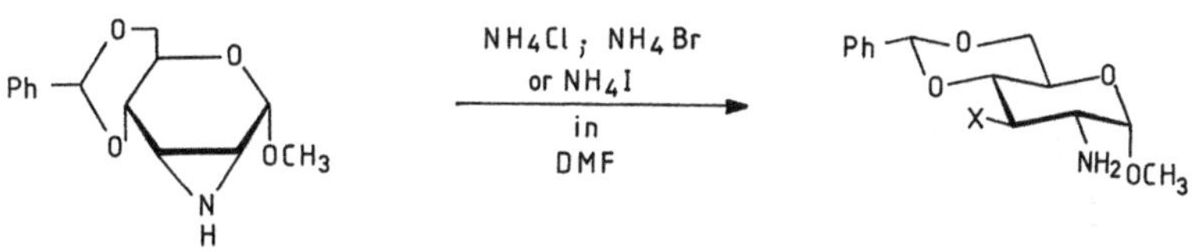

325

326 X = Cl (40%)

327 X = Br (52%)

328 X = I (54%)

Scheme 65

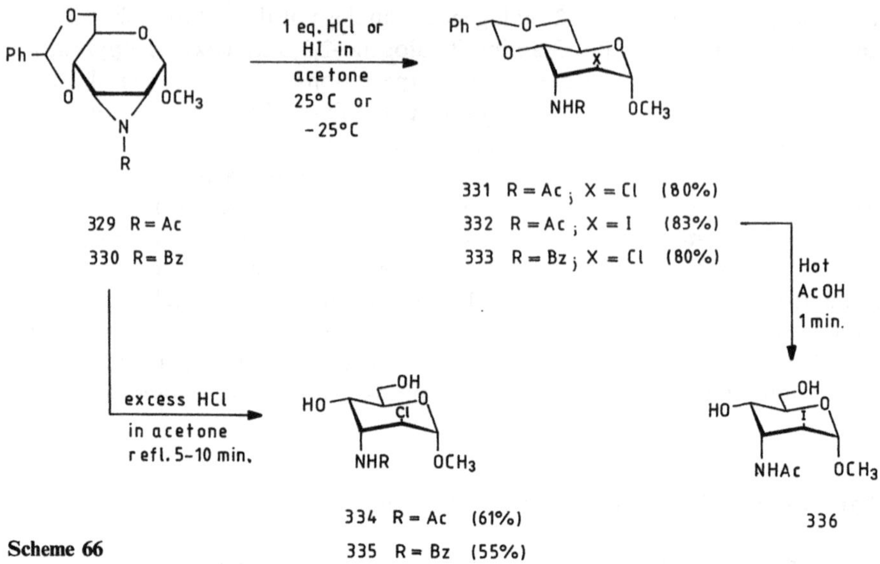

Scheme 66

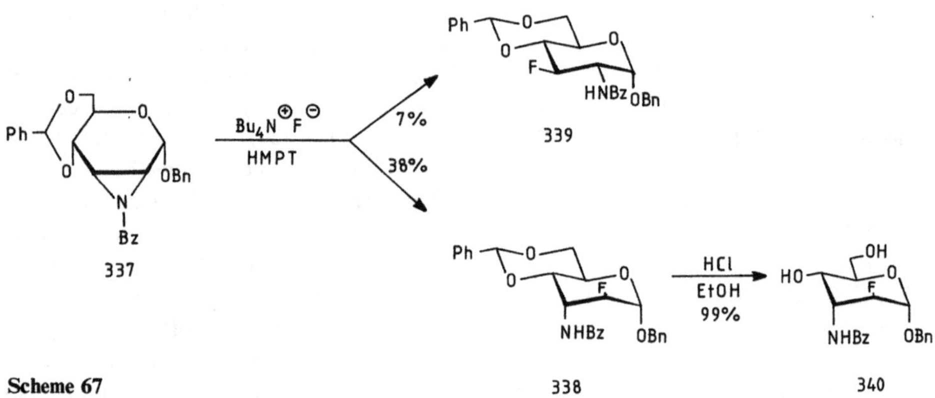

Scheme 67

dideoxy-α-D-allopyranoside (**337**) with tetrabutylammonium fluoride, similarly resulted (396) in a 38% yield of the 2-fluoroaltropyranoside **338** and 7% 3-fluoroglucoside **339** (Scheme 67). Mild acid hydrolysis of the acetal protecting group of **338** then led quantitatively to benzyl 3-benzamido-2-fluoro-2,3-dideoxy-α-D-altropyranoside (**340**), a key intermediate of (s)-2-fluoro-L-daunosamine. Methyl 4,6-O-benzylidene-3-benzamido-2,3-dideoxy-2-fluoro-α-D-altropyranoside (**343**), another popular compound in strategies for the preparation of (s)-2-fluoro-L-daunosamine, was directly synthesized from methyl 4,6-O-benzylidene-2-benzamido-3-O-methanesulfonyl- (**341**) (397) or-3-*p*-toluenesulfonyl-2-deoxy-α-D-glucopyranoside (**342**) (394, 398) without isolation of the intermediary benzoyalziridine **330** (Scheme 68). Thus, treatment of **341** or **342** with anhydrous tetrabutylammonium fluoride in hot acetonitrile, in hexamethylphosphoric triamide, or in oxolane gave a 40–75% yield of **343**. Baer et al. (398) reported the formation of ca. a 6% yield of methyl 4,6-O-benzylidene-2-benzamido-3-fluoro-2,3-dideoxy-α-D-glucopyranoside by-product. Picq and Anker

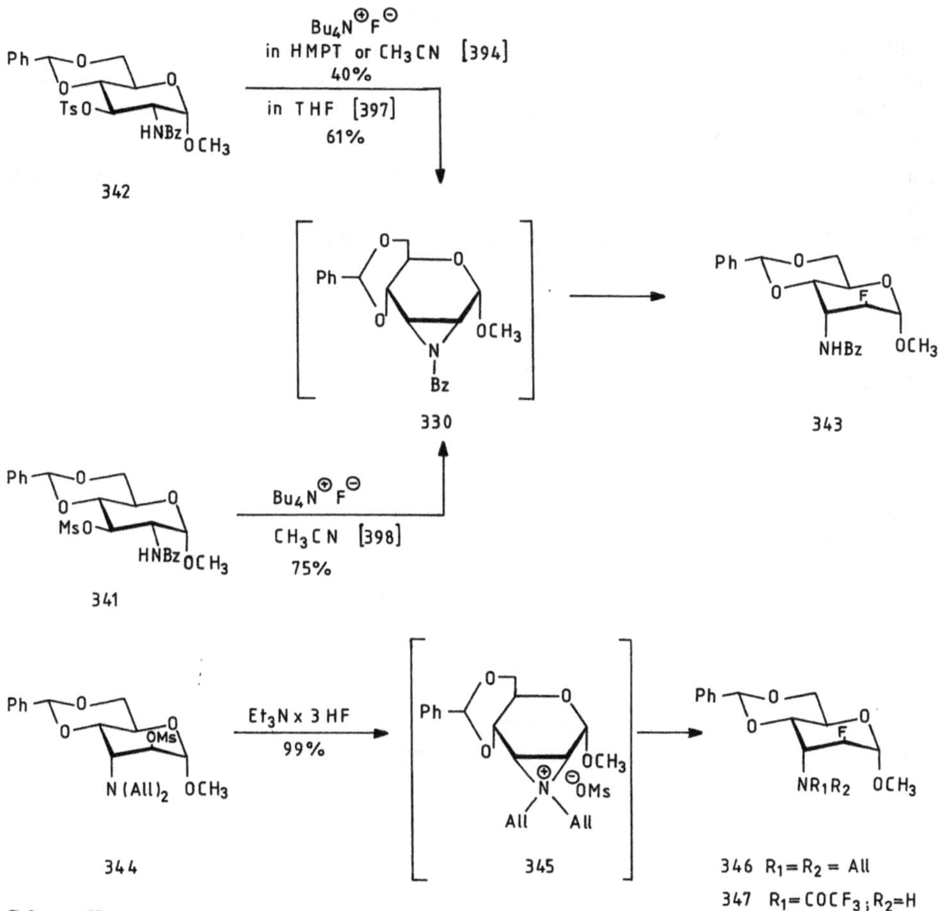

Scheme 68

(399) carried out the reaction of methyl 4,6-O-benzylidene-3-diallylamino-2-O-methanesulfonyl-α-D-altropyranoside **(344)** with the triethylamine-hydrogen fluoride complex. Because of the neighbouring-group participation involving the diallyl-aziridinium ion **345**, the fluorine was introduced with complete retention of the configuration at C-2, and the diallylamino-fluorosugar **346** was obtained. Reductive N-deallylation and subsequent trifluoroacetylation then gave methyl 4,6-O-benzylidene-2-fluoro-3-trifluoroacetamido-α-D-altropyranoside **(347)** (Scheme 68). The 2-fluoro-3-aminohexose derivatives prepared by these methods described have been used for the synthesis of (s)-2-fluoro-L-daunosamine (**348**, 3-amino-2-fluoro-2,3,6-trideoxy-L-galactose) and its derivatives. As a general strategy, compounds **340, 343**, and **347** are converted into the respective 6-bromo- or 6-iodo-6-deoxy analogs (Sect. 2.7), and then the C-6-deoxy function is generated with a simultaneous change in the chirality at C-5 (Sect. 2.7.4). Thus, the 4,6-diol **340** is first transformed (396) into the 6-iodo compound **555** (Sect. 2.7.1, Scheme 110), and the β-L-benzyl glycoside **(636)** of **348** is then prepared via the hex-5-enopyranoside **630** (Table 12 entry 4). The 6-bromo-6-deoxysugars **585** and **586** derived (397, 398) from **343** and

347 (Table 8 entries 1 and 2) are similarly converted into 2-fluoro-L-daunosamine **(348)** and its derivatives **633–635** and **638** (Table 12 entries 2 and 3). In addition, from the 3-amino-2-fluoroaltropyranoside **343** methyl (s)-2-fluoro-α-D-ristosaminide **(349)** could also be synthesized (398) (Table 8 entry 1). The reductive dehalogenation (Sect. 2.6.3.) at C-2 of the chloro compounds **333** (395) and **334** (276) and the 6-iodohexopyranoside **336** (276) (Schemes 101 and 102) offers an easy access to the 3-amino-2,3-dideoxy-α-D-*ribo*-hexopyranoside series. Key intermediates such as **535** and **536**, for conversion into derivatives of D-ristosamine **(20)** and L-daunosamine **(12)** could be so prepared.

348 349

Scheme 69

2.4.2 Syntheses from 2-hydroxy- and 2-O-sulfonyl derivatives

Lukacs and coworkers (400) treated benzyl 3-azido-4,6-O-benzylidene-3-deoxy-α-D-altropyranoside **(350)** with diethylaminosulfur trifluoride (DAST) in hot benzene to produce a mixture of three reaction products **(351, 352**, and **353)** that was then separated by means of column chromatography (Scheme 70). One of the major

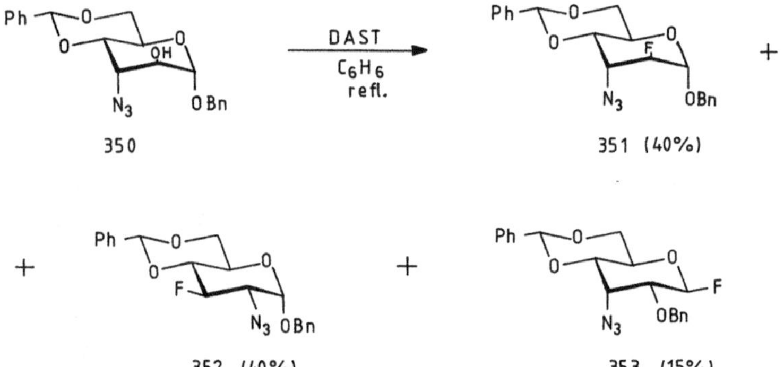

Scheme 70

products (40%), benzyl 3-azido-4,6-O-benzylidene-2,3-dideoxy-2-fluoro-α-D-altropyranoside **(351)** was formed by a stereospecific attack of the fluoride ion at carbon C-2 from the less hindered upper side of the pyranoside ring. An α approach by the nucleophile is severely inhibited by the two neighbouring bulky *axial* substituents at C-1 and C-3. The formation of the 3-fluoro and 1-fluoro compounds (**352** and **353**) can be explained by an S_Ni process and neighbouring-group participation, respectively. Similar effects cause the treatment of the α-D-gluco- and β-D-idopyranose configurational isomers of **350** with DAST to result (402), likewise, in a high proportion of 2-O-benzyl-glycosylfluorides as the benzyloxy-migration products.

An analogous intramolecular rearrangement, mediated by DAST, occurs in a furanoid-system as has been reported by Hasegawa et al. (403). The reaction of methyl or benzyl 3-azido-5-O-benzoyl-3,6-dideoxy-α-L-talofuranoside (**354** or **355**) with DAST led to the migration of the anomeric methoxy or benzyloxy group to position C-2 and produced yields of 50.3% and 79%, respectively, of 3-azido-5-O-benzoyl-2-O-methyl- (**356**) and 2-O-benzyl- (**357**) -3,6-dideoxy-β-L-galactofuranosyl fluoride (Scheme 71).

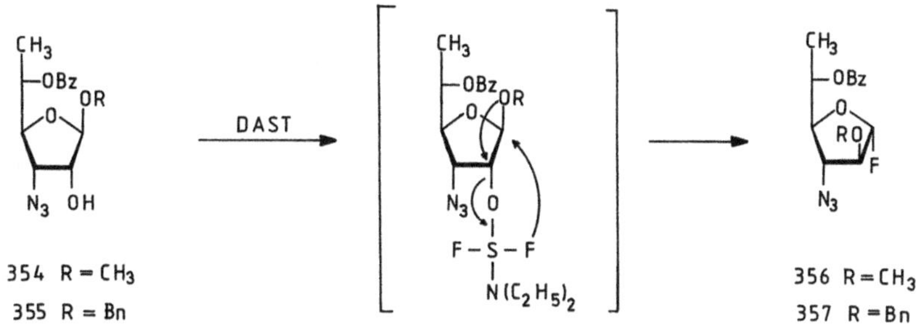

Scheme 71

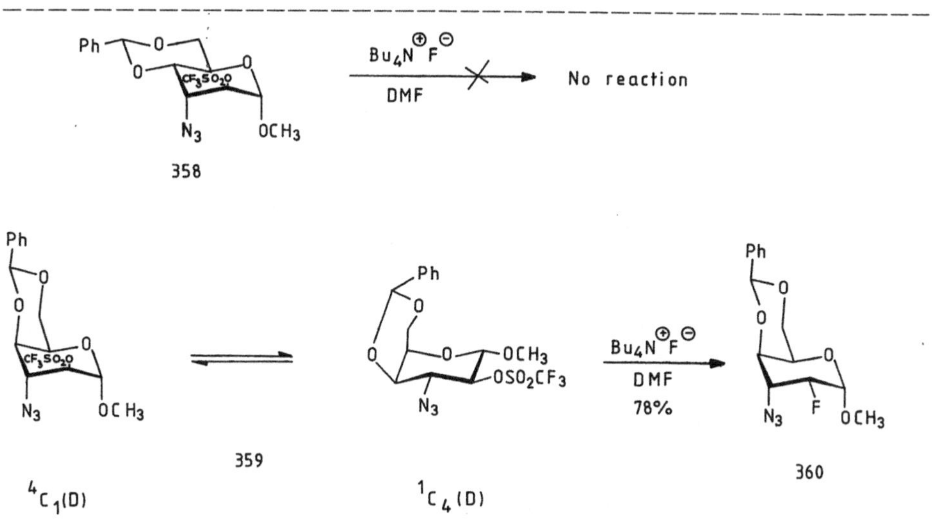

Scheme 72

The S_N2 displacement (401) of the triflate function in the 3-azido-2-O-trifluoromethanesulfonyl-altropyranoside **358** with tetrabutylammonium fluoride (Scheme 72) was unsuccessful because of the severe steric hindrance. On the other hand, an S_N2-like displacement reaction of the α-D-ido stereoisomer **359** with the same nucleophile readily produced the 2-fluoro-α-D-gulo sugar **360**, because of the ability of **359** to adopt an alternative $^1C_4$(D) conformation as demonstrated by means of ^{1}H-NMR measurements (401). In a similar manner, replacement of the O-2 trifluoromethanesulfonyl group of benzyl 3-azido-4,6-O-benzylidene-3-deoxy-2-O-trifluoromethanesulfonyl-α-D-gluco- and β-D-idopyranosides using cesium fluoride proceeded smoothly to afford (402) the corresponding 2-fluorosugars in 60–70% yield with inversion of the configuration at C-2. In contrast to these latter results, methyl 3-azido-5-O-benzoyl-3,6-dideoxy-2-O-trifluoromethanesulfonyl-α-L-talofuranoside gave only the elimination product methyl 3-azido-5-O-benzoyl-2,3,6-trideoxy-α-L-*threo*-hex-2-enofuranoside when treated (403) with tetrabutylammonium fluoride.

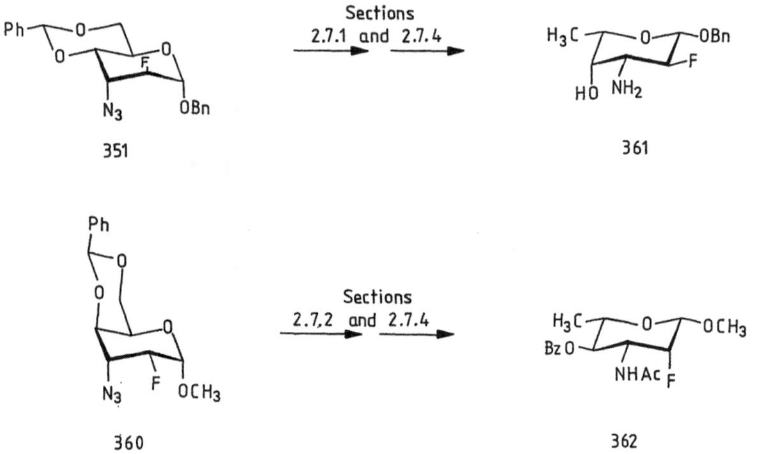

Scheme 73

The 3-azido-2-fluorosugars **351** and **360** were transformed by the cleavage of the benzylidene acetal function into the respective 6-iodo-6-deoxy- **(626)** and 6-bromo-6-deoxy- **(587)** derivatives. Subsequent C-6 dehydrohalogenation (Sect. 2.7.4), followed by saturation of the exocyclic double bond of the resulting **627** and **631** by simultaneous reduction of the azide function (Table 12 entries 1 and 5) then led to benzyl (s)-2-fluoro-β-L-daunosaminide **(361)** (400) and methyl 3-acetamido-4-O-benzoyl-2,3,6-trideoxy-2-fluoro-β-L-mannopyranoside **(362)** (401), respectively. The latter is a protected 2-fluoro derivative of L-acosamine (Scheme 73).

2.4.3 Simultaneous introduction of the 3-amino and 2-halogeno substituents

An elegant method of simultaneously introducing an amino function on C-3 and a halogen atom (bromine or iodine) on C-2 of a carbohydrate *via* the 2-halogeno-oxazoline or oxazolidine derivatives **253, 254, 259, ±263,** and **366** derived by means of the bromonium or iodonium ion induced cyclization (344–349, 351) of the O-4 imidate esters **252, 258** and **±262**, is described in Sect. 2.2.5. The treatment of oxazolines **253** and **254** with hydrochloric acid in aqueous methanol at room temperature results in (345) methyl 3-amino-2-bromo- **(363)** and -2-iodo-2,3,6-trideoxy-α-L-altropyranoside **(364)**, the (R)-2-bromo- and iodo analogs of methyl α-L-ristosaminide (Scheme 74).

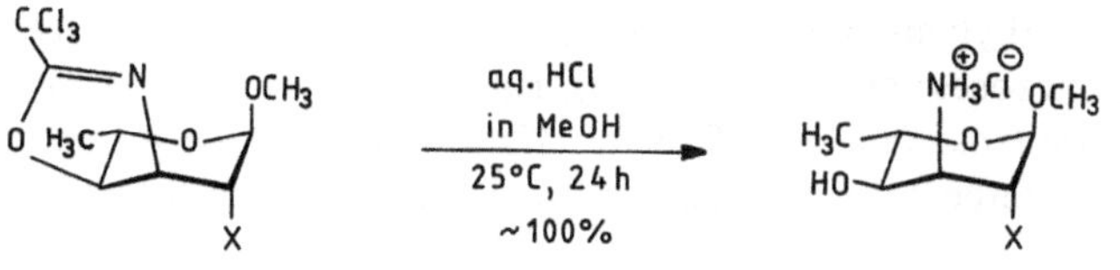

253 X = Br
254 X = I

363 X = Br
364 X = I

Scheme 74

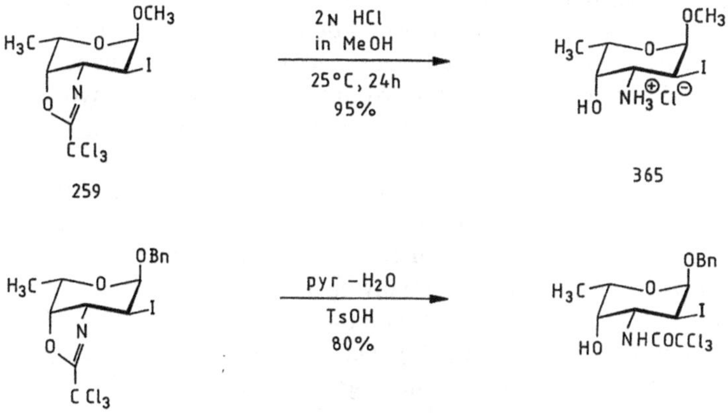

259 → 365

366 → 367

Scheme 75

The reaction of the 2-iodo-oxazoline **259** with hydrochloric acid in methanol led to a similar formation of methyl (s)-2-iodo-α-L-daunosaminide **(365)** in 95% yield (348). When the corresponding benzyl glycoside **(366)** of **259** was kept (351) in pyridine-water in the presence of *p*-toluenesulphonic acid catalyst, the cleavage of the oxazoline ring was not accompanied by N-deacylation and benzyl 2-iodo-3-trichloroacetamido-2,3,6-trideoxy-α-L-galactopyranoside **(367)** was produced (Scheme 75). Methyl (s)-2-bromo-N,O-dibenzoyl-α-D,L-daunosaminide (±**264**) has been prepared (347) from the racemic imidate ester ±**262** as shown in Scheme 52. Reductive dehalogenation at C-2 of the 2-bromo- and 2-iodo-aminogylcosides ±**264, 363, 364,** and **367** using tributyltin hydride gave the corresponding 2-deoxysugars (±**12, 141, 260**, and **538**) as described in Sects. 2.2.5 and 2.6.3 (Schemes 52 and 103–105).

2.5 Methodologies for the preparation of 3-C-methyl branched-chain 3-amino- and 3-nitrosugars of antibiotic substances

As shown in Chap. 1 and Sect. 2.2.4, polydeoxy sugars carrying the $CH_3-\overset{|}{\underset{|}{C}}-NR$ branch (with amino or nitro functions) (60) have been found in nature as building blocks of antibiotic substances. These three branched-chain aminohexoses;

L-vancosamine **(15)**, N,N-dimethylvancosamine **(16)**, and 3-epi-L-vancosamine **(21)** (Table 1), and the four nitrohexoses; evernitrose **(240)**, rubranitrose **(241)**, kijanose (tetronitrose, **242**), and decilonitrose **(243)** (cf. Scheme 47) have in many cases been synthesized via their unnatural stereoisomers and according, basically, to the following five methods.

2.5.1 Cyclization of dialdehyde derivatives with nitroethane

The base-catalyzed Baer-Fischer cyclization (299) (Scheme 42) of "sugar dialdehydes" (see Sect. 2.2.4) with two-carbon nitro compounds (nitroethane, nitroethanol, ethyl nitroacetate) gives rise (300, 301) to mixtures of 3-deoxy-3-nitro-3-C-substituted aldohexopyranoside steroisomers. Upon cyclization (404) using nitroethane and subsequent acetylation and purification, L′-methoxy-L-methyl-diglycolic aldehyde **368** produced a 19% yield of crystalline methyl 2,4-di-O-acetyl-3,6-dideoxy-3-C-methyl-3-nitro-α-L-glucopyranoside **(369)** as the only product isolable in pure form (374) (Scheme 76). The nitro sugar **369** was transformed into 3-acetamido-1,2,4-tri-O-acetyl-3,6-dideoxy-3-C-methyl-α-L-glucopyranose **371** in 56% overall yield via sequential mild acid hydrolysis, acetylation, reduction of the nitro function of **370**, and N-acetylation. An alternative approach involving the reduction of **369** over Raney nickel followed by N-acetylation to give the 3-acetamido-2,4-diol **372**, provided a significantly lower (~30%) yield (374). The branched-chain aminosugars **371** and **372** were converted by the Brimacombe group (Sects. 2.6.1 and 2.6.2) into their 2-deoxy analogs **416** (181, 405) (Scheme 95) and also into **420** and **517** (181, 374, 405) (Scheme 92), which are substituted analogs of 4-epi-L-vancosamine. The 3-C-methyl-3-aminohexoses **371** and **372** were also used as starting materials in a synthesis (406) of methyl N-acetyl-α-L-vancosaminide **(681**, Sect. 2.8.1, Table 15 entry 4).

The nitroethane cyclization method is also restricted (similarly to that for

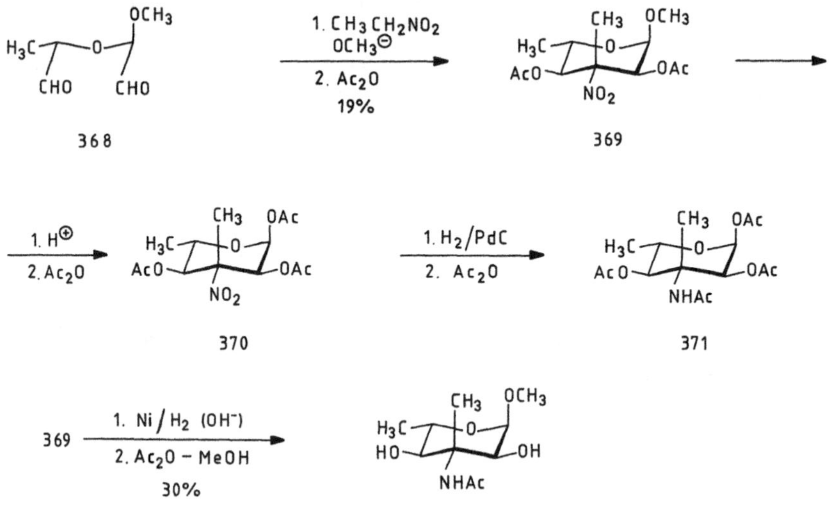

Scheme 76

nitromethane, cf. Sect. 2.2.4) from the preparative point of view to the preparation of 3-nitro- and 3-amino-3-C-methyl branched hexoses bearing *equatorial* nitrogen functions (the vancosamine and evernitrose derivatives).

2.5.2 Cyanomesylation of hexopyranosid-3-ulose derivatives

One route to the C-methyl-branched aminosugars is based on the cyanomesylation of protected ketosugars, followed by reduction with lithium aluminum hydride to form a spiro-aziridine and the opening of the ring of this latter compound by catalytic hydrogenation over Raney nickel (Scheme 77). The foregoing method was introduced to carbohydrate chemistry by Bourgeois (407) using 1,2:5,6-di-O-isopropylidene-α-D-*ribo*-hexopyranosid-3-ulose as the starting material and cyanide ion as the other reagent. Using such a reaction sequence, Bourgeois (407) prepared 3-acetamido-5-O-acatyl-3,6-dideoxy-1,2-isopropylidene-3-C-methyl-α-D-glucofuranose, a protected derivative of 4-epi-2-hydroxy-D-vancosamine, in 1976.

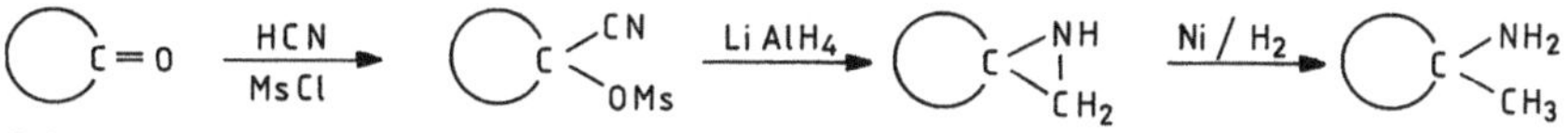

Scheme 77

Extensive efforts to synthesize L-vancosamine and the other structurally related branched-chain 3-amino- and 3-nitrosugars that occur in antibiotics, have been made by the groups of Brimacombe, Lukacs, and Yoshimura. They have tried to adapt the cyanomesylation method to the conversion of the 2-deoxy- and 2,6-dideoxy-hexopyranosid-3-uloses in both the D- and the L-series.

The Lukacs group (177, 409) showed that the cyanomesylation of ketose derivatives can be conducted under either kinetic or thermodynamic control. The steric outcomes, i.e. the configurations of the branch point of the C-3 cyanomesylates formed from such 3-ulopyranosides under conditions of kinetic (hydrogen cyanide in pyridine followed by mesylation) and thermodynamic control (potassium cyanide and sodium hydrogen carbonate in aqueous dichloromethane and subsequent mesylation) are shown in Table 5. Under kinetic control the cyanohydrin, the initial product of the reaction, is most likely produced (186) by an *equatorial* attack of the cyanide ion at the carbonyl group — unless the latter is sterically hindered from this direction. This initial product can be equilibrated, a to a varying extent, to form a mixture of the C-3 epimeric compunds prior to the mesylation, which stabilizes the configuration of the cyanohydrin.

Yoshimura et al. (186) suggested that for the cyanomesylates produced from the α-D- **(100)** and -α-L-*erythro*- **(104** and **106)** ulosides (see Table 5 entries 1–3) the configuration of the initial cyanohydrin made by an *equatorial* attack is presumably fixed through hydrogen-bonding between the *axially* disposed C-3 hydroxyl and anomeric methoxy groups. Thus, under kinetic control **100, 104,** and **106** yield high proportions (64%, 80%, and 87%, respectively) of the cyanomesylates with the *ribo* configuration **(373, 375**, and **377)**. Because no such stabilizing effect operates on the cyanohydrins of the β-*erythro* ketones **379** and **382** (Table 5 entries 4 and 5), they form C-3 epimeric mixtures of the *arabino*- **(380** and **383)** and *ribo*- **(381** and **384)** cyanomesylates containing the *arabino* stereoisomers **380** and **383** as the major kinetic products.

Table 5. Cyanomesylation of 2-deoxy-hexopyranoside-3-uloses[a]

Entry	Starting hexopyranosid-3-ulose		Kinetic product		Thermodynamic product	
	Structure	Prepn.	Structure	Ref.	Structure	Ref.
1	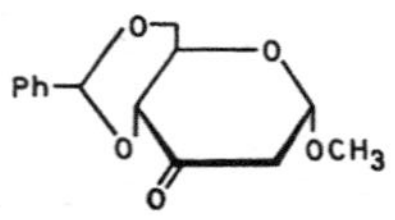 100 α-D-erythro	167	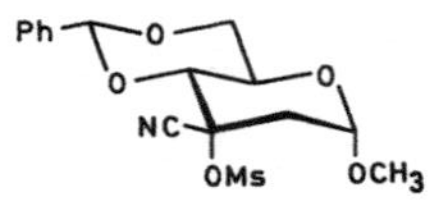 373 α-D-ribo	177 180 183	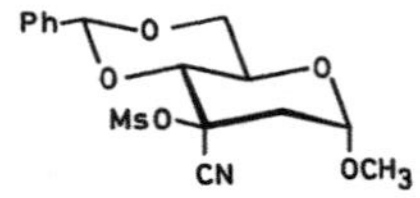 374 α-D-arabino	177 183
2 3	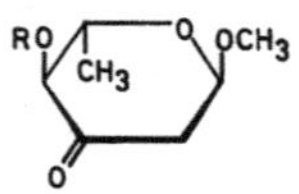 104 R=Me 106 R=MOM α-L-erythro	169 170, 171	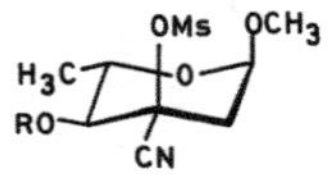 375 R=Me 377 R=MOM α-L-ribo	180, 186 170, 183, 185	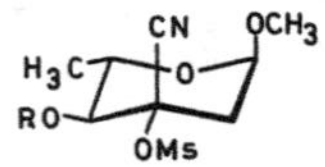 376 R=Me 378 R=MOM α-L-arabino	186 170, 183, 185
4 5	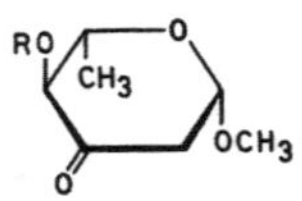 379 R=Bn 382 R=Me β-L-erythro	326 186	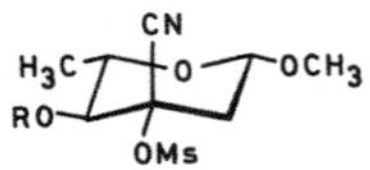 380 R=Bn 383 R=Me β-L-arabino	326 180, 186	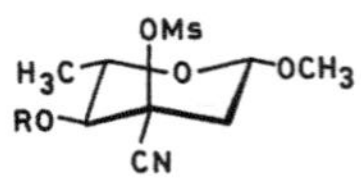 381 R=Bn 384 R=Me β-L-ribo	326 180, 186
6	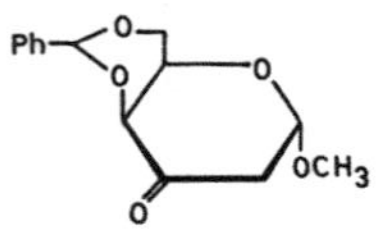 385 α-D-threo	408	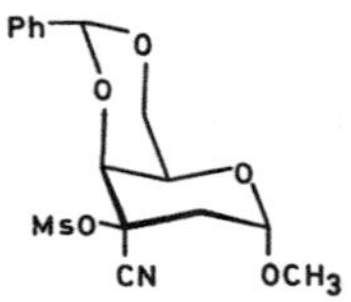 386 α-D-lyxo	409	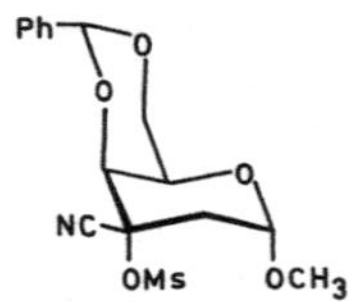 387 α-D-xylo	409

	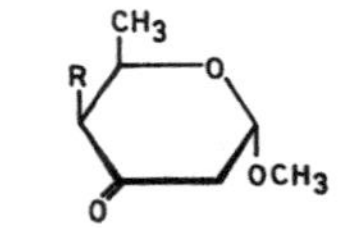 		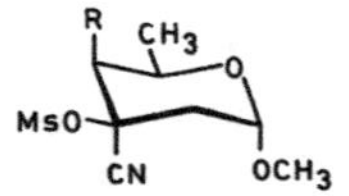 		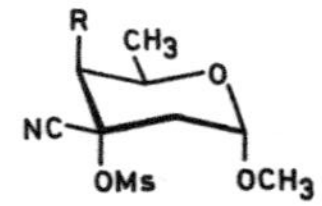 	
7	388 R=OMe	410	389 R=OMe	410	390 R=OMe	410
8	391 R=$NHCOOCH_3$	410	392 R=$NHCOOCH_3$	410	393 R=$NHCOOCH_3$	410
	α-D-threo		α-D-lyxo		α-D-xylo	
	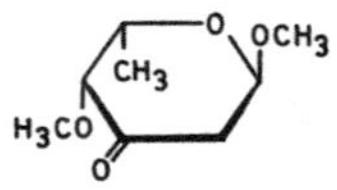 		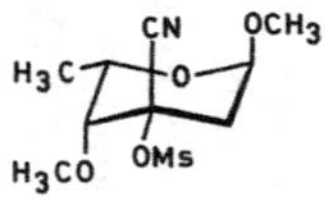 		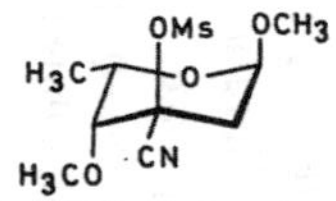 	
9	394 α-L-threo	194	395 α-L-lyxo	194	396 α-L-xylo	194
	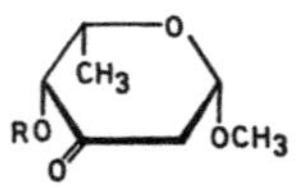 		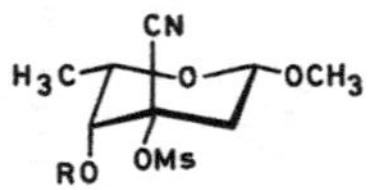 		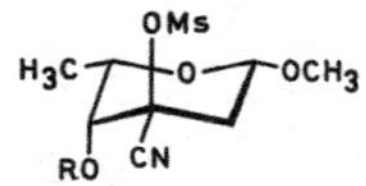 	
10	397 R=Me	193	398 R=Me	193	399 R=Me	193
11	400 R=Bz	192	401 R=Bz	192	402 R=Bz	192
12	403 R=MEM	192	404 R=MEM	192	405 R=MEM	192
13	406 R=Bn	192	407 R=Bn	192	408 R=Bn	192
	β-L-threo		β-L-lyxo		β-L-xylo	

[a] For transformation reactions leading to 3-amino-3-C-methylhexoses see Table 6

Because the D-*threo*- **(385, 388,** and **391)** and L-*threo*-ketones **(394, 397, 400, 403,** and **406)** always give *lyxo* cyanomesylates as the kinetic products regardless of the anomeric configuration (see Table 5, entries 6–13), an initial approach of the cyanide ion to the carbonyl group from the less hindered *axial* direction is definitely anticipated. Experimental data (194, 410) show, however, that even under kinetic control of the cyanohydrin formation, the α-*threo* ketones **385, 388, 391,** and **394** (Table 5 entries 6–9) quickly give considerable quantities of the equilibrated thermodynamic α-*xylo* cyanohydrins, in which the *axial* C-3 hydroxyl group might be hydrogen-bonded to the *axial* anomeric methoxy group, as is the case (186) for the α-*erythro* isomers (Table 5 entries 1–3). In contrast, no such quick equilibration has been observed (192, 193) with the cyanohydrins of the corresponding β-*threo* ketones **397, 400, 403,** and **406.** This allows the convenient preparation of the β-L-*lyxo*-**(398, 401, 404,** and **407)** and β-L-*xylo*- **(399, 402, 405,** and **408)** cyanomesylates, respectively, under conditions of kinetic and thermodynamic control (see Table 5 entries 10–13). The development of the conformational equilibria (409) and the hitherto unexplored stabilizing effects (194) might also be responsible for the aforementioned apparent difference between the steric outcome of the cyanomesylation reaction of the *erythro*- and *threo* hexopyranosid-3-uloses. Nevertheless, these differences allow proper choices of both the starting materials and the reaction conditions for the syntheses of cyanomesylates of deoxyhexopyranosid-3-uloses possessing the required stereochemistry at C-3.

Upon treatment with lithium aluminum hydride, the C-3 cyanomesylates are readily converted into the C-3 spiroaziridines via an intramolecular S_N2 mechanism, as shown by Brimacombe and his coworkers (411, 412). The subsequent hydrogenolytic ring cleavage of these spiroaziridines over Raney nickel then leads to C-3 methyl-branched aminosugars in high yields. As a result of the extensive studies of the Lukacs group (177) the configuration of the branch-point of such branched-chain carbohydrates can be reasonably determined by means of ^{13}C-NMR methods.

The application of the above sequence of reactions to the cyanomesylate **374** with α-D-*arabino* configuration produced (177, 380) methyl 3-acetamido-4,6-O-benzylidene-3-C-methyl-2,3-dideoxy-α-D-*ribo*-hexopyranoside **(410)** via the aziridine **409** (Scheme 78). The protected amine **410** was transformed by Lukacs et al. into O-benzoyl-N-trifluoroacetyl-L-vancosamine **(301)** (380) and methyl N-acetyl-β-L-vacosaminide **(411)** (177) via the 6-bromo-6-deoxy derivatives **591** and **590**, respectively, (Table 13

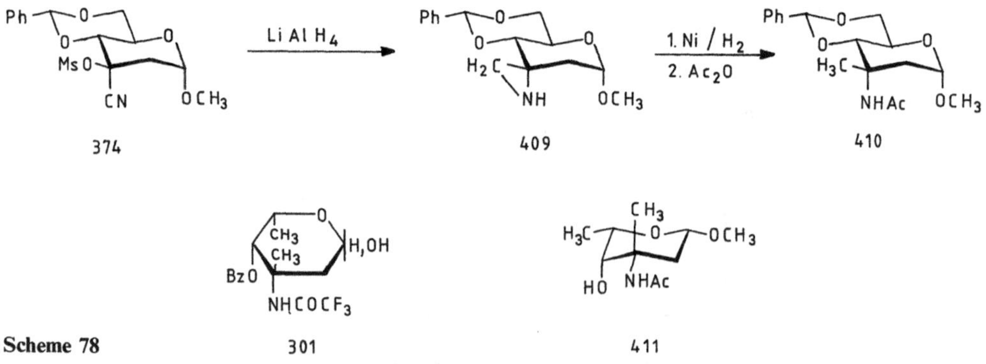

Scheme 78

entries 1–2) by means of deoxygenation at C-6 with a concomitant inversion of the configuration at C-5 (Sect. 2.7.4).

An essentially similar procedure with the α-D-*ribo*-cyanomesylate **373** led to the preparation (177, 180) of methyl 3-amino-4,6-O-benzylidene-3-C-methyl-2,3-dideoxy-α-D-*arabino*-hexopyranoside **(412)**. Sequential deoxygenation at the primary position (Sect. 2.7.2, Table 9 entry 3), removal of the 4-O-benzoyl group, and O-methylation gave the 3-C-methyl amino-glycoside **413** (Scheme 79). The amino function of this latter compound was oxidized with *m*-chloroperbenzoic acid to obtain (180) the unnatural D enantiomer (**414**, 3-C,4-O-dimethyl-3-nitro-2,3,6-trideoxy-D-*arabino*-hexopyranose) of evernitrose **(240)** after mild acid hydrolysis.

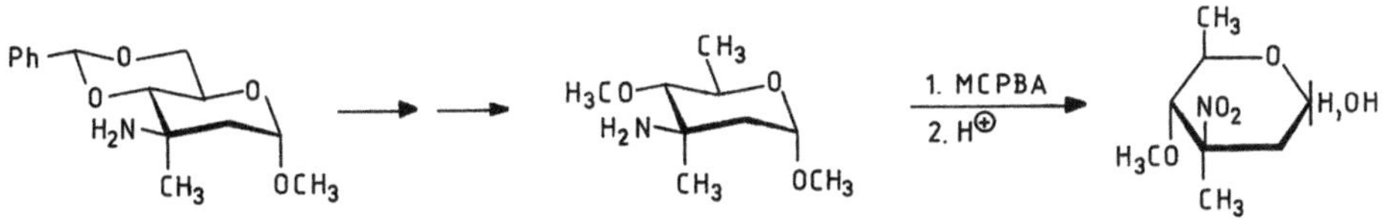

412 413 414

Scheme 79

This general reaction sequence allows the preparation of numerous 3-amino- and 3-nitro-3-C-methyl-hexose derivatives occuring in antibiotics and their unnatural stereoisomers, as well. Table 6 gives a summary of the related work and indicates the starting cyanomesylate, the intermediary C-3 spiroaziridines, and the target 3-amino- or 3-nitro-3-C-methyl-branched sugar derivatives. The results show that each of the eight stereoisomeric 3-amino- (and) 3-nitro-3-C-methyl-2,3,6-trideoxy-D- and L-hexoses (or derivatives thereof) has been prepared by the cyanomesylation procedure and subsequent standard transformation in a well-established, straightforward way and — in most cases — in a stereocontrolled manner. For these reasons the cyanomesylate route to 3-C-methyl-branched amino- and nitrosugars has become the most popular choice for the production of such carbohydrate derivatives.

Thus, among such functionalized sugars present in antibiotic substances, a derivative **(442)** of vancosamine **(15)** was synthesized (194) (Table 6 entry 12) from the cyanomesylate **396** with the α-L-*xylo* configuration via the spiroaziridine **441**. Protected analogs **(424, 440, 446, 448,** and **449;** see Table 6 entries 4, 11, 14, and 15) of the C-3 epimer **(21)** of vancosamine have been obtained (185, 192, 194) from the cyanomesylates **378, 395, 404,** and **407** via the epimines **421, 439, 445,** and **447,** respectively. In the work summarized in Table 6 entry 4, **424** was produced (185) by the stereoselective reduction of an intermediary ulose **(692)** derived from the O-4 deprotected analog **(690)** of **422** (see Sect. 2.8.3, Table 17 entry 2).

For the preparation of the evernitrose derivatives **417** and **431** the corresponding aminosugars **416** and **430** were applied (180, 186, 326) (Table 6 entries 1 and 7). Rubranitrose **(241)** was synthesized (410) (Table 6 entry 9), via the spiroaziridine **435**, by the oxidation of the aminoglycoside **436**. The α- **(425)** (413) and β-methyl glycoside **(428)** (326) of decilonitrose **(243)** and its β-1-O-acetyl-4-O-methyl derivative **434** (186) were prepared using the reaction sequence outlined in Table 6, entries 5, 6, and 8. The use of the cyanomesylation route to synthesize (410) methyl α-kijanoside (**437**, methyl α-tetronitroside) and its C-3 epimer **(438)** from a mixture of **392** and **393** is shown in entry 10 (Table 6).

Table 6. Conversion of cyanomesylates into 3-amino- and 3-nitro-3-C-methyl sugar derivatives

Entry	Starting cyanohydrin	Intermediary C-3 spiroaziridine	Aminosugar from ring opening	Target branched-chain sugar	Ref.
1	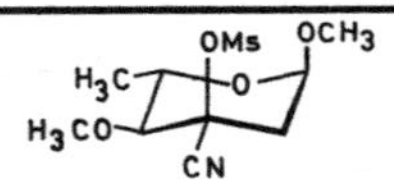 375	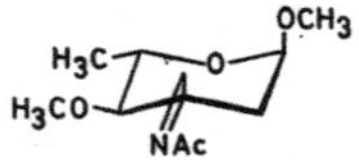 415	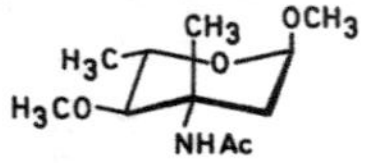 416	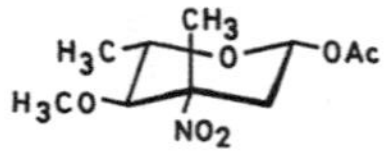 417	180 186
2	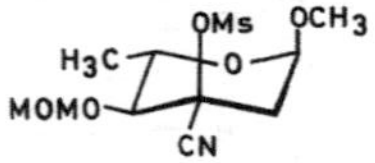 377	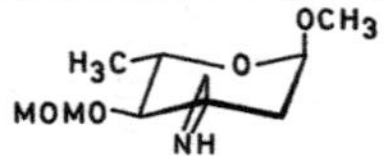 418[a]	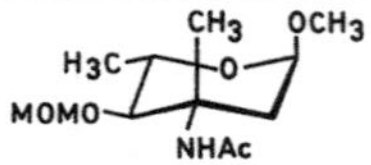 419	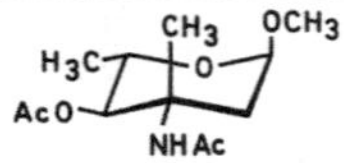 420	183
3				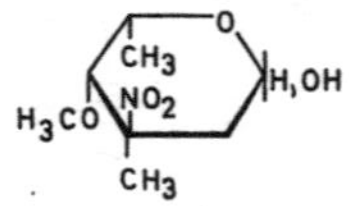 423[b]	194
4	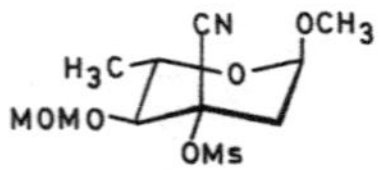 378	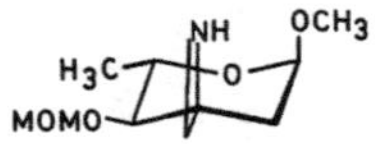 421[a]	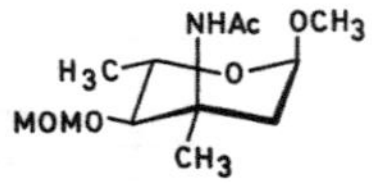 422	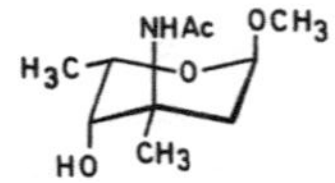 424[b]	185
5				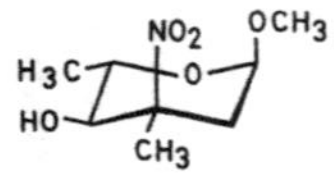 425	413
6	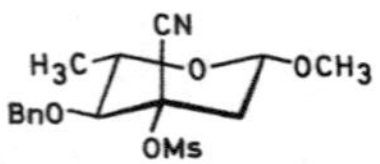 380	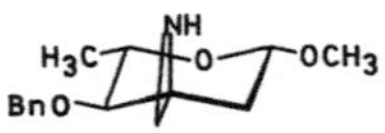 426[a]	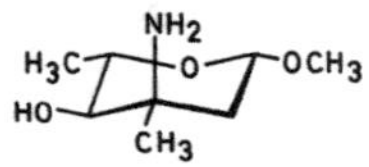 427	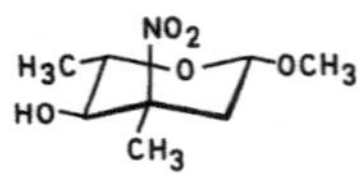 428	326
7	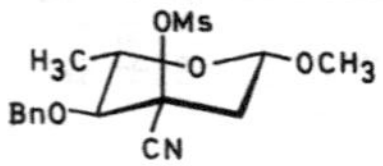 381	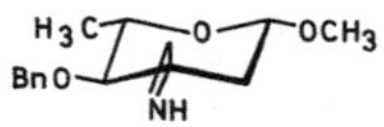 429[a]	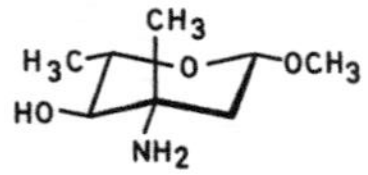 430	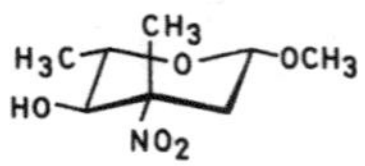 431	326

8	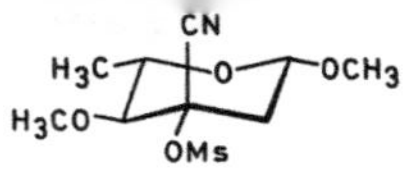 383[c]	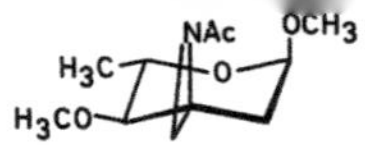 432	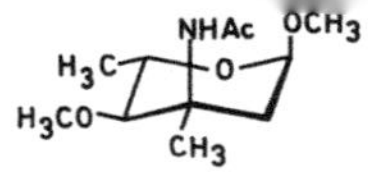 433	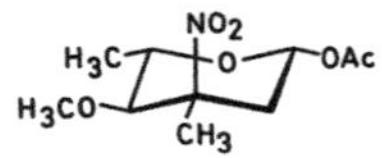 434	186
9	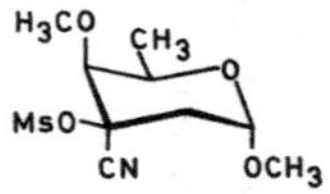 389	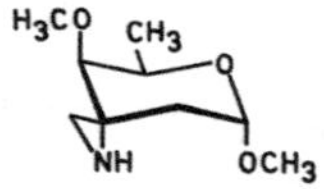 435	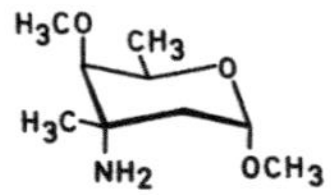 436	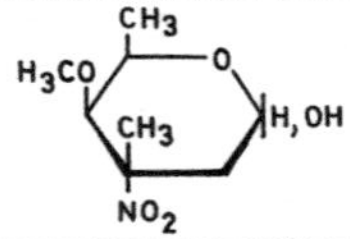 241	410
10	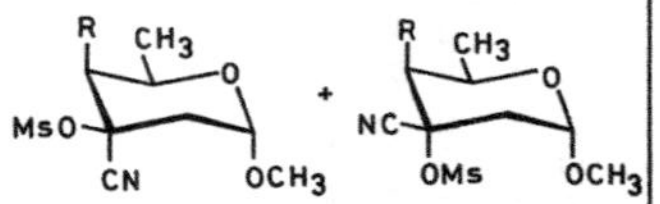 R = -NH-C(=O)-OCH3 392 393	Unseparable mixture of C-3 epimeric spiroaziridines	Unseparable mixture of C-3 epimeric 3-amino-3-C-methyl compounds	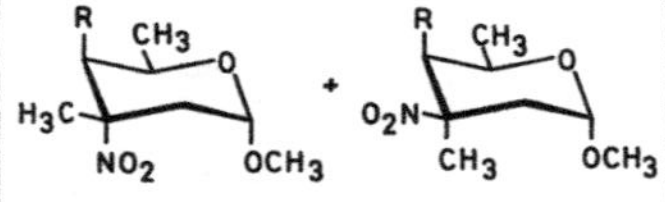 R = -NH-C(=O)-OCH3 437 438	410
11	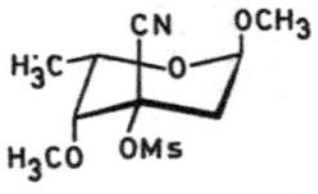 395	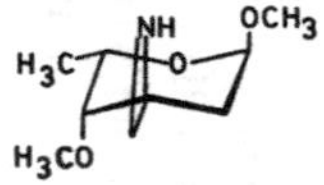 439[a]	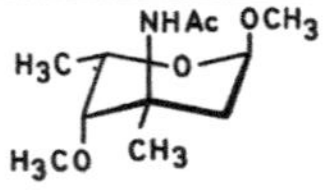 440	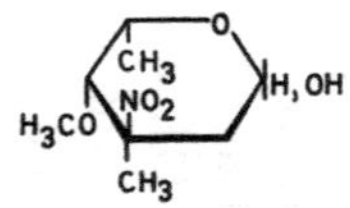 423	194
12	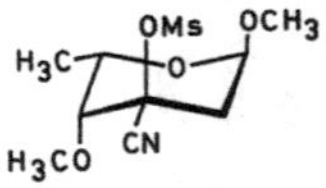 396	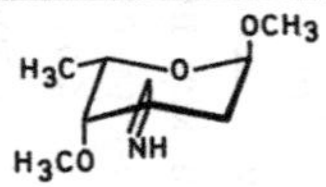 441[a]	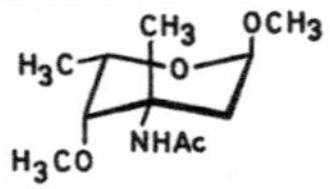 442	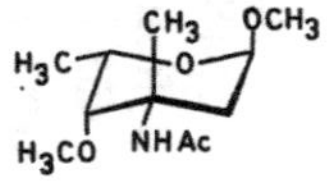 442	194
13	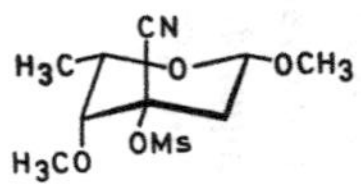 398	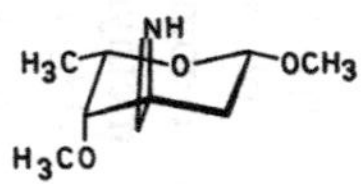 443	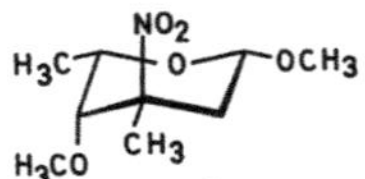 444[d]	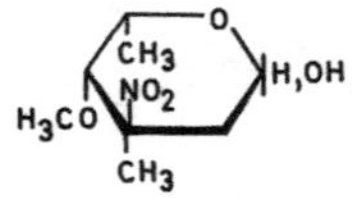 423	193
14	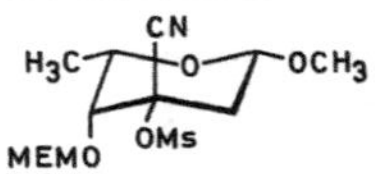 404	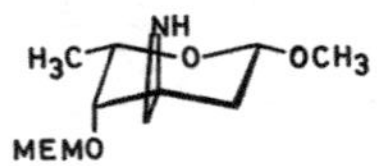 445	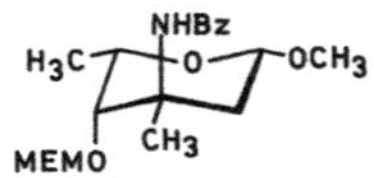 446	Attempts for the removal of the MEM protecting group of 446 remained unsuccessful	192
15	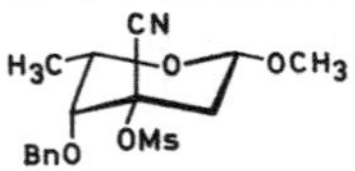 407	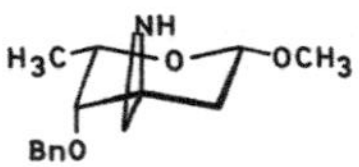 447	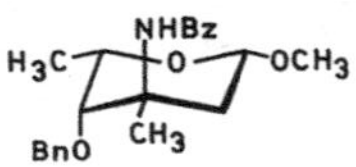 448	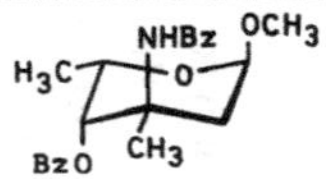 449[e]	192

See p. 76 for footnotes

2.5.3 The spirooxirane route to 3-C-methyl-branched amino- and nitrosugars

The third procedure for the development of a $CH_3-\overset{|}{\underset{|}{C}}-NH_2$ branch in carbohydrates has recently been introduced by the Brimacombe group (182, 414). The method involves the nucleophilic ring opening of the spiro-oxirane derivative **450** (129) with azide ion to obtain a primary alkyl azide **451** (Scheme 80). The subsequent hydrogenolysis of the azido-mesylate **452**, derived from **451**, over a platinum catalyst proceeds through the primary alkyl amine **453** and the spiroaziridine **454**, furnishing the methyl-branched amine **455** with inversion of the configuration at the tertiary center. An advantage of this procedure over the cyanomesylation route (Scheme 77) is that there is no need to isolate the intermediary spiroaziridine **454**, and the reduction of **452** into **455** can thus be accomplished (182) as a one-pot operation.

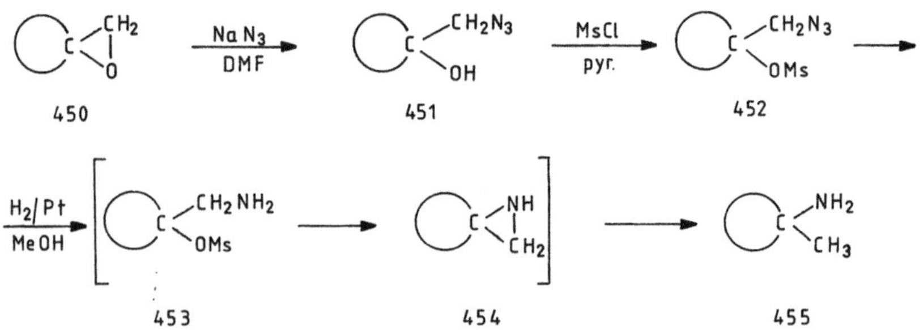

Scheme 80

The application of this methodology to methyl 3,3′-anhydro-[4,6-O-benzylidene-2-deoxy-3-C-(hydroxymethyl)-α-D-*arabino*-hexopyranoside] **(456)** (415) led to the preparation (182, 414, 416) of methyl 3-acetamido-3-C-methyl-2,3,6-trideoxy-α-D-*xylo*-hexopyranoside **(457)**, a common key intermediate in the preparation of both D-rubranitrose **(241)** and methyl α-D-kijanoside **(437)** (Fig. 8).

Cleavage of the epoxide ring of **456** with sodium azide gave the azido-methyl alcohol **458**, which was mesylated to obtain a 76% yield of the azido-mesylate **459** and a 16% yield of the elimination product **460**. When **459** was subjected sequentially to hydrogenolysis over Adams catalyst, N-acetylation, and chromatographic separation, methyl 3-acetamido-4,6-O-benzylidene-3-C-methyl-2-deoxy-α-D-ribo-hexopyranoside **(410)** was isolated in 54% overall yield. Deoxygenation of this latter compound at C-6 via the corresponding C-6 bromo derivative **590** (see Table 9 entry 1, Sect. 2.7.2) and the removal of the 4-O-benzoyl group allowed the preparation of the trideoxy aminosugar **461**. Inversion of the configuration of the C-4 asymmetric center of **461** to obtain **457** was achieved via the oxidation and stereoselective

---

Table 6 (continued)

[a] Not isolated

[b] Via reduction of the intermediary C-4 ulose, see Sect. 2.8.3

[c] A 1:1.4 equilibrated mixture of **383** and **384** was used

[d] The amino compound was in situ oxidized into the nitro derivative

[e] Under neutral conditions of 4-O-benzoylation an unexpected anomerization occured

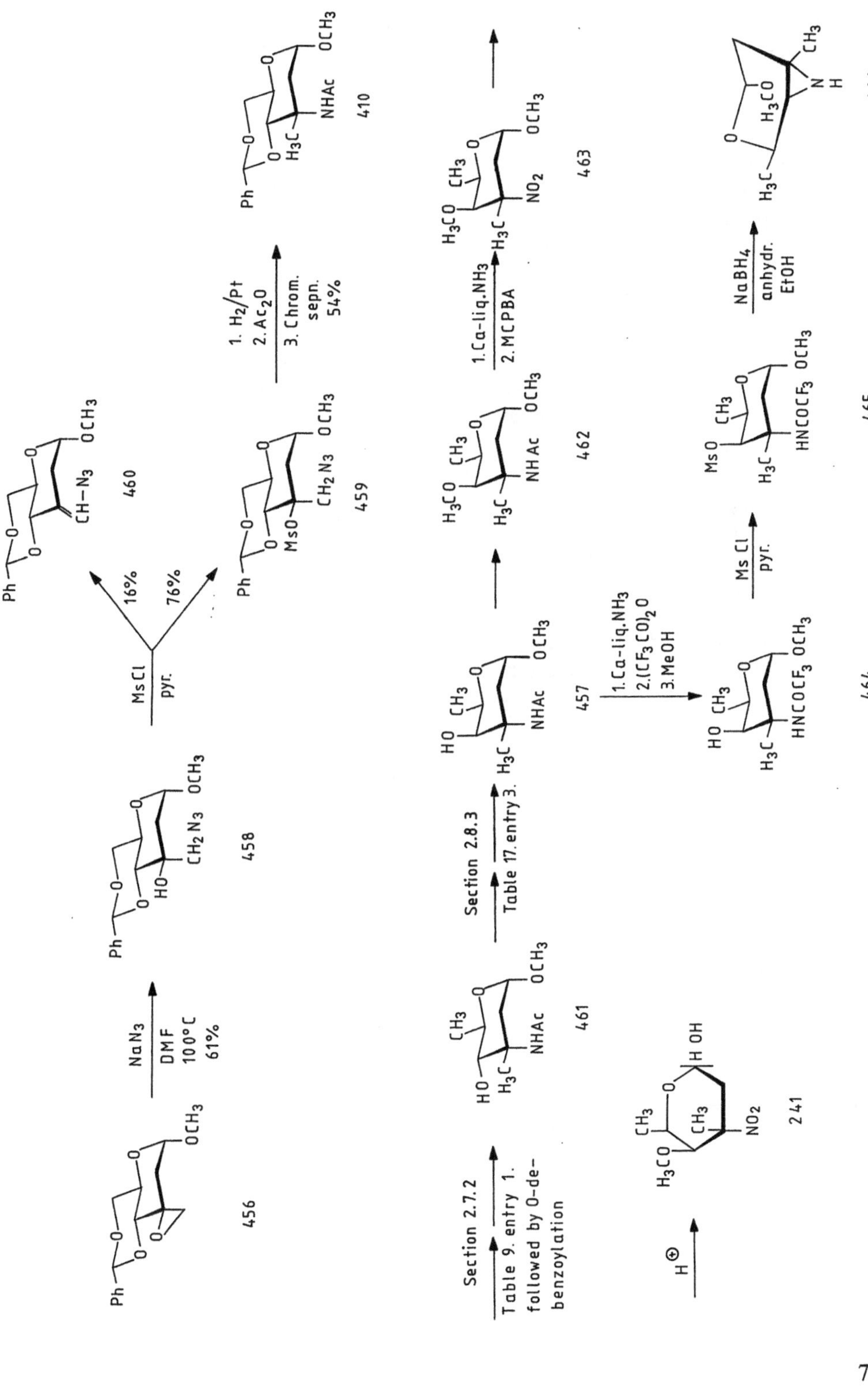

Fig. 8. Synthesis of D-rubranitrose **(241)** and precursors to methyl α-D-kijanoside

reduction of the resulting C-4 ulose **693** as discussed in Sect. 2.8.3 (Table 17 entry 3). O-Methylation of **457** yielded the 4-O-methyl ether **462**, which upon N-deacylation and oxidation using *m*-chloroperbenzoic acid produced methyl α-D-rubranitroside **463**. Subsequent acid hydrolysis then gave (182) the natural D-rubranitrose (**241**, Fig. 8).

For the synthesis of methyl α-D-kijanoside (**437**, Scheme 83) the acetamidosugar **457** was transformed (182, 416) into the corresponding N-trifluoroacetyl analog **464** (Fig. 8), which carries a more suitable N-protecting group for the generation of a 3,4-aziridine function under reductive conditions (vide infra). An additional route to **464** from the N-trifluoroacetyl analog of **410** proved to be essentially inconvenient because of difficulties with the C-4 inversion step (cf. Sect. 2.8.3, Table 17 entry 4). Mesylation of **464** gave the 4-O-mesyl ester **465**. Treatment of this latter substance with sodium borohydride in ethanol resulted in the reductive splitting off of the N-trifluoroacetyl group from **465** followed by the intramolecular displacement of the O-4-mesyl function by the *trans*-oriented amino group at C-3 to produce the desired 3,4-epiminohexopyranoside **466**.

This aziridine **(466)** was also synthesized in an independent way by the Yoshii group (417). Starting with D-mannose, the reaction proceeded via methyl α-D-mycaroside **(467)** (Scheme 81). The 3,4-di-O-benzoate **(468)** of **467** was reacted with N,N-thiocarbonyldiimidazole. The resulting thiocarbonate was not isolated but treated with hot dimethyl trimethylsilyl phosphite to obtain the 3,4-unsaturated sugar **469** in 77% yield. The reaction of this latter substance with iodine isocyanate and then with sodium methoxide in methanol produced the iodo-carbamate **470**, which upon reflux in aqueous methanol containing potassium hydroxide gave the aziridine **466**.

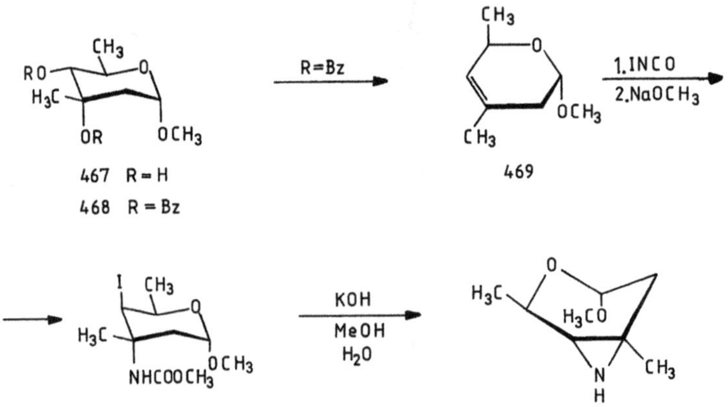

467 R = H
468 R = Bz
469
470
466

Scheme 81

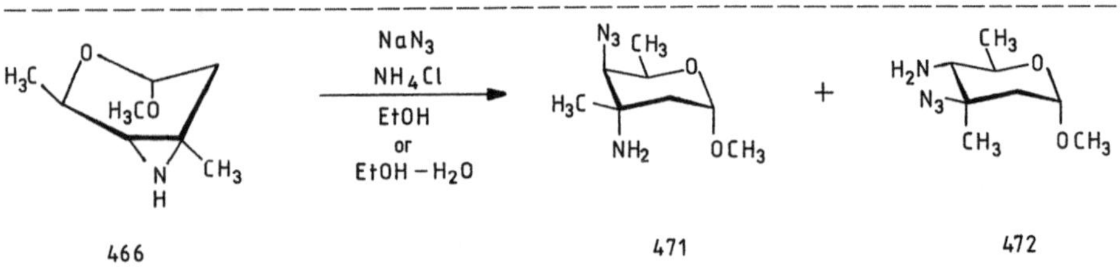

466
471
472

Scheme 82

Cleavage of the epimine ring of **466** (Scheme 82) was achieved with sodium azide in the presence of ammonium chloride in refluxing aqueous ethanol (182, 416) or in anhydrous ethanol (417) to form a separable mixture of methyl 3-amino-4-azido-2,3,4,6-tetradeoxy-3-C-methyl-α-D-*xylo*-hexopyranoside **(471)** and the corresponding *arabino* regio-isomer **472**. The ratio of **471**:**472** was found by Brimacombe et al. to be 4:1 (182, 416). The Japanese group (417) had previously reported a ratio of 3:2. The observed dissimilar proportions of the two ring-opened products might be attributable to the different reaction conditions. The desired 4-azido compound **471** was then transformed (417) into methyl α-D-kijanoside (methyl α-D-tetronitroside, **437**) by standard procedures **(471 → 473 → 474 → 475 → 437,** Scheme 83), of which the chemoselective removal of the benzyloxycarbonyl group in the presence of the C-4 methoxycarbonyl function **(474 → 475)** is particularly noteworthy for those working in related fields.

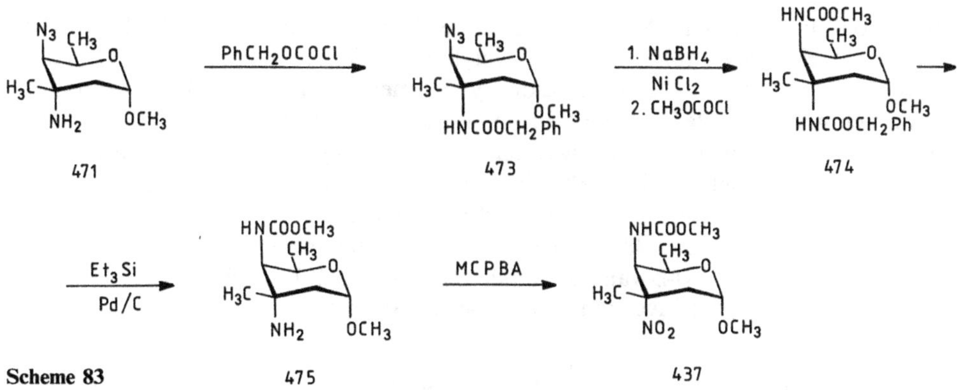

Scheme 83

2.5.4 Addition of iodine azide to C-3 methylene sugars

The addition of iodine azide (generated in situ from sodium azide and iodine monochloride) to 3-C-methylene sugars **(476)** results in (181, 418) a primary-iodo azide **(477)**, which transforms into a spiro-aziridine **(454)** upon treatment with lithium aluminum hydride (Scheme 84). As shown in Sect. 2.5.2, such aziridines can be readily hydrogenolyzed into 3-amino-3-C-methyl sugars **(455)** over a Raney nickel catalyst. The application (181, 418) of this procedure to methyl 4,6-O-benzylidene-2,3-dideoxy-3-C-methylene-α-D-*erythro*-hexopyranoside **(478,** prepared from ulose **100)** gave — in the first step — the adduct **479** (Scheme 85) in 24% yield as the only isolable product. The reaction of **478** with mercuric azide (412) did not lead to an assignable adduct. The treatment of **479** with lithium aluminum hydride in ether produced the spiroaziridine **480**, whose structure was identified by X-ray measure-

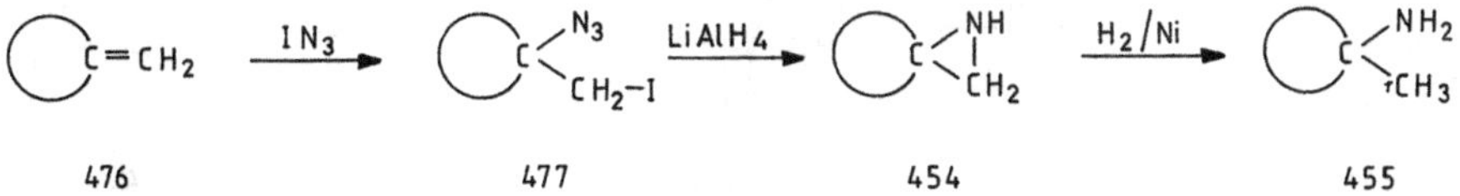

Scheme 84

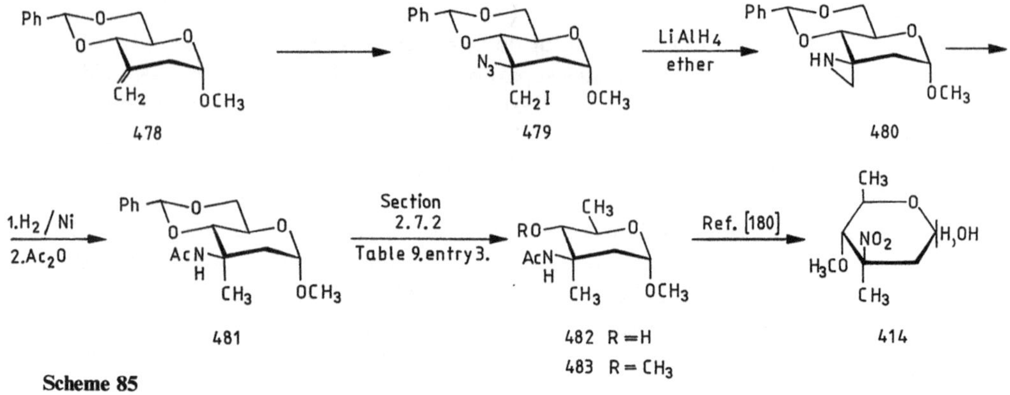

Scheme 85

ments on its crystalline N-acetyl derivative. The hydrogenolysis of **480** over a Raney nickel catalyst followed by N-acetylation afforded a 40% yield of methyl 3-acetamido-4,6-O-benzylidene-2,3-dideoxy-3-C-methyl-α-D-*arabino*-hexopyranoside **481** which was converted *via* the 6-bromo-6-deoxy analog **592** (Sect. 2.7.2, Table 9 entry 3) into the trideoxyamino sugar **482**. Careful methylation of this latter product provided the 4-O-methyl ether **483**, which can be converted into D-evernitrose **(414)** by the method of Yoshimura et al. (180) as shown in Sect. 2.5.2 (Scheme 79). The trideoxyamino sugar **482** also served as the starting material in a synthesis (183) of methyl N-acetyl-D-vancosaminide (**682**, Sect. 2.8.1, Table 15 entry 5), conducted by means of the inversion of the configuration at C-4.

2.5.5 Introduction of the $CH_3-\overset{|}{\underset{|}{C}}-NH_2$ branching by means of [3.3]-sigmatropic rearrangement

As mentioned in Sect. 2.2.5, the high-yielding route for the regio- and stereoselective synthesis of non-branched 3-amino-2,3,6-trideoxyhexoses carrying *cis*-oriented C-3 amino and C-4 hydroxyl functions, has recently been extended by

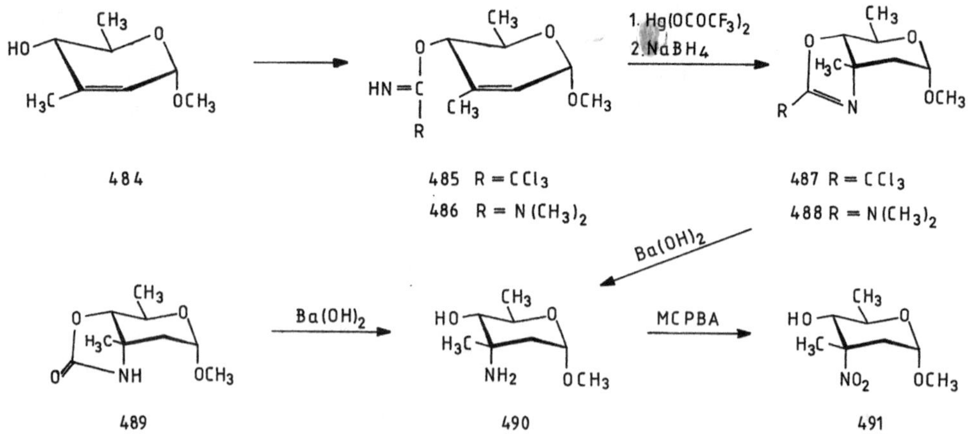

Scheme 86

Giuliano et al. (352, 419) to produce structurally related branched-chain 3-amino-sugar derivatives in both the D- and the L-series. The α-D-*erythro*-hex-2-enopyranoside **484**, available in five steps (420) from the hexopyranosid-3-ulose **100**, was converted (352) into the allylic imidates **485** and **486** by treatment with sodium hydride and trichloroacetonitrile (in ether) or dimethylcyanamide (without solvent) in 83% and 90% yields, respectively (Scheme 86). Mercury trifluoroacetate-mediated cyclization of **485** and **486**, followed by demercuration with sodium borohydride yielded the oxazolines **487** and **488**; the former is contaminated with both the respective dichloro-methyl analog and the urethane **489**. Although this latter mixture could be completely transformed upon mild acid hydrolysis into **489** and this in turn converted into the 3-aminosugar **490** by hydrolysis using barium hydroxide, a more straightforward way involved the direct alkaline hydrolysis of **488** into **490** with the same reagent (in 41% overall yield from **484**). Treatment of the aminoalcohol **490** with *m*-chloroperbenzoic acid then gave (352) methyl α-D-decilonitroside **(491)**, a potential precursor of D-rubranitrose **(241)**.

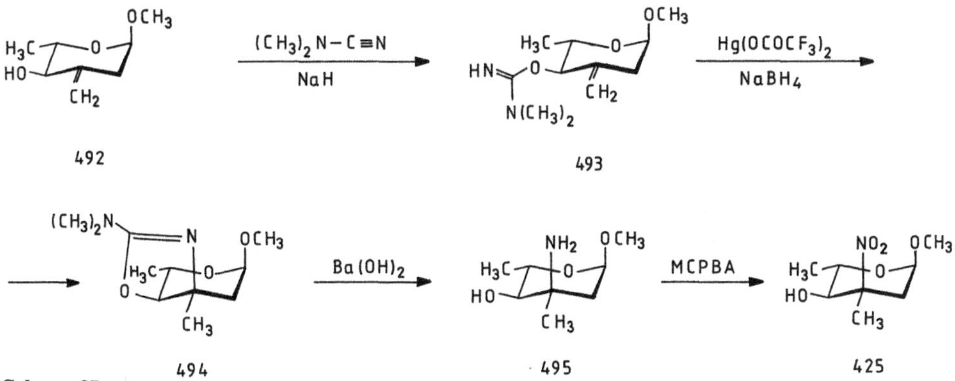

492 493 494 495 425

Scheme 87

Based on the recognition that the allylic imidates derived from dimethylcyanamide are more stable and resistant under the conditions of reductive demercuration with sodium borohydride than the respective trichloroacetimidates which suffer partial reduction, a stereoselective synthesis (Scheme 87) of methyl α-L-decilonitroside **(425)** was performed (419) via the 5-O-dimethylamidino derivative **493** obtained from methyl 2,3,6-trideoxy-C-methylene-α-L-*erythro*-hexopyranoside **(492)** (421). Thus, a strategy essentially similar to that outlined above for **486** yielded the target nitrosugar **425** in a total of nine steps starting with 3,4-di-O-acetyl-L-rhamnal. In contrast to these results, the anomeric trichloroacetimidate **497**, prepared from 4-O-benzoyl-3-C-methyl-2,3,6-trideoxy-α,β-D-*erythro*-hex-2-enopyranose **(496)**, does not undergo the expected sigmatropic rearrangement to yield the branched-chain aminoglycals **498**, although a successful attempt (353) has been reported. Detailed studies (422) have demonstrated, however, that when the reaction mixture of the allylic imidate **497** (prepared at −10 °C) was allowed to warm up to ambient temperature, it spontaneously rearranged into anomers of the trichloroacetyl glycosylamines **499** in 80% yield. At a slightly elevated temperature the furanose isomers **500** (ca. 47% yield) were observed in addition to this anomeric mixture **499** (ca. 41% yield), (Scheme 88).

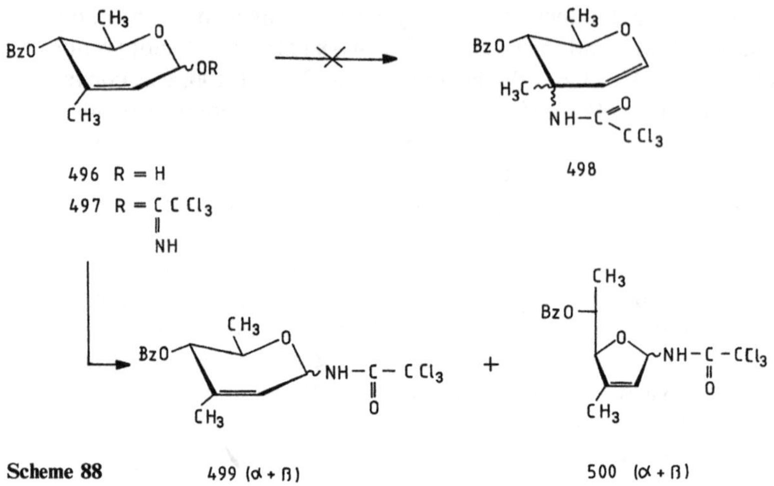

Scheme 88

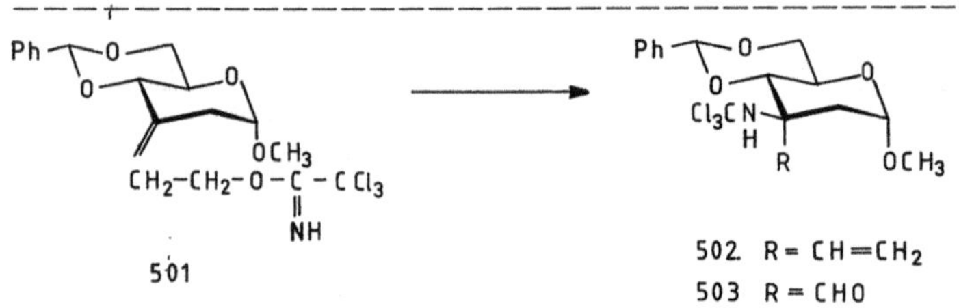

Scheme 89

On the other hand, [3.3]-sigmatropic rearrangement of the trichloroacetimidate **501** with an exocyclic allylic alcohol moiety readily proceeded (354) under thermal conditions (165 °C, 6 h) to give a 62% yield of the branched-chain aminosugar **502** with D-*arabino* configuration and not the D-*ribo* diastereoisomer, as previously supposed (353). The corresponding 3-C-formyl derivative **503** was obtained upon subsequent ozonolysis (Scheme 89).

In order to synthesize (354) D-vancosamine derivatives with the utilization of this method, compound **502** was deprotected in alkaline medium to give **504**, which was tosylated to produce **505** (in 40% overall yield). This latter substance was ozonized and the resulting C-formyl sugar **506** was reduced with sodium borohydride. The hydroxymethyl group of **507** so obtained was deoxygenated via the lithium aluminum hydride reduction of the intermediary tosylate **508**, and yielded methyl 4,6-O-benzylidene-2,3-dideoxy-3-C-methyl-3-tosylamino-α-D-*arabino*-hexopyranoside **(509)**, a potential precursor of D-vancosamine derivatives (Scheme 90).

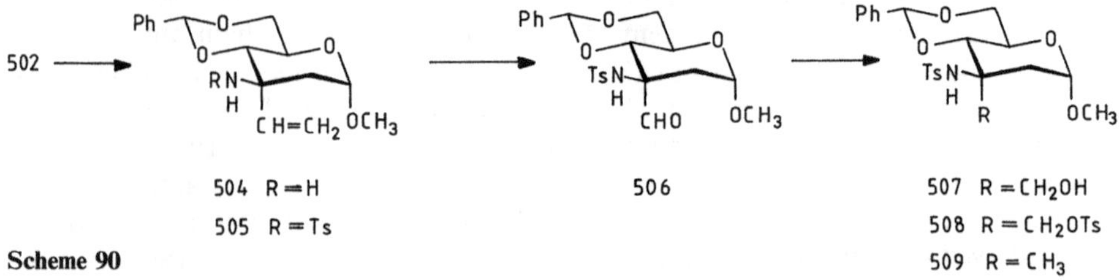

Scheme 90

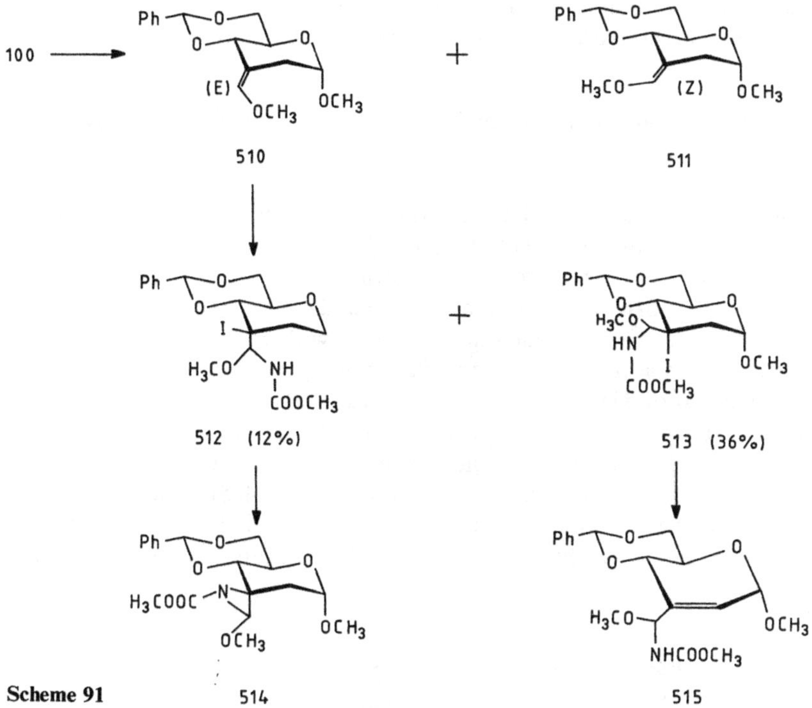

Scheme 91

The exocyclic *E*- **(510)** and *Z*- **(511)** enol ethers, derived from the ulose **100** by Wittig-olefination using methoxymethylenetriphenylphosphorane, have been used by Dyong et al. (423) to synthesize a novel type of branched-chain aminosugars (Scheme 91). The addition of silver isocyanate and iodine onto the double bond of **510** and subsequent methanolysis gave the epimeric 3-deoxy-3-iodo-3′-aminals **512** and **513**. Alkaline treatment of these two compounds then led to the substituted spiroaziridine **514** and the allylic aminal **515**, respectively.

2.6 Subsequent generation of the 2-deoxy functionality of unbranched- and branched-chain 3-aminosugars

Until the mid-1970s not much interest was devoted to the deoxygenation of 3-amino-3-deoxyhexosesat C-2 because such a strategy did not appear to be advantageous in the light of the pioneering studies in related fields by the groups of Brimacombe and Umezawa. The 2-deoxysugar precursors of 3-amino-2,3-dideoxy- and 3-amino-2,3,6-trideoxyhexoses were apparently more conveniently available by means of well-established methods (Sect. 2.1). In the majority of cases 2-deoxyhexoses were the starting materials of choice, and the corresponding methodologies as outlined and exemplified throughout the previous chapters, were used. However, the introduction of novel procedures (see Sects. 2.2.5 and 2.4) than led directly to aminohexoses with a reducible functionality at C-2, offering the alternative of designing routes to form daunosamine-type unbranched and branched-chain hexoses via the subsequent C-2 deoxygenation of carbohydrates already carrying the C-3 amino group.

2.6.1 Deoxygenation of 3-aminohexoses at C-2 according to the Fischer's glycal procedure

One of the earliest approaches to the deoxygenation of 3-aminosugars at position C-2 is based on the classical Fischer's glycal method. Derivatives of 4-epivancosamine (precursors to L-evernitrose) were prepared by Brimacombe et al. (181, 374, 405) from the acetaminoglycal **516** obtained in moderate yield (48%) from the aminosugar **371** (Sect. 2.5.1, Scheme 76) by sequential treatment with hydrogen bromide and the zinc-copper couple in acetic acid (Scheme 92). Attempts to adapt this procedure for the nitro analog **370** of **371** remained unsuccessful (374). The boron trifluoride-catalyzed addition of methanol to **516** produced a 3:1 mixture of the 2-deoxy-α-**(420)** and β- **(517)** methyl glycosides, and the pure components could be isolated only by means of a tedious chromatographic separation (181, 405). The α-glycoside **420** was then transformed into the corresponding 4-O-methyl ether **416**, a key intermediate in independent syntheses (180, 186) of 1-β-O-acetyl-L-evernitrose (**417**, see Table 6 entry 1).

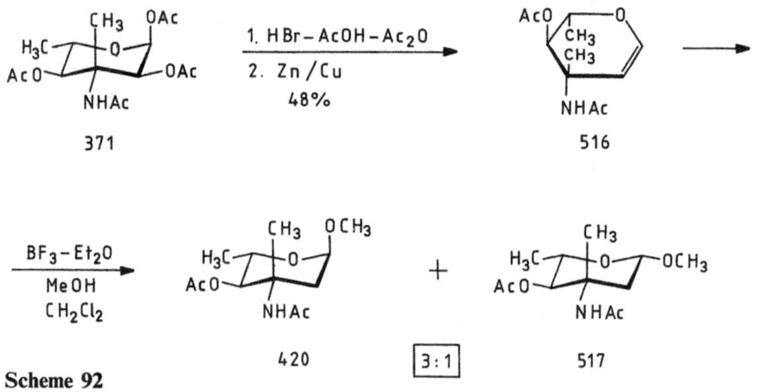

Scheme 92

3-Amino-3-deoxy-D-glucose (kanosamine) was converted into the aminoglycal **296** by essentially the same operation (Scheme 93). Acid-catalyzed methanol addition produced (372) a high yield of methyl 3-acetamido-4,6-di-O-acetyl-2,3-dideoxy-α-D-*arabino*-hexopyranoside **(518)**, a precursor in the total synthesis of negamycin derivatives (Sect. 4.5.2). It has been mentioned in Sect. 2.2.6 that no allylic rearrangement proceeded for such protected 3-aminoglycals upon the addition of alcohols under Lewis or mineral acid catalysis (cf. also Sect. 2.1.1). In an alternative (99) to such a glycosylation with protected aminoglycals, **296** was reacted with alcohols in the presence of N-bromosuccinimide, to produce the corresponding 3-acetamido-2-bromo-2-deoxy-α-D-mannopyranosides (*trans-diaxial* addition products) in a reaction similar to that for 3,4,6-tri-O-acetyl-D-glucal **(30)**. In contrast to **30**, however, the yields decreased as the steric bulk of the alcohol increased. When methanol was used, **296** produced 3-acetamido-4,6-di-O-acetyl-2-bromo-2,3-dideoxy-α-D-mannopyranoside

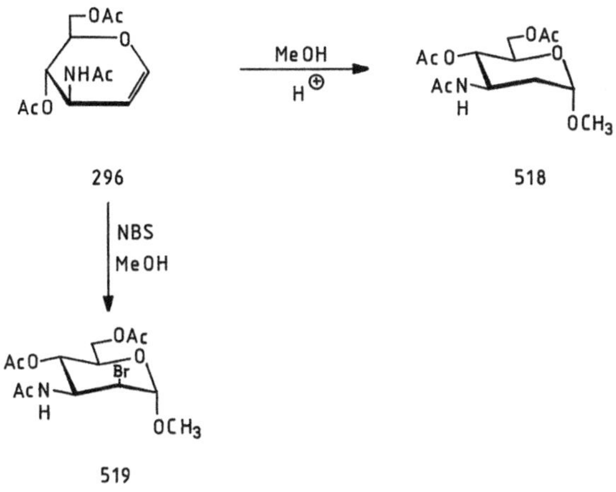

Scheme 93

(519) exclusively. Reductive dehalogenation of this latter product then gave the 2-deoxyglycoside **518** (Scheme 93). Additional examples of such C-2 dehalogenations are given in Sect. 2.6.3. In one of the procedures leading to L-daunosamine **(12)**, Tanaka (373) also applied the classical glycal route for the deoxygenation of the protected aminosugar **520** (Scheme 94). Methoxymercuration and subsequent reductive demercuration (Sect. 2.1.1) were used to introduce the anomeric methoxy group into the glycal **521**, which was then hydrolized with hydrochloric acid to obtain the hyhrochloride of L-daunosamine **(12)**. Details for the preparation of the starting aminosugar **520** are given in Sect. 2.7.4 (Scheme 115).

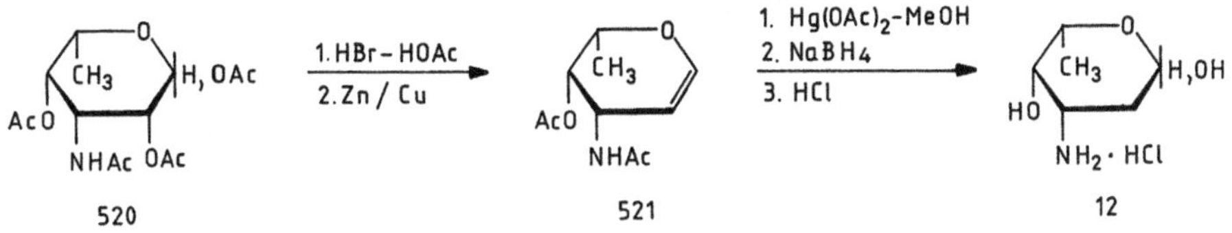

Scheme 94

2.6.2 Deoxygenation by means of the reductive removal of C-2 thiolester- and thioether functions

The Barton method (206), reported for the deoxygenation of neutral sugar derivatives (Sect. 2.1.5), was soon extended to 3-amino-3-deoxysugars in order to obtain the corresponding 2-deoxy analogs. However, the first attempts, carried out by the Brimacombe group, brought disappointingly low yields (181, 405) of the desired 3-amino-2,3-dideoxy compounds. In the course of these studies the branched-chain aminosugar **372** was transformed (181, 405) by sequential monotosylation at O-2 under phase-transfer conditions into **522**; methylation at O-4 gave **(523)**, and detosylation produced methyl 3-acetamido-3,6-dideoxy-3-C,4-O-dimethyl-α-L-glucopyranoside **(524)**. Treatment of this latter substance with N,N-dimethyl-α-chlorobenzylidene ammonium chloride (generated in situ) and hydrogen sulfide (206) produced only a 27% yield of the 2-thiobenzoate ester **525**, which upon reduction with tributyltin hydride in refluxing toluene gave the 2-deoxyglycoside **416** in as low as 37% yield (Scheme 95).

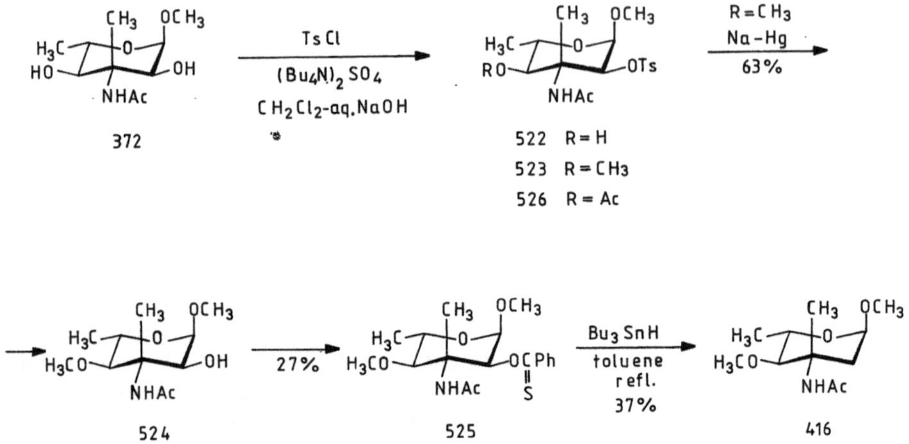

Scheme 95

Brimacombe and coworkers also examined (405) whether the same radical deoxygenation with an *axially* disposed 2-O-thiobenzoyl ester function would provide a more efficient access to the 2-deoxy series. In order to obtain the C-2 epimer **(528)** of **525**, the monotosylate **522** was acetylated to form **526**, which was subjected to inversion of the configuration at C-2 via nucleophilic displacement with acetate ion (Scheme 96). The resulting mannose derivative **527** with OH-2 free was converted into the 2-O-thiobenzoate **528** in 31% yield. However, the reduction of this ester (as shown above for **525**) gave only a 30% yield of methyl 3-acetamido-4-O-

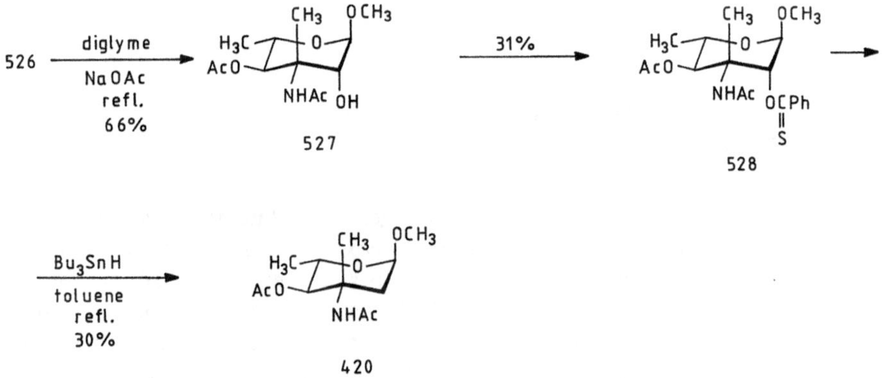

Scheme 96

acetyl-3-C-methyl-2,3,6-trideoxy-α-L-*arabino*-hexopyranoside **(420)**. These experimental findings clearly show that both the glycal procedure (Sect. 2.6.1) and the cyanomesylation route (Sect. 2.5.2, Table 6 entries 1 and 2) are apparently more efficient for the preparation of the branched-chain 2-deoxyglycosides **416** and **420**.

On the other hand, a successful attempt to improve the Barton deoxygenation method in the case of protected 3-amino-3-deoxyhexopyranosides has most recently been reported (383) (Scheme 97).

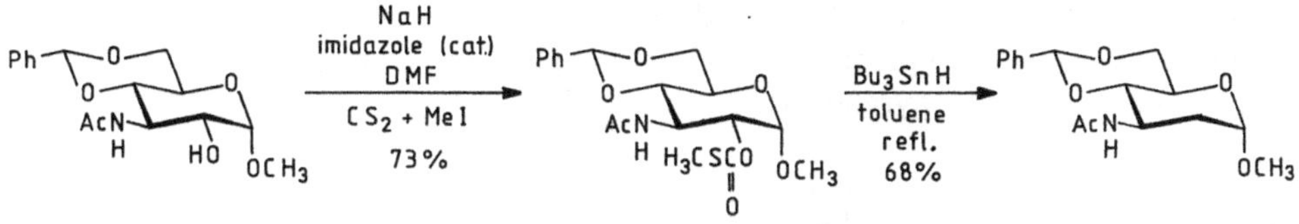

529 530 197

Scheme 97

Methyl 3-acetamido-4,6-O-benzylidene-3-deoxy-α-D-glucopyranoside **(529)**, prepared from **308** (see Sect. 2.2.7, Scheme 59) was sequentially treated with sodium hydride and a catalytic amount of imidazole in DMF and then with carbon disulfide and methyl iodide in a one-pot operation, to obtain the dithiocarbonate ester **530** in 73% yield. Reduction of this latter compound using tributyltin hydride in refluxing toluene yielded a readily separable 8:2 mixture of the 2-deoxy compound **197** and the starting **529**, from which the desired aminoglycoside **197** could be obtained in 68% yield by means of fractional crystallization. We believe that this procedure for the preparation of the α-*arabino*-glycoside **197** may not be as effective as that leading through the 2-deoxy-3-O-mesylate **144** and the α-D-*arabino*-azide **146** (see Sect. 2.1, Scheme 27).

A method of much greater utility was reported by Picq et al. (424) in 1985. They found that the secondary mesylate group of methyl glycopyranosides, being in a *trans* relationship with a diallylamino function, can be readily reduced using sodium borohydride into deoxygenated aminoglycosides in an intramolecular participation process involving an aziridinium ion intermediate (cf. Sect. 2.4.1, Scheme 68).

In the case of the aziridinium ion **345** produced from methyl 4,6-O-benzylidene-3-diallylamino-2-O-methanesulfonyl-α-D-altropyranoside **(344)**, the attack of the hydride nucleophile occurs exclusively at C-2 to yield the α-D-*ribo* diallylamine **531**

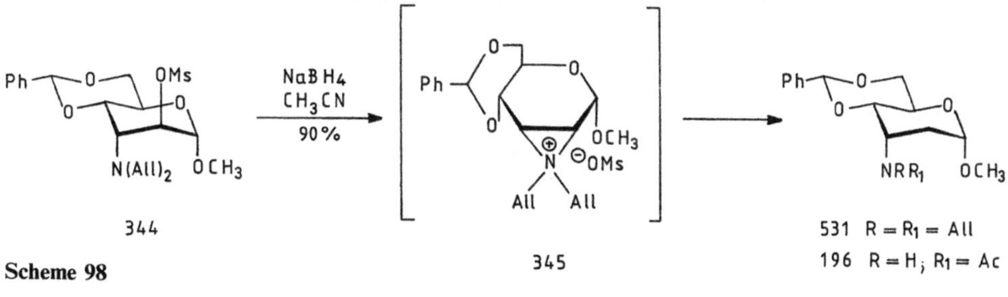

344

345

531 R = R1 = All
196 R = H; R1 = Ac

Scheme 98

(Scheme 98). This aminosugar was transformed into the N-acetyl derivative **196**, which was converted (425) by Horton's procedure (172) into L-daunosamine **(12)** via the 6-bromosugar **565** and the hex-5-enopyranoside **613** (Sect. 2.7.4, Table 11 entry 3). L-Daunosamine **(12)** and its methyl α,β-furanoside glycoside **533** were obtained by Yamaguchi et al. (248) (Scheme 99) by the reductive removal of the 2-S-benzyl ether function of methyl 3-amino-2-S-benzyl-3,6-dideoxy-2-thio-α,β-L-galactofuranoside (**532**, see Sect. 2.7.4, Scheme 116) upon desulfurization using Raney nickel. An analogous desulfurization, accompanied by the reductive cleavage of the benzylidene acetal function of methyl 3-acetamido-4,6-O-benzylidene-2-benzylthio-3-deoxy-α-D-gluco (426) and altro-pyranoside (388), was used by Christensen and Goodman to prepare methyl 3-acetamido-2,3-dideoxy-α-D-*arabino*- **(534)** and α-D-*ribo*- **(535)** hexopyranosides (Scheme 100).

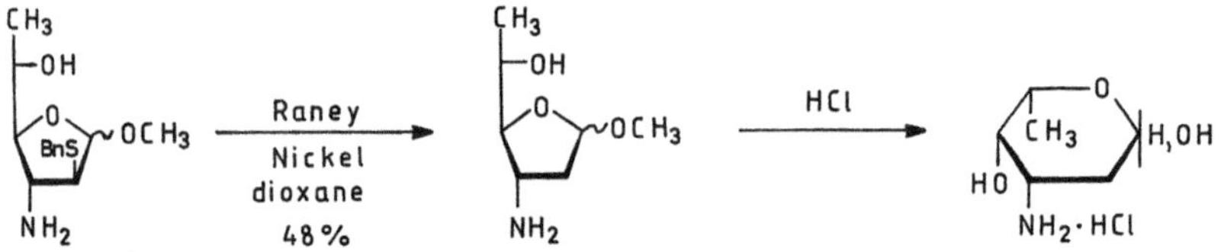

532 533 12

Scheme 99

534 535

Scheme 100

2.6.3 Reductive dehalogenation of 3-amino-2-halogeno-2,3-dideoxyhexose derivatives

The C-2 halogeno derivatives of alkyl 3-amino-2,3-dideoxyhexopyranosides are readily converted into the corresponding 2-deoxyglycosides by reductive dehalogenation. In such a manner the 2-chloro- **(334)** and 2-iodo- **(336)** 3-acetamidoglycosides, prepared as shown in Sect. 2.4.1 via ring opening of 2,3-epiminohexopyranosides (Scheme 66), were hydrogenolyzed (276) over Raney nickel (in 60–70% yield) (Scheme 101) to form methyl 3-acetamido-2,3-dideoxy-α-D-*ribo*-hexopyranoside

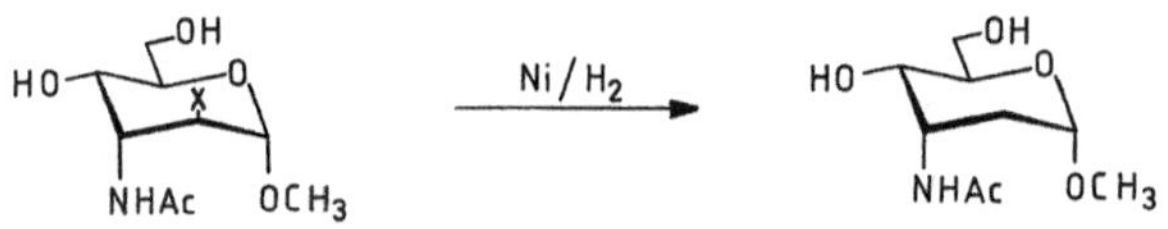

334 X=Cl

336 X=I

535

Scheme 101

(535), a potential precursor of D-ristosamine **(20)** and L-daunosamine (**12**; cf. Sect. 2.7.4).

A key step in the synthesis (395) of methyl 3-N-benzoyl-4-O-benzoyl-β-L-daunosaminide **(537)** is the dehalogenation of compound **333** (Sect. 2.4.1, Scheme 66) using one equivalent of tributyltin hydride and an excess of sodium borohydride in refluxing ethanol under UV-irradiation to produce **536** in 80% yield (Scheme 102). This aminosugar was then converted into the desired daunosamine glycoside **537** by deoxygenation at C-6 with a concomitant inversion of the configuration at C-5 (Sects. 2.7.2 and 2.7.4) through the 6-bromo-6-deoxy derivative **567** (Table 7 entry 7) and the hex-5-enopyranoside **615** (Table 11 entry 5).

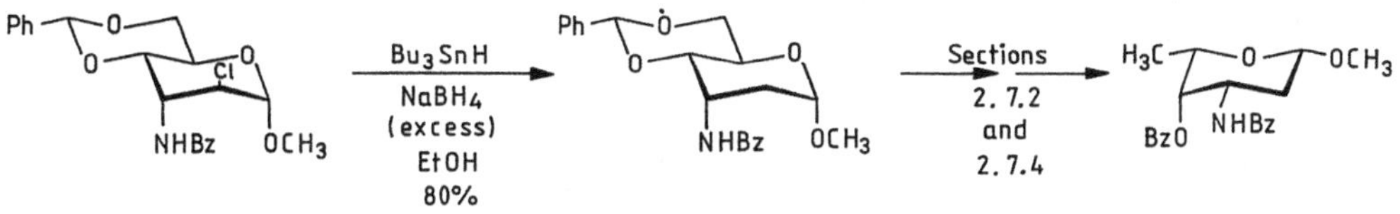

333 536 537

Scheme 102

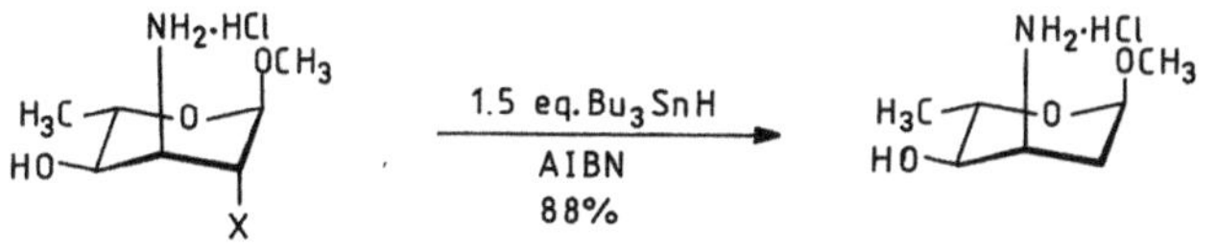

363 X = Br
364 X = I
141

Scheme 103

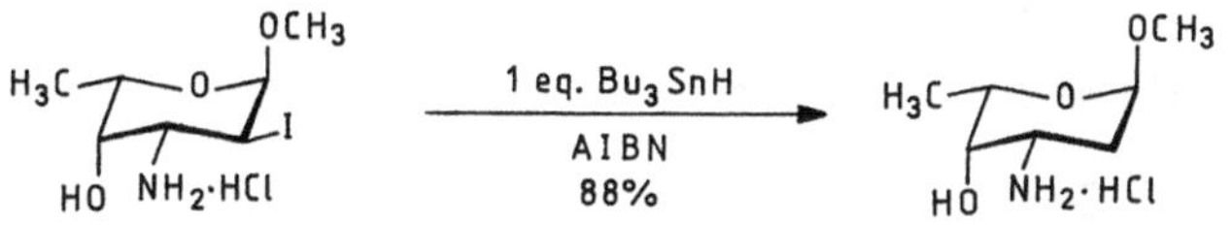

365 260

Scheme 104

Cleavage of the C_2-halogen bond of the 3-amino-2-halogenohexopyranosides **363–365**, **367**, and ±**264** — prepared by means of the bromonium and iodonium ion-induced cyclization of allylic trichloroacetimidates (see Sects. 2.2.5 and 2.4.3) — has also been utilized in alternative syntheses (345, 347, 348, 351) of L-ristosamine and L-daunosamine derivatives (Schemes 103–105). Treatment of the methyl 2-bromo- **(363)** and 2-iodo- **(364)** -α-L-altropyranosides with 1.5 equivalents of tributyltin hydride in the presence of azobis(isobutyronitrile) produced (345) methyl α-L-ristosaminide **(141)** in 88% yield (Scheme 103). The (s)-2-iodoglycoside **365** was converted (348) into methyl α-L-daunosaminide (**260**, Scheme 104) in a similar manner. D,L-Daunosamine **(±12)** was obtained (347) from the racemic 2-bromosugar ±**264** upon analogous radical reduction followed by the removal of the protecting groups (Sect. 2.2.5, Scheme 52).

When benzyl 2-iodo-3-trichloroacetamido-2,3,6-trideoxy-α-L-galactopyranoside **(367)** was allowed (351) to react with six equivalents of tributyltin hydride in benzene at room temperature for ten days, a 79% yield of the completely dehalogenated compound benzyl N-acetyl-α-L-daunosaminide **(538)** was obtained (Scheme 105). With the same reagent, **367** transformed into the de-iodinated dichloroacetamidosugar **539** in only 12 hours. At the same time, catalytic hydrogenation of **367** over various catalysts led (351) to the replacement of a single chlorine atom of the acylamide moiety with hydrogen even upon pressure hydrogenation. Reductive de-iodination of various alkyl 3-azido-2-iodo-hexopyranosides using tributyltin hydride was achieved, without a concomitant reduction of the azide function, to obtain (351) the corresponding 3-azido-2,3-dideoxysugar derivatives. This knowledge has been used (351) in syntheses of daunosamine-containing oligosaccharide derivatives.

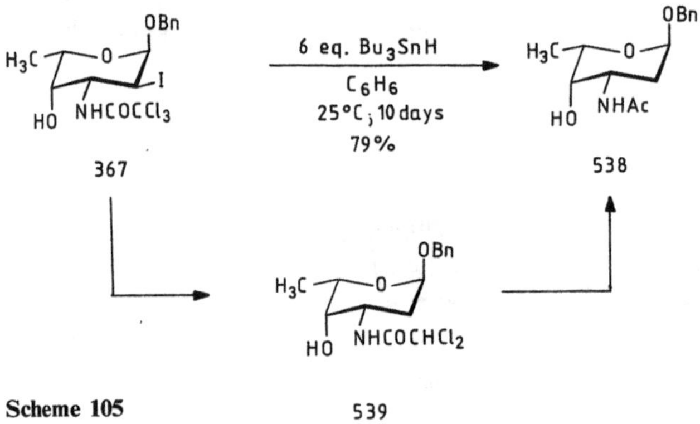

Scheme 105

2.7 Synthetic strategies for the preparation of the 6-deoxy analogs of 3-amino-3-deoxyhexopyranosides

2.7.1 Deoxygenation at C-6 via C-6 sulfonate ester derivatives

As mentioned in Sect. 2.2.1, the primary sulfonate ester group of hexopyranosides can be readily reduced to a terminal deoxy function using lithium aluminum hydride (140, 227, 256, 427) or lithium triethyl borohydride (428). By using lithium aluminum hydride, Overend et al. (228) transformed the 6-O-tosylate **540**, derived from **201** (Sect. 2.2.3, Scheme 36) by means of acid hydrolysis of the acetal moiety followed by tosylation, into methyl 3-methylamino-2,3,6-trideoxy-α-D-*xylo*-hexopyranoside **(541)** in 47% yield (Scheme 106).

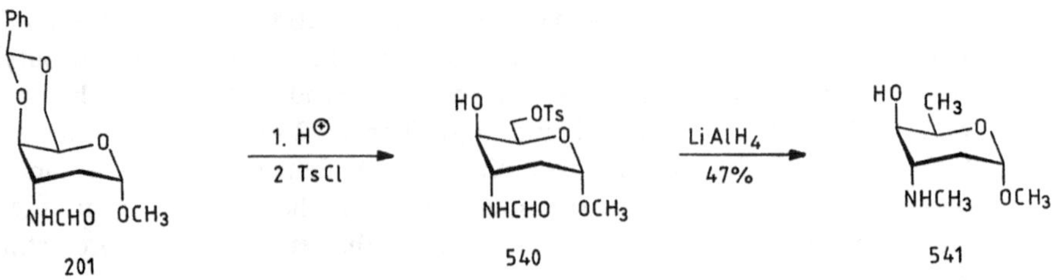

Scheme 106

The reason for the moderate yields observed for this and similar transformations is that the displacement of such sulfonates by the hydride nucleophile, — expectedly resulting in the cleavage of the terminal C-O bond, is frequently accompanied by varying amounts of concurrent O-S bond scission (ester hydrolysis). Hydrogenolysis over Raney nickel results only in the hydrolysis of the sulfonate ester group, and the unchanged starting materials were usually recovered when palladium-on-carbon catalysts were employed (429). Sodium borohydride has also been reported to reduce primary O-mesylates (430) and O-trifluoromethanesulfonates (431) of carbohydrates under mild conditions; in most cases without significant concurrent ester hydrolysis.

In practice, however, primary tosylates are usually transformed into the 6-iodo-6-deoxy analogs which upon catalytic hydrogenation or reduction with sodium borohydride in DMSO (432) can be readily converted into 6-deoxyhexopyranosides. Deoxygenation of 6-deoxy-6-iodohexopyranosides carrying a C-2 azide function can be achieved, without concomitant reduction of the azide, using sodium cyanoborohydride in hexamethyl phosphoric triamide (433, 434). The superior chemoselectivity of this reagent system for the reductive removal of primary and secondary iodo- and bromo-, as well as primary tosylate functions from non-carbohydrate substances, leaving other substituents (such as chloro, nitro, cyano, keto and epoxide groups) and unsaturated bonds unaffected, has been demonstrated by numerous examples (435).

In an effort (Scheme 107) to prepare the first synthetic daunosamine derivative, enatiomerically related to that of the natural L-sugar, Richardson (229, 230) converted methyl 4,6-O-benzylidene-3-azido-2,3-dideoxy-α-D-*arabino*-hexopyranoside; (**146**, Sect. 2.2.1, Scheme 27) into the acetamide **197** from which the 6-O-tosylate was prepared by the sequential removal of the benzylidene acetal function and the selective tosylation of the terminal hydroxyl group. The resulting 6-O-sulfonate readily underwent displacement with iodine in acetone, to produce the 6-deoxy-6-iodo compound **543** in 75% yield. This was subjected to reductive dehalogenation over Raney nickel to give an 86% yield of methyl 3-acetamido-2,3,6-trideoxy-α-D-*arabino*-hexopyranoside (**544**, methyl N-acetyl-α-D-acosaminide). Inversion of the configuration of this latter aminosugar at C-4 was then effected (229–231) as shown in Sect. 2.8.1 to obtain N-benzoyl-D-daunosamine **(671)** via the *arabino*-4-O-methanesulfonate **658** (Table 14 entry 6).

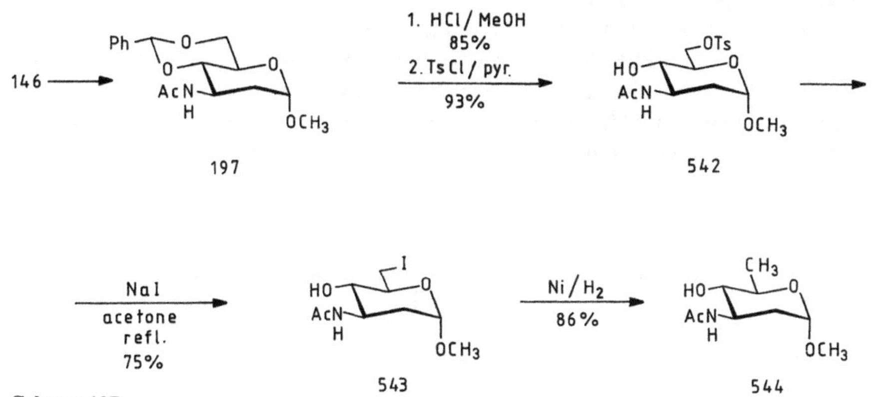

Scheme 107

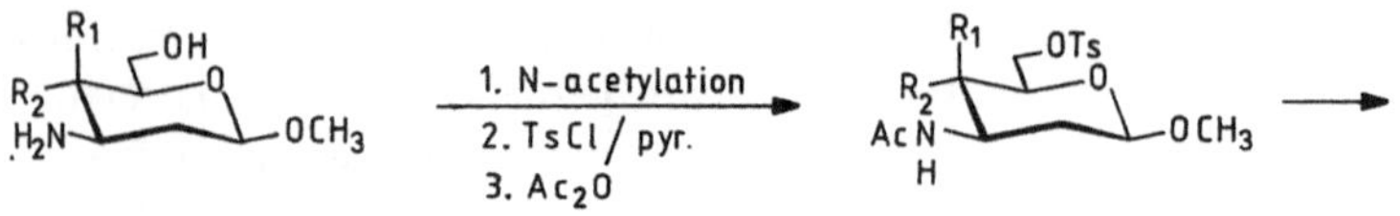

229 R_1=H; R_2=OH

230 R_1=OH; R_2=H

547 R_1=H; R_2=OAc

548 R_1=OAc; R_2=H

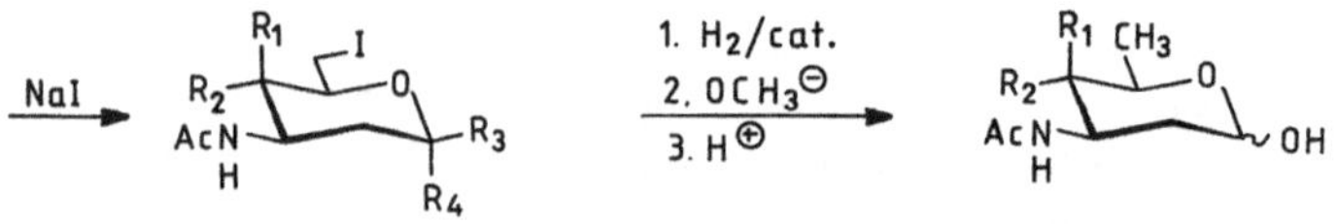

549 R_1=H; R_2=OAc; R_3,R_4=H,OCH_3

550 R_1=OAc; R_2=R_3=H; R_4=OCH_3

551 R_1=R_4=H; R_2=OAc; R_3=OCH_3

545 R_1=OH; R_2=H

546 R_1=H; R_2=OH

Scheme 108

In an alternative synthesis of N-acetyl-D-daunosamine **(545)** and N-acetyl-D-acosamine **(546)**, Baer and associates (318) also employed (Scheme 108) the 6-O-tosylate → 6-iodoglycoside → 6-deoxyglycoside route to deoxygenate the *arabino*- **(229)** and *lyxo*- **(230)** 3-amino-2,3-dideoxy-β-D-hexopyranosides into the target aminodeoxy sugars. Thus, **229** and **230** (obtained by nitromethane cyclization and subsequent Schmidt-Rutz reaction (Sect. 2.2.4, Scheme 44) were N-acetylated, mono-tosylated at O-6, and O-acetylated to produce the acetylated primary tosyloxy compounds **547** and **548**, which were treated with sodium iodide. The reaction of the *arabino* isomer **547** was performed in butanone, and an anomeric α,β-mixture **(549)** of the 6-deoxy-6-iodoglycosides was isolated. Iodination of the *lyxo* compound **548**, carried out in a sealed tube with the same reagent and solvent, led to the α-methyl 6-iodoglycoside **550** in rather low yield (20%). The formation of the α,β-mixture **549** was explained (318) by anomerization of the extremely acid-labile (436) 2-deoxy-β-glycoside **547** by the action of the traces of acid liberated during the displacement reaction. On the other hand, when iodination of **547** was carried out in the presence of sodium hydrogen carbonate (as acid acceptor), the 6-iodo-β-glycoside **551** was produced in excellent yield (93%). Catalytic hydrogenation of the iodo-compounds **550** and **551**, followed by O-deacylation and mild acid hydrolysis then furnished N-acetyl-D-daunosamine **(545)** and N-acetyl-D-acosamine **(546)**.

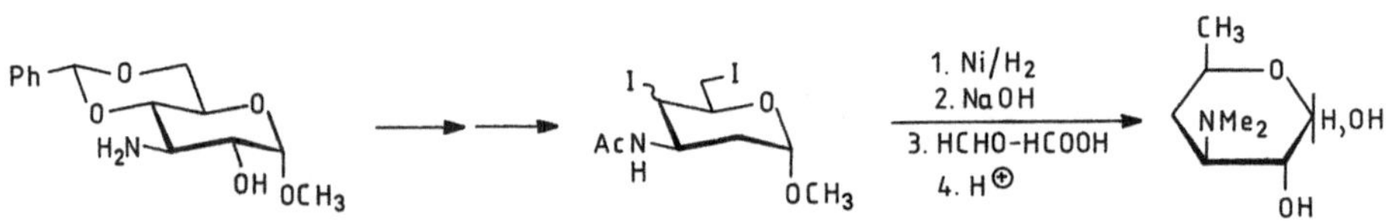

308 552 553

Scheme 109

The *lyxo* sugar **545** was also obtained (318) by means of the inversion of the configuration of the 6-deoxysugar **650** (derived from **551**) at C-4, as shown in Sect. 2.8.1 (Table 14 entry 7).

In a similar way, Richardson (384) converted the aminosugar **308** (Sect. 2.2.7, Scheme 59) into the 4,6-di-iodo compound **552** by sequential N-acetylation, acetal hydrolysis, mesylation, and replacement of the 4-O- and 6-O-mesyloxy functions with iodine (Scheme 109). Removal of the halogen atoms by catalytic hydrogenation, followed by reductive methylation of the liberated amino function and acid hydrolysis of the glycosidic moiety, produced desosamine (**553**, 3-dimethylamino-3,4,6-trideoxy-D-*xylo*-hexopyranose), a common, basic, nitrogen-containing component of the macrolide antibiotics erythromycin, narbomycin, picromycin, methymycin, and oleandomycin. Khan and Hough (437, 438) applied a similar procedure for conversion of methyl D-glucosaminide derivatives into the corresponding 6-deoxy analogs.

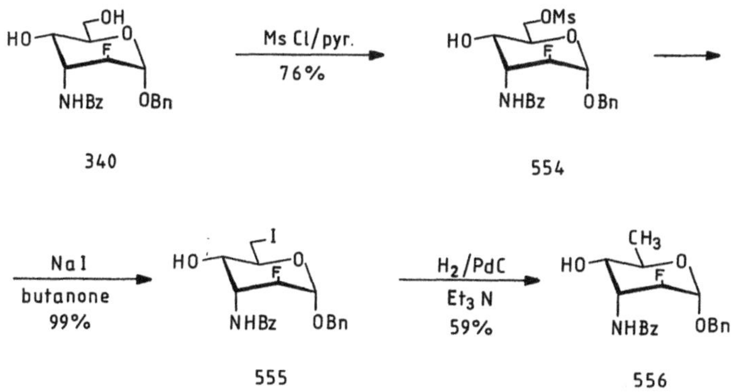

Scheme 110

The synthesis of benzyl 3-benzamido-2,3,6-trideoxy-2-fluoro-α-D-altropyranoside **(556)** from the 4,6-diol **340** (Sect. 2.4.1, Scheme 67) was carried out according to an analogous strategy (396) (Scheme 110). The unimolar mesylation of **340** gave the 6-O-mesylate **554** (76%), which was transformed into the 6-deoxy-6-iodo analog **555**. Catalytic hydrogenation of this latter substance over palladium-on-carbon in the presence of triethylamine produced a 59% yield of **556**, a protected (s)-2-fluoro derivative of D-ristosamine **(20)**. An essentially similar procedure was applied (400) in the preparation of the 3-azido analog **(626)** of **555** from the corresponding 4,6-O-benzylidene acetal (**351**, Scheme 70). Conversion of **626** into benzyl (s)-2-fluoro-β-L-daunosaminide **(361)** via the C-5 methylene derivative **627** is discussed in Sect. 2.7.4 (Table 12 entry 1). The transformation of the trifluoromethanesulfonate esters of protected neutral carbohydrates (involving 6-O-triflates) into the respective halo-deoxy sugars was easily achieved (439) using tetrabutylammonium halogenides under very mild conditions. This reaction might also find use in the synthesis of antibiotic aminodeoxy sugars.

2.7.2 C-6 Deoxygenation of 3-azido- and 3-amino-2,3-dideoxyhexopyranosides with the utilization of the ring opening reactions of 4,6-O-benzylidene acetals

Following pioneering studies of the action of N-bromosuccinimide on acyclic (440) and cyclic (441) acetals of benzaldehyde, this reaction was introduced to carbohydrate chemistry by Hanessian (159) and Hullar et al. (442) in 1966 for the acetal ring-cleavage of 4,6-O-benzylidenehexopyranosides. As a result of extensive studies, primarily by the Hanessian group (160–162, 443), this process — resulting in 4-O-benzoyl-6-bromo-6-deoxyhexopyranosides with high regioselectivity — has emerged as one of the most widely used procedures for the deoxygenation of functionalized carbohydrates at C-6 by the dehalogenation of the 6-bromo derivatives produced.

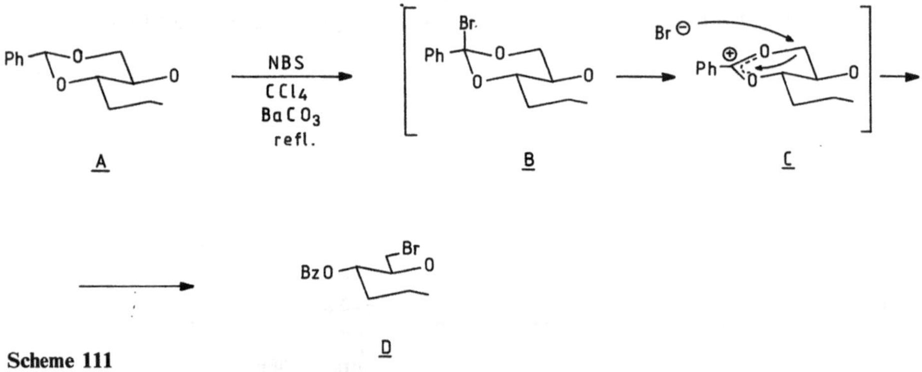

Scheme 111

The mechanism of the reaction is shown in Scheme 111. The bromine atom produced from the reagent replaces the benzylic hydrogen of the benzylidene acetal *(A)* in a radical and/or ionic reaction (160, 443), to result in the formation of an unstable geminal bromoacetal *(B)*. This latter product may dissociate into a 2-phenyl-1,3-dioxolenium ion *(C)* and bromide ion which attacks at carbon C-6 to yield the 4-O-benzoyl-6-bromo-derivative *(D)*.

The scope and limitations of the Hanessian-Hullar reaction have been circumstantially demonstrated (160–162, 443) and thoroughly reviewed (444–446); therefore these are not the subjects of this chapter. It is to be emphasized, however, that except for benzyl and allyl substituents, the procedure works well in the presence of most functional groups employed in carbohydrate chemistry, including nitrogen-carrying moieties (azide, alkyl- or acylamides) also. Applications of this acetal ring cleavage method to di-O-benzylidene-hexopyranosides have been outlined in Sect. 2.1.4.

Since 1966 the Hanessian-Hullar process has found extensive utility in numerous syntheses (443–446) of various 6-bromo- and 6-deoxyhexopyranosides and related oligosaccharides having biological importance but this work is not treated in this chapter either. An important step in the systemic layout of antibiotic aminodeoxy sugar molecules, the extension of this method to 3-azido- and 3-amino-4,6-O-benzylidene-2,3-dideoxyhexopyranosides in order to provide access to deoxygenation at C-6, is discussed in detail in light of the hitherto published results. The conversion of 4,6-O-benzylidene-2,3-dideoxy-D-glycopyranosides carrying a C-3 nitrogen substituent and their 2-halogeno- and 3-C-methyl-branched derivatives into the corresponding 6-bromo-6-deoxyhexopyranosides are summarized, in Tables 7, 8, and 9, along with the methods employed for their subsequent dehalogenation to obtain the desired 6-deoxy-D-hexose analogs.

The ring-opening of the benzylidene acetal function of the *α-arabino* compounds **146**, **197**, **557**, and **558** (Table 7 entries 1–4) using N-bromosuccinimide in refluxing carbon tetrachloride in the presence of barium carbonate gave the 6-bromo compounds **561–564** in 75–100% yields. Reductive dehalogenation of **561** and **562** (Table 7 entries 1 and 2) readily yielded the methyl α-D-acosaminide derivatives **571–573**, and **544**, and removal of the protecting groups of aminoglycoside **573** resulted in (174) D-acosamine hydrochloride **(10)**, the aminosugar component of sporaviridin (12). The bromosugars **561** and **563** (Table 7 entries 1 and 3) were also utilized in syntheses of the highly functionalized carbohydrates **577** (234) and **579** (233) (precursors of thienamycin and azaprostaglandin derivatives) and in the preparation (447, 448) of the azidocyclohexanone **578**, as shown in Sects. 4.3.2, 4.3.3 and 4.7.

From the α- **(565)** and β-*ribo*- **(568)**-6-bromo compounds (Table 7 entries 5 and 8) derived from the acetals **196** and **152**, D-ristosamine hydrochloride **(20)** was synthesized via the 6-deoxy-α- **(574)** (174, 175) and β- **(575** and **576)** glycosides (114, 239, 240). Hydrogenation of methyl 3-azido-4-O-benzoyl-6-bromo-2,3,6-trideoxy-β-D-*ribo*-hexopyranoside **(568)** to produce **575** (Table 7 entry 8) represented (114, 239) the first example of the simultaneous generation of the 3-amino- and 6-deoxy functions of di- or trideoxy-3-aminohexoses from 3-azido-6-bromosugar derivatives. Such synthetic strategy has been widely utilized (231, 234, 240, 451, 452) since then.

Conversion of the benzylidene azide **152** into the 3-acylamino derivatives **559** and **560** allowed the preparation (448) of the 6-bromo compounds **569** and **570** (Table 7 entries 9 and 10), which were transformed into the aminocyclohexanones **583** and **584** via the corresponding C-5 methylene compounds, as discussed in Sects. 2.7.4 and 4.7. Note that in the case of the methyl β-D-*ribo* 3-acylaminoglycosides **569**, **570**, and **575** (Table 7 entries 8, 9, and 10) the $^1C_4$(D) conformer with an *equatorially* disposed C-3 function is favored in solution. The 6-bromo-aminoglycoside **567** (Table 7 entry 7) was employed (450) in the preparation (Sect. 4.7) of a unique carbocyclic analog **(581)** of L-daunosamine **(12)**.

The 6-bromo-6-deoxyglycosides with the α-D-*arabino* **(564)** and α-D-*ribo* configurations (**566** and **567**) also served as the starting materials to prepare 3-amino-2,3,6-trideoxyhexoses **179** (449), **580** (178, 179), and **537** (395) in the L-series (Table 7 entries 4, 6 and 7), via the appropriate hex-5-enopyranosides **612**, **614**, and **615**, respectively (see Sect. 2.7.4, Table 11 entries 2, 4, and 5). An analogous reaction sequence, used with the 6-bromo-2-fluoro-3-aminoglycosides **586** and **587** derived from the acetals **347** and **360** (Table 8 entries 2 and 3), led to 2-fluoro-daunosamine **(348)** (399) and the 2-fluoro-α-L-acosaminide derivative **362** (401), respectively, by means of the hydrogenation of the C-5 methylene compounds **629** and **631** (see Sect. 2.7.4, Table 12 entries 3 and 5). Additional examples of such conversions resulting in 6-deoxysugar derivatives with concomitant inversion of the configuration at C-5 are discussed in Sect. 2.7.4.

Protected C-5 epimers (**461** and **595**) of L-vancosamine **(15)** have been synthesized (177, 182, 414) by the reductive dehalogenation of the 6-bromo-3-C-methyl aminoglycosides **590** and **591**, respectively, (Table 9 entries 1 and 2). The natural D-rubranitrose **(241)** was also prepared (182, 414) from the 6-bromo intermediate **591**

Table 7. Preparation and conversion of 3-amino- and 3-azido-6-bromo-2,3,6-trideoxy-D-hexopyranosides into 6-deoxy analogs

Entry	Starting sugar (Section)	Benzylidene acetal	6-Bromo sugar	Yield (%)	Ref.	6-Deoxy-D-sugar	Method	Yield (%)	Ref.	Target compound	Method	Ref.
1	144 (2.2.1) Scheme 27	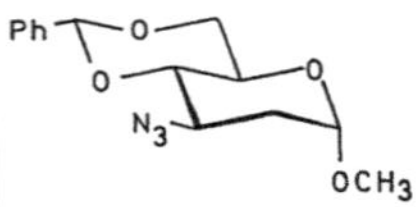 146	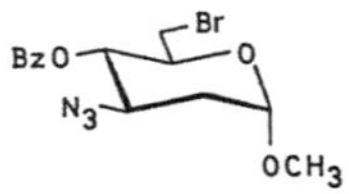 561	75 75 76 ~100	231 232 234 238	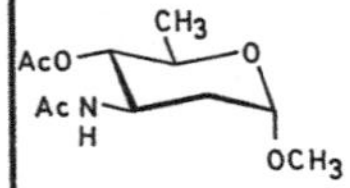 571	i. $LiAlH_4$ ii. acetyl.	68	231	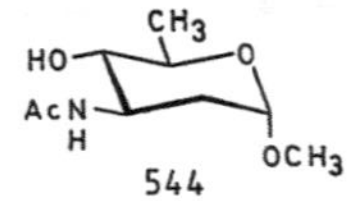 544	$OCH_3^{\ominus}$	231
										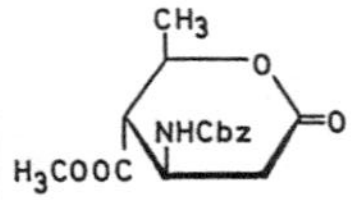 577	Section 4.3.3 Scheme 183	234
						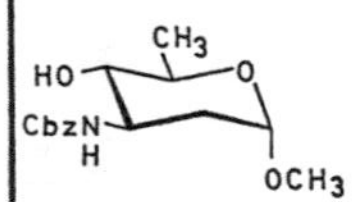 572	i. Ni/H_2 Et_3N ii. CbzCl	52	234	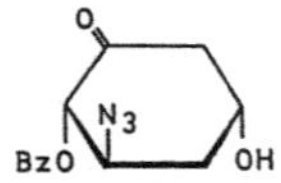 578	Section 4.7 Scheme 199	447 448
2	146 (2.2.1) Scheme 27 or 189 or 191 (2.2.3) Table 3	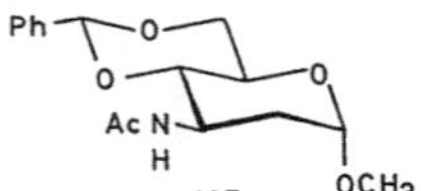 197	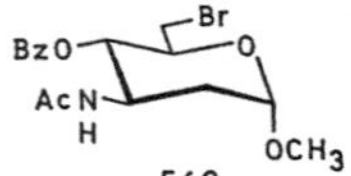 562	99	174	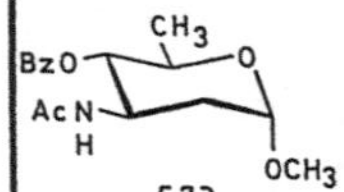 573	Ni/H_2 Et_3N	99	174	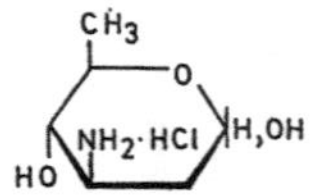 10	i. $OCH_3^{\ominus}$ ii. $Ba(OH)_2$ iii. HCl	174
3	146 (2.2.1) Scheme 27	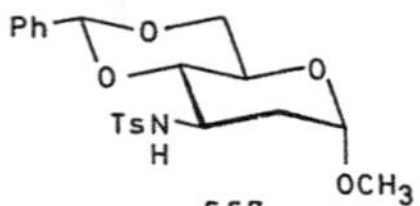 557	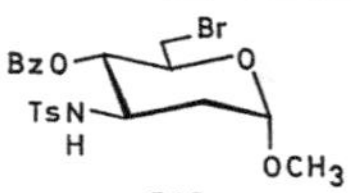 563	92	233	—	—	—	—	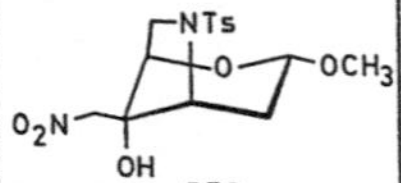 579	Section 4.3.2 Scheme 180	233
4	198 (2.2.3) Table 3	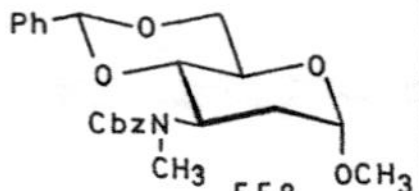 558	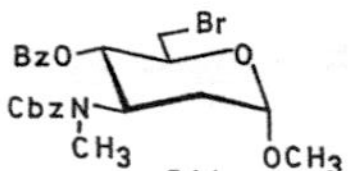 564	75	449	—	—	—	—	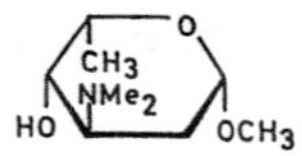 179	Section 2.7.4 Table 11 entry 2	449
5	189 (2.2.3) Table 3 531 (2.6.2) Scheme 98	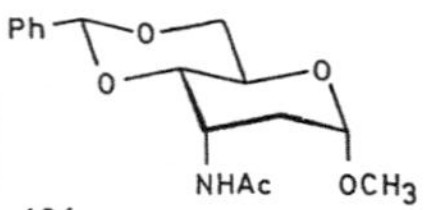 196	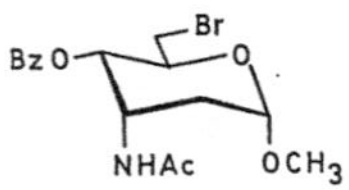 565	70	172 175 425	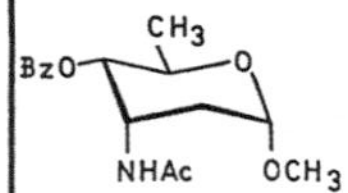 574	Ni/H_2 Et_3N	91	175	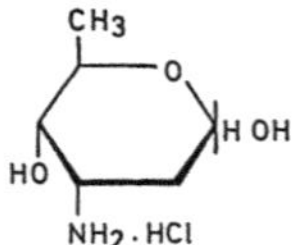 20	i. $OCH_3^{\ominus}$ ii. $Ba(OH)_2$ iii. HCl	174 175

6	189 (2.2.3) Table3	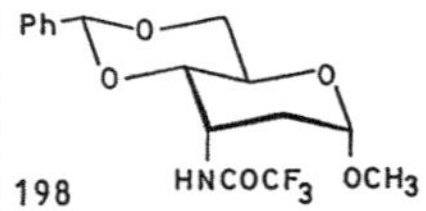 198	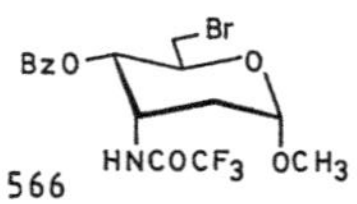 566	73 88	178 179	—	—	—	—	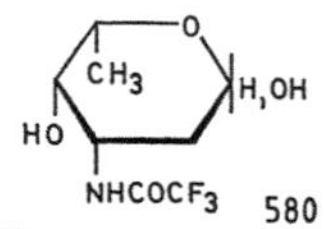 580	Section 2.7.4 Table 11 entry 4	178 179
7	333 (2.6.3) Scheme 102	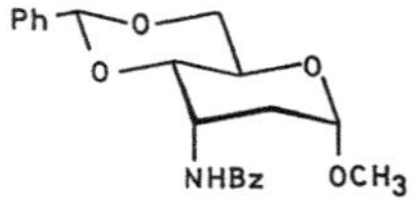 536	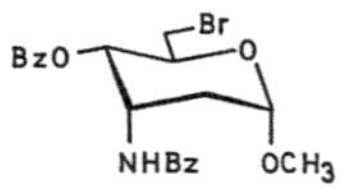 567	76 80	395 450	—	—	—	—	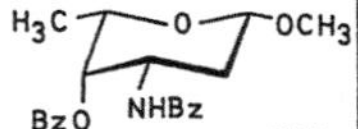 537	Section 2.7.4 Table 11 entry 5	395
										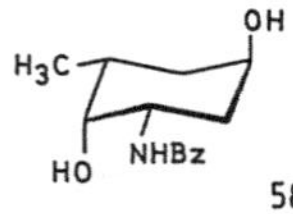 581	Section 4.7 Scheme 201	450
8	150 (2.2.1) Scheme 28	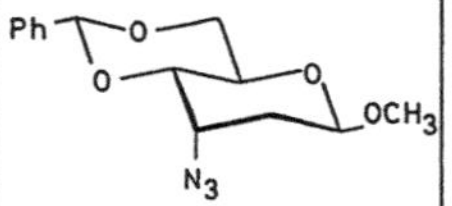 152	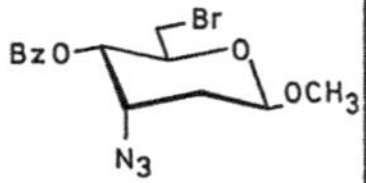 568	90	114 239	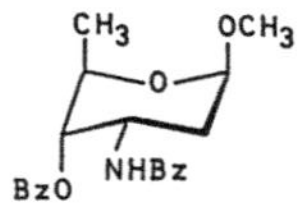 575	i. Ni/H_2 Et_3N ii. BzCl	90	114 239	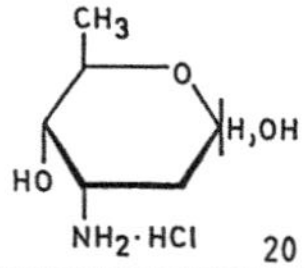 20	i. $OCH_3^{\ominus}$ ii. $Ba(OH)_2$ iii. HCl	114 239
											HCl	240
				69	240	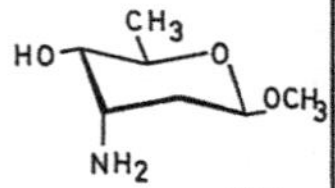 576	i. $OCH_3^{\ominus}$ ii Pd C/ H_2 $BaCO_3$	not isolated	240	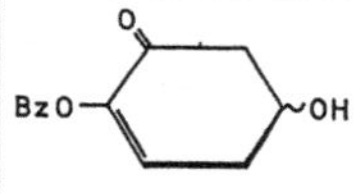 582	Section 4.7 Scheme 199	447 448
9	152 (2.2.1) Scheme 28	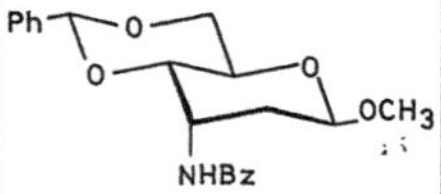 559	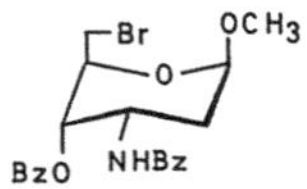 569	78	448	—	—	—	—	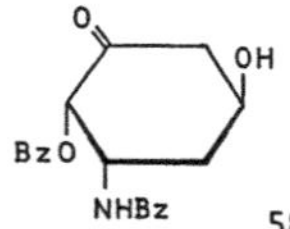 583	Section 4.7 Scheme 200	448
10	152 (2.2.1) Scheme 28	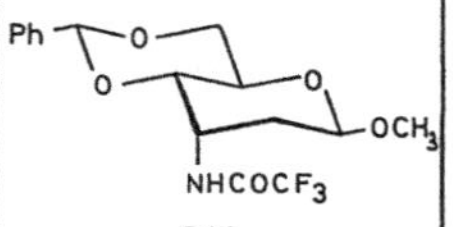 560	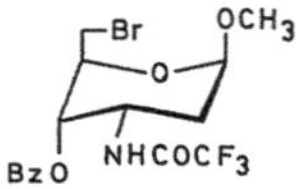 570	90	448	—	—	—	—	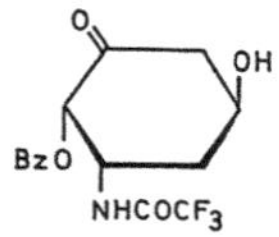 584	Section 4.7 Scheme 200	448

Table 8. Preparation and conversion of 3-amino and 3-azido-6-bromo-2-fluoro-2,3,6-trideoxy-D-hexopyranosides into 6-heoxy analogs

Entry	Starting sugar (Section)	Benzylidene acetal	6-Bromo sugar	Yield (%)	Ref.	6-Deoxy-D-sugar	Method	Yield (%)	Ref.	Target compound	Method	Ref.
1	341 or 342 (2.4.1) Scheme 68	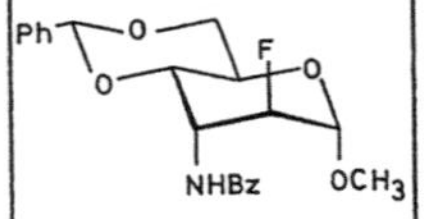 343	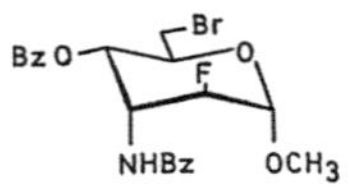 585	41 83 82	396 397 398	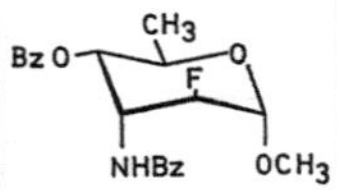 588	Ni/H_2 Bu_3SnH AIBN	~100 98.5	397 398	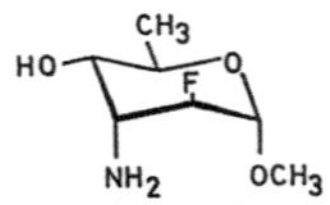 349	aq. NaOH	397 398
2	344 (2.4.1) Scheme 68	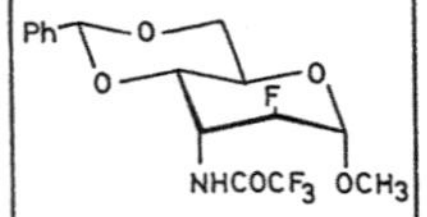 347	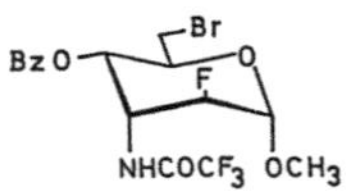 586	94	399	—	—	—	—	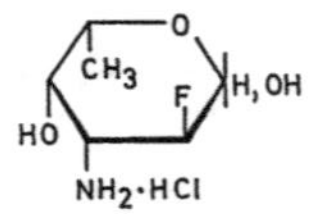 348	Section 2.7.4 Table 12 entry 3	399
3	359 (2.4.2) Scheme 72	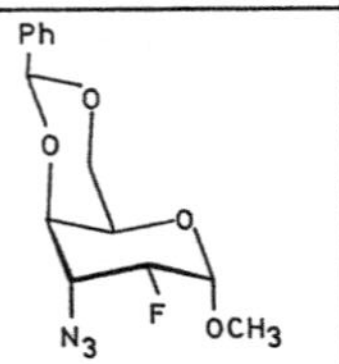 360	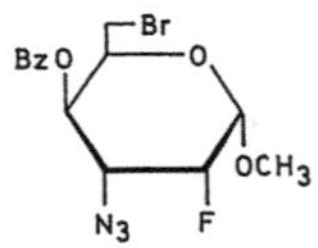 587 [a]	56	401	—	—	—	—	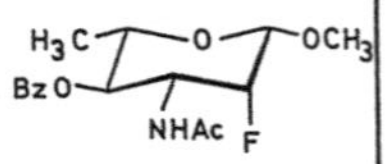 362	Section 2.7.4 Table 12 entry 5	401

[a] Exists in conformational equilibrium ($^4C_1 \rightleftharpoons {}^1C_4$)

Table 9. Preparation and conversion of 3-amino-6-bromo-3-C-methyl-2,3,6-trideoxy-D- hexopyranosides into 6-deoxy analogs

Entry	Starting sugar (Section)	Benzylidene acetal	6-Bromo sugar	Yield (%)	Ref.	6-Deoxy-D-sugar	Method	Yield (%)	Ref.	Target compound	Method	Ref.
1	409 (2.5.2) Scheme 78 or 456 (2.5.3) Fig. 8	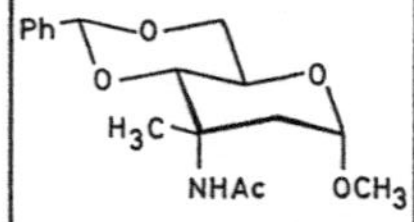 410	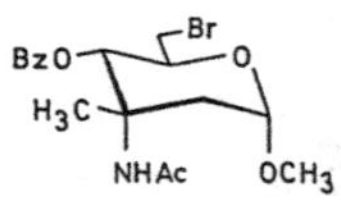 590	89 a	177 414	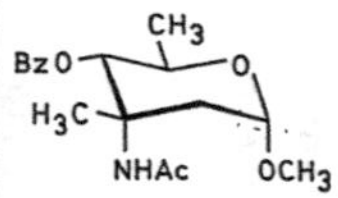 593	Ni / H_2	~100	177	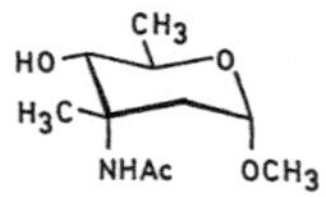 461	$OCH_3^{\ominus}$	177 414
2	409 (2.5.2) Scheme 78	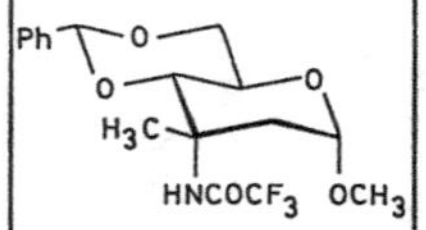 589	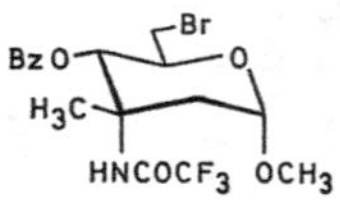 591	75 90 75	182 380 414	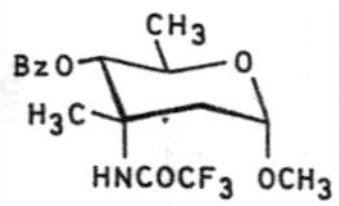 594	i. NaI ii. PdC/H_2	85 89	182 414	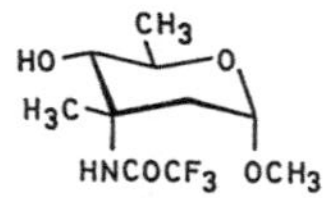 595	$OCH_3^{\ominus}$	182 414
										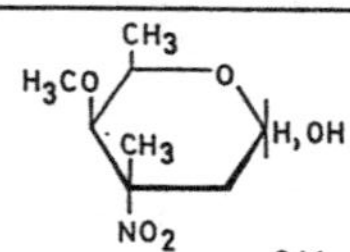 241	See also Section 2.5.3 Fig. 8	
3	412 (2.5.2) Scheme 79 or 480 (2.5.4) Scheme 85	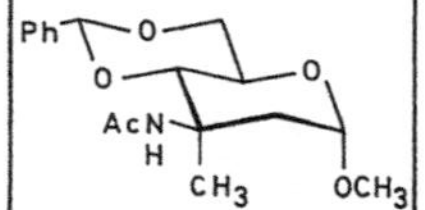 481	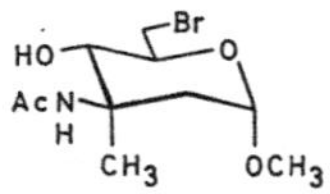 592	60 86	180 181 418	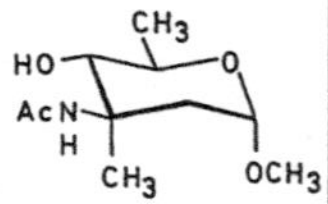 482	Ni / H_2 Ni / H_2 Et_3N	95 37	180 181 418	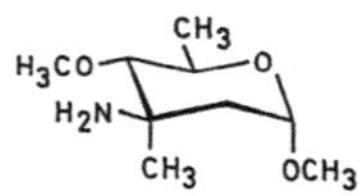 413 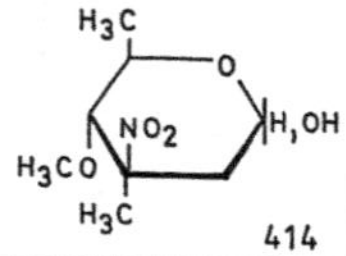 414	See also Section 2.5.2 Schemes 78 and 79	180 181 418

[a] No yield is reported

(Table 9 entry 2). A procedure (Sect. 2.5.4, Scheme 85) leading to D-evernitrose **(414)**, the unnatural enantiomer of the nitrosugar **240**, was based (180, 181, 418) on methyl 3-acetamido-3-C-methyl-2,3,6-trideoxy-α-D-*arabino*-hexopyranoside **(482)**, which can be obtained by the route **481** → **592** → **482** (Table 9 entry 3).

In addition to the Hanessian-Hullar reaction, several other methods have been developed for the cleavage of 1,3-dioxane-type carbohydrate benzylidene acetals (446, 453–455). As shown by Lipták and coworkers (454–457), the hydrogenolysis of 4,6-O-benzylidene acetals of the hexopyranosides using the lithium aluminum hydride–aluminum chloride ($LiAlH_4$–$AlCl_3$) reagent gives (in most cases) the corresponding 4-O-benzyl ether derivative with OH-6 free as the major product accompanied by various quantities of the respective O-4 unsubstituted 6-O-benzyl derivative. Extensive research work by the Lipták group has led to a precise exploration (454, 456, 457) of the factors affecting the direction of the splitting of the 4,6-O-benzylidene acetal function of various carbohydrates. The cleavage of the type of acetals previously mentioned with sodium cyanoborohydride–hydrogen chloride exhibits (458, 459) a regiospecifity opposite to that generally observed for $LiAlH_4$–$AlCl_3$ reductions and yields major products with a 6-O-benzyl group and a free C-4 hydroxyl function. Because both of these acetal-ring cleavage reactions are compatible (458–461) with the presence of acylated amino functions in the 2-acylamido-4,6-O-benzylidene-hexopyranosides, these procedures can apparently be used in the syntheses of 3-amino-polydeoxyhexose derivatives of antibiotic substances. Some additional novel procedures (446, 455, 462–464) for the opening of acetal rings may also find application in this field.

2.7.3 Miscellaneous methods for the preparation and dehalogenation of 6-halogeno-3-amino- and 3-azidohexopyranosides

Although the Hanessian-Hullar procedure (Sect. 2.7.2) for the direct conversion of various 4,6-O-benzylidene-hexopyranosides into the corresponding 6-bromo-6-deoxy analogs in high yield is apparently the most popular route to such halogenosugars, numerous additional processes (204, 465–474) have been worked out for the introduction of a halogen atom into the primary position of a carbohydrate. Of these methods, however, only a single one (465, 466) has so far been adopted (184, 351) for the generation and subsequent removal of the C-6 halogen function of 3-amino- or 3-azido-2,3-dideoxyhexopyranosides.

Thus, in a synthesis (Scheme 112) of methyl N-trifluoroacetyl-α-L-ristosaminide **(597)**, Arcamone and coworkers (184) treated methyl 3-trifluoroacetamido-2,3-dideoxy-α-D-*ribo*-hexopyranoside **(203)**, obtained from L-**157** as shown in Sect. 2.2.3 (Scheme 37) with N-bromosuccinimide in DMF in the presence of triphenylphosphine,

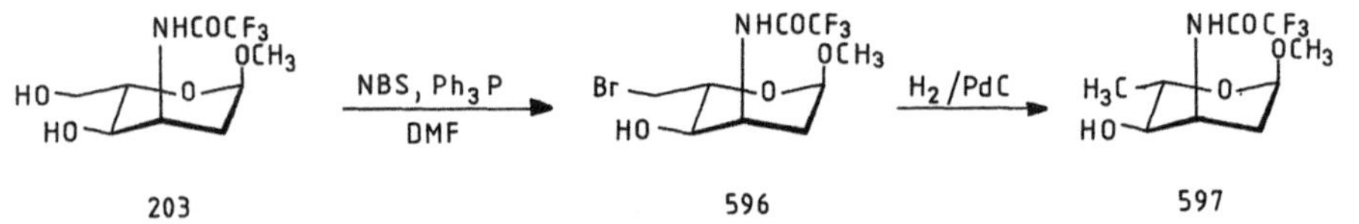

Scheme 112

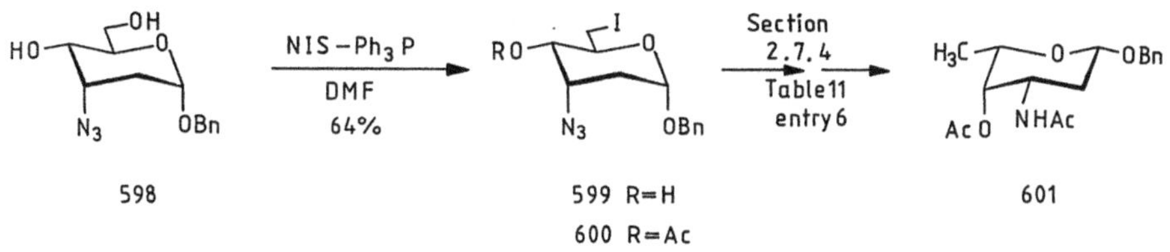

Scheme 113

and obtained the 6-bromo compound **596**. Removal of the bromo substituent was then achieved by catalytic hydrogenation to yield the ristosamine glycoside **597** which was used for glycosylation of anthracyclinones. In a modified version of this halogenation reaction (Scheme 113), the 3-azidodiol **598** was treated (351) with N-iodosuccinimide-triphenylphosphine to produce the 3-azido-6-iodoglycoside **599** in 64% yield. Subsequent acetylation gave the 4-O-acetate **600**, which was converted to benzyl N,O-diacetyl-β-L-daunosaminide **(601)** through the 3-azido-hex-5-enopyranoside **617**, as shown in Sect. 2.7.4 (Table 11 entry 6).

2.7.4 Preparation of the 6-deoxy analogs of C-3 nitrogen-substituted 2,3-dideoxyhexopyranosides with concomitant inversion of the configuration at C-5

Due to the high cost and limited availability of L-sugar starting materials, functionalized 6-deoxy-L-hexoses — including aminodeoxy derivatives — are often synthesized from their C-5 epimeric diastereoisomers, which are readily obtainable from commercial D-sugars by means of the required transformation reactions (substitution, deoxygenation, etc.). For the purpose of such C-5 configurational inversions, the most widely applied strategy involves the conversion of a suitably substituted 6-bromo- or 6-iodo-D-sugar with the desired stereochemistry (see Sects. 2.7.1–2.7.3) into the corresponding deoxy-hex-5-enopyranoside. This dehydrohalogenation reaction can be achieved using sodium alkoxides (475), silver fluoride in pyridine (476, 477), 1,5-diazabicyclo-[5,4,0]-undec-5-ene (478) in various solvents (HMPA, benzene, acetonitrile, DMF, DMSO), or sodium hydride in DMF (479). Subsequent saturation of the exocyclic double bond of these C-5 methylene compounds then results in the development of the respective 6-deoxy derivatives. The L-configuration of the new asymmetric centre at C-5 can be reasonably predicted and generated in light of a generalizable recognition (vide infra). As an example, taken from the pioneering results in this field, Helferich and Himmen (476) described the transformation of methyl 2,3,4-tri-O-acetyl-6-deoxy-6-iodo-α-D-glucopyranoside **(602)** into methyl 2,3,4-tri-O-acetyl-6-deoxy-α-D-*xylo*-hex-5-enopyranoside **(603)** by treatment with silver fluoride in pyridine. Catalytic hydrogenation (480) of **603** gave methyl 2,3,4-tri-O-acetyl-6-deoxy-β-L-idopyranoside **(604)** in quantitative yield (Scheme 114).

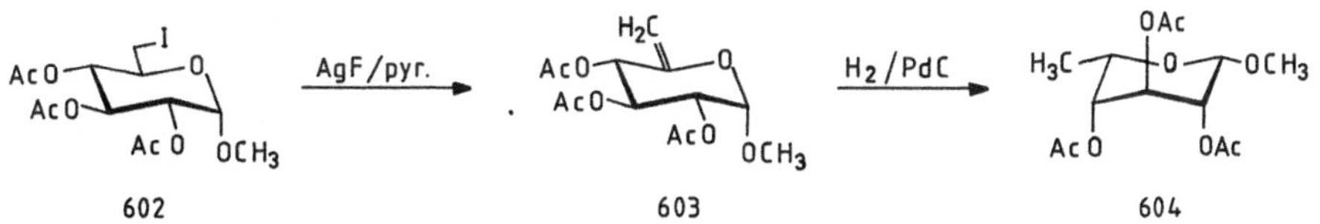

Scheme 114

Table 10. Formation of 3,6-iminohexopyranosides from 6-bromo-6-deoxyhexopyranosides

562, 563, 592, 605   606 – 609

6-Bromo-deoxysugar					3,6-Iminopyranoside					Reagent	Yield %	Ref.
No.	R	R_1	R_2	R_3	No.	R	R_1	R_2	R_3			
605	OAc	H	$COOCH_3$	Ac	**606**	OAc	H	$COOCH_3$	Ac	AgF/pyr	80	480
562	H	H	Ac	Bz	**607**	H	H	Ac	Bz	AgF/pyr	83	232
563	H	H	$pCH_3C_6H_4SO_2$	Bz	**608**	H	H	$pCH_3C_6H_4SO_2$	H	$NaOCH_3$	64	233
592	H	CH_3	Ac	Bz	**609**	H	CH_3	Ac	Bz	AgF/pyr	82	180

Analogous hex-5-enopyranoside derivatives are also readily produced from protected 2-amino-2,6-dideoxy-6-halogeno-hexopyranosides (438, 480–482) and from the structurally related regioisomeric 3-amino- or 3-azidosugars upon treatment with silver fluoride or DBU (see data in Tables 11–13). However, the failure of such a dehydrohalogenation reaction was experienced in the cases of the 3-acetamido-6-bromo-3,6-dideoxy-α-D-glucopyranoside **605** (480) and the 3-acylamino-6-bromo-2,3,6-trideoxy-α-D-*arabino*-hexopyranosides **562** (232), **563** (233), and **592** (180) when the intramolecular nucleophilic attack of the nitrogen atom of the 3-acylamino moiety at C-6 in the sterically favored conformational arrangement resulted in the formation of the 3,6-imino-hexopyranosides **606–609** as the exclusive products (Table 10).

The production of structurally related bicyclo-pyrrolidine derivatives from suitably protected carbohydrate 3-amino-6-O-sulfonate precursors was also reported by zu Reckendorf (483), Brimacombe (411, 412), Bourgeois (407), and Baer et al. (310). Nevertheless, these difficulties in obtaining 3-aminodeoxy-hex-5-enopyranosides with the α-D-*xylo*- and the α-D-*threo* stereochemistry can be easily avoided by the application of either 3-phthalimidohexopyranosides (480) or another strategy for N-disubstitution (449) (Table 11 entry 2). An even more convenient route (231, 232) involves the dehydrobromination of C-3 azidoglycosides, such as **561**, followed by either chemoselective reduction of the azido group using lithium aluminum hydride (232) (to give compound **611**) or the simultaneous generation (231, 232) of the 3-amino- and C-5 methyl functions by catalytic hydrogenation (see Table 11 entry 1).

Based on numerous reports about the saturation of the double bond in diversely substituted deoxy-hex-5-enopyranosides, it was recognized (398, 480, 484, 485) that catalytic addition of hydrogen proceeds preferentially at the face of the pyranoside ring opposite the anomeric function. Regardless of the nature and stereochemical arrangement of the substituents and practically irrespective[3] of the applied catalyst (398, 486, 492), hydrogenation of α-D-hex-5-enopyranosides (193, 480–482,

[3] In a few instances, see for example Refs. (486) and (492) the product ratio was somewhat influenced by the choice of the catalyst.

484, 486–490, 492) gives exclusively or predominantly 6-deoxy-β-L-glycosides with high stereoselectivity. The exo-methylene sugars with the β-D-configuration (438, 484, 485, 489, 491, 493) transform into the corresponding 6-deoxyhexose derivatives of the β-D-series with analogous stereoselectivity. Related results for 3-amino- and 3-azido-2,3,6-trideoxy-hex-5-enopyranosides and their 2-halogeno- and 3-C-methyl analogs are presented in Tables 11–13. The only exceptions in this field were reported by Descotes et al. (489) and Brockhaus and associates (491). The catalytic addition of hydrogen to methyl 2,3,4-tri-O-acetyl-β-L-*arabino*-hex-5-enopyranoside gave (489) a 1:1 mixture of the corresponding 6-deoxy-β-L-altro- and α-D-galactopyranoside derivatives. Also, when methyl 4-acetamido-2,3-di-O-acetyl-α-D-*xylo*-hex-5-enopyranoside was hydrogenated an 82% yield of methyl 4-acetamido-2,3-di-O-acetyl-4,6-dideoxy-α-D-glucopyranoside was obtained (491).

In a limited number of cases, difficulties were encountered with the 4-O-benzoyl substituent (232, 397, 482) of the hex-5-enopyranoside, but these could be readily overcome by prior O-deacylation (see Table 12 entries 2–4). When methyl 2,3-O-isopropylidene-4-O-methanesulfonyl-6-deoxy-α-D-*lyxo*-hex-5-enopyranoside was treated (494) with lithium aluminum hydride, the loss of the O-4-substituent resulted in a 4,5-unsaturated compound: methyl 2,3-O-isopropylidene-4,6-dideoxy-α-D-*erythro*-hex-4-enopyranoside.

As demonstrated by the data summarized in Tables 11–13, numerous L-aminodeoxy hexose components of antibiotic substances, as well as their structurally related stereoisomers, have been prepared by the foregoing convenient step over from the D- into the L-carbohydrate series. Thus, L-daunosamine **(12)** (172, 425) and N-trifluoroacetyl-L-daunosamine **(580)** (178, 179) (Table 11 entries 3 and 4) and their substituted derivatives **537**, **601**, **623**, and **624** were synthesized via the stereoselective hydrogenation of the 3-acylamino- **(613–615)** and 3-azido- **(617)**-α-D-*erythro*-hex-5-enopyranosides (Table 11 entries 3, 4, 5, and 6). Similar reductions of the β-D-*erythro*-hex-5-enopyranosides **618** and **619** gave (239, 485) exclusively methyl N,O-dibenzoyl-β-D-ristosaminide **575** (Table 11 entries 7 and 8).

Derivatives (**132**, **179**, **621**, **622**, and **625**) of the C-3 epimer (L-*xylo*) of L-daunosamine were obtained (Table 11 entries 1 and 2) by the reduction of the α-D-*threo*-hex-5-enopyranosides **610–612**. As mentioned in Sect. 2.7.2, the 3-azido- (**610** and **618**) and 3-amino- (**616**, **619**, and **620**) C-5-methylene sugars also served as starting materials for the preparation of the functionalized cyclohexanones **578**, **582–584** (Table 11 entries 1, 7–9) and for *pseudo*-α-L-daunosamine (**581**, Table 11 entry 5) (450). These conversions are described in Sect. 4.7.

(s)-2-Fluoro-L-daunosamine **(348)** and its several N- or O- or both substituted analogs **(361, 633–638)** have also been synthesized (396–400) by a strategy involving the generation and dehydrobromination of the 6-bromo- (**585** and **586**) and 6-iodo-sugars **(626, 555)**, followed by catalytic hydrogenation of the resulting hex-5-enopyranosides **627–630** (Table 12 entries 1–4). Such a procedure was also employed for the preparation (401) of a 2-fluoro-L-acosamine derivative **362** from the 6-bromo-sugar **587** (Table 12 entry 5).

The first stereospecific synthesis (177) of a derivative **(411)** of L-vancosamine **(15)** was also based on a hex-5-enopyranoside intermediate (**639**, Table 13 entry 1), which was derived from the benzylidene acetal **410** (Sect. 2.5.2, Scheme 78) via the 6-bromo-6-deoxy compound **590**. Similar utilization of the C-5 unsaturated sugar

Table 11. Synthesis and reduction of C-3 nitrogen-substituted deoxy-hex-5-enopyranosides

Entry	Starting sugar (Section)	Starting 6-Halogeno sugar	Hex-5-enopyranoside	Method	Yield (%)	Ref.	Reduction product D (%)	Reduction product L (%)	Method	Ratio D:L	Ref.	Target compound	Method	Ref.
1	146 (2.7.2) Table 7 entry 1	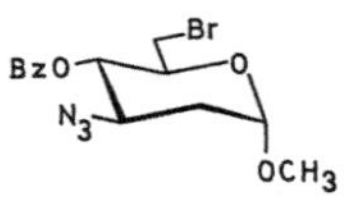 561	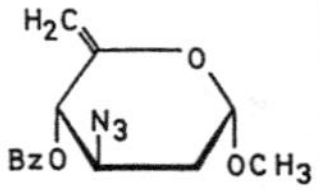 610	AgF/pyr	79	231	—	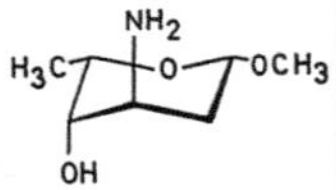 132 (98)	i. $OCH_3^{\ominus}$ ii. Ni/H_2	0:1	231	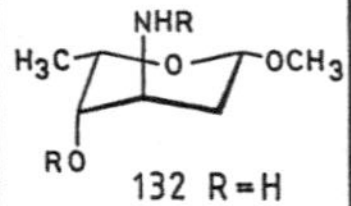 132 R = H 622 R = Ac	acetylation	231
												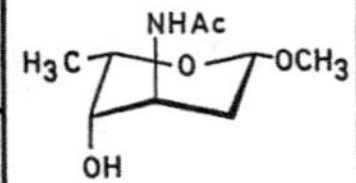 625	$OCH_3^{\ominus}$	231 232
				AgF/pyr	96	232	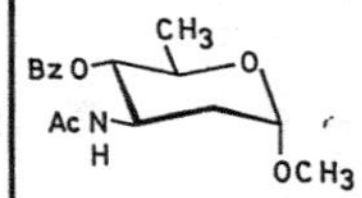 573 (6.3)	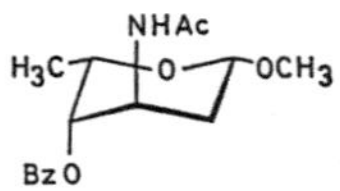 621 (40)	i. PdC/H_2 ii. acetylation	1:6	232	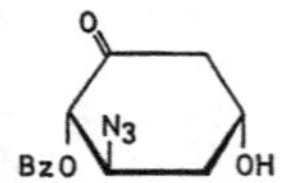 578	Section 4.7 Scheme 199	447 448
			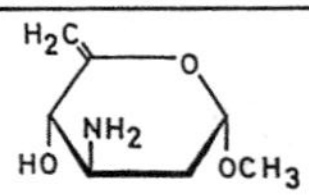 611	i. AgF/pyr ii. NaOH iii. $LiAlH_4$	96 91 84	232	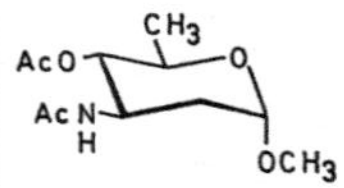 571 (4.5)	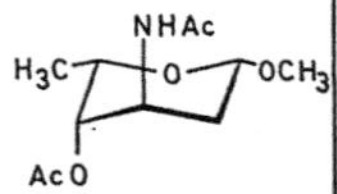 622 (83)	i PdC/H_2 ii. acetylation	1:18	232	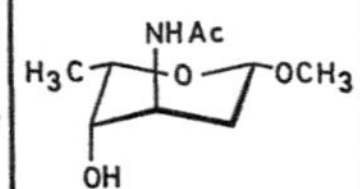 625	$OCH_3^{\ominus}$	232
2	558 (2.7.2) Table 7 entry 4	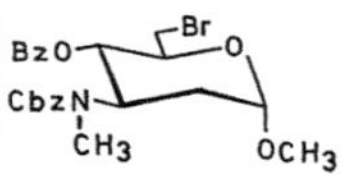 564	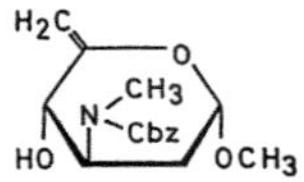 612	AgF/pyr	71	449	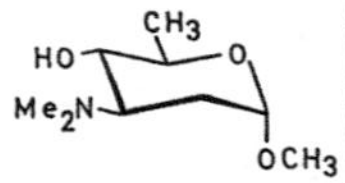 136[a]	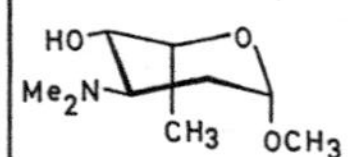 179 (16)	i. $LiAlH_4$ (88 %) ii. PdC/H_2	3:7	449	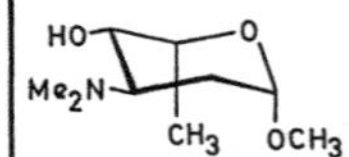 179	—	449

3	196 (2.7.2) Table 7 entry 5	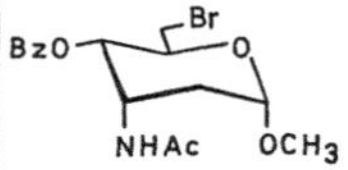 565	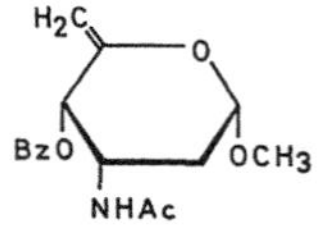 613	AgF/pyr	~100	172	—	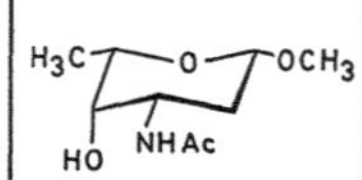 623 (100)	i. $OCH_3^{\ominus}$ ii. Pd/ $BaSO_4$ H_2	0:1	172	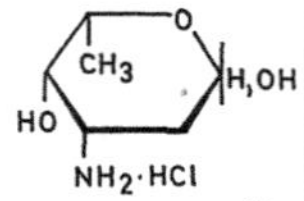 12	i. $Ba(OH)_2$ ii. HCl	172 425
4	198 (2.7.2) Table 7 entry 6	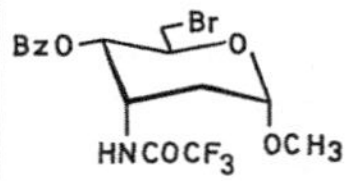 566	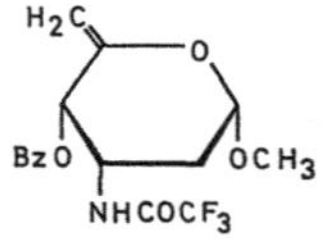 614	DBU-HMPT DBU-C_6H_6	90 75	178 179	—	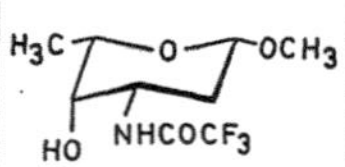 624[b]	i. $OCH_3^{\ominus}$ ii. Pd/ $BaSO_4$ H_2 i. $OCH_3^{\ominus}$ ii. PdC/H_2	0:1 0:1	178 179	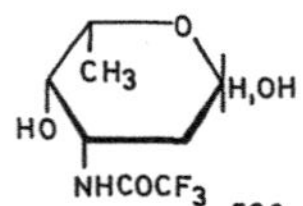 580	20% aq. HOAc	178 179
5	536 (2.7.2) Table 7 entry 7	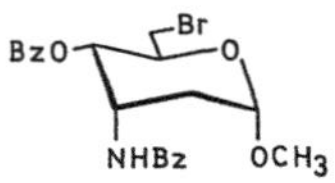 567	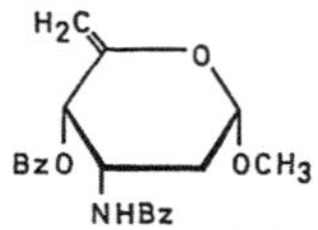 615	DBU-HMPT	97	395	—	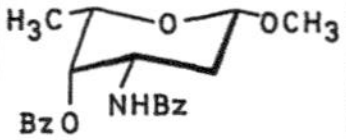 537 (85)	PdC/H_2	0:1	395	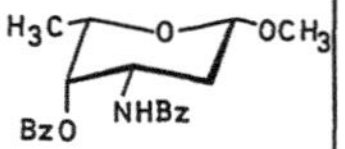 537	—	395
			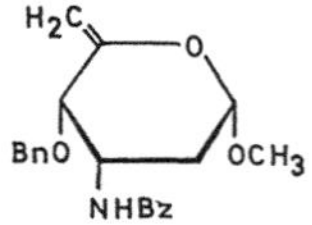 616	i. $OCH_3^{\ominus}$ ii. BnCl iii. AgF/pyr	80	450	—	—	—	—	—	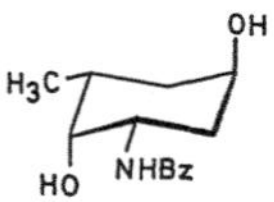 581	Section 4.7 Scheme 201	450
6	598 (2.7.3) Scheme 113	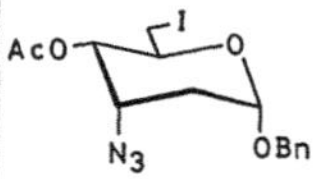 600	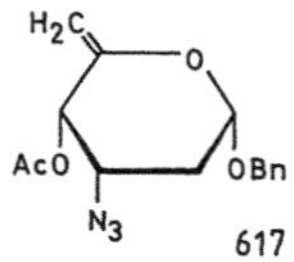 617	DBU-C_6H_6	97	351	—	 601 (56)	i. PdC/H_2 ii. acetylation	0:1	351	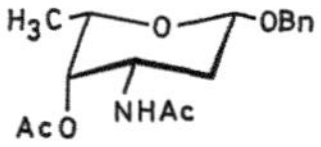 601	—	351
7	152 (2.7.2) Table 7 entry 8	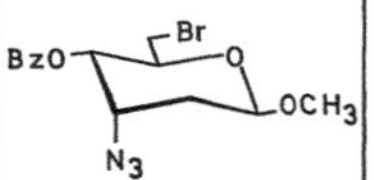 568	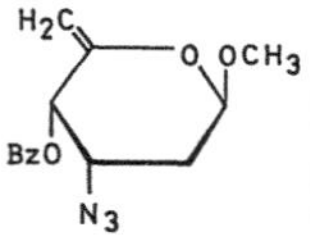 618	AgF/pyr DBU-HMPT	55 70	114 239 c	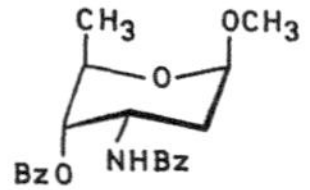 575 (92)	—	i. $OCH_3^{\ominus}$ ii. Pd/ $BaSO_4$ H_2 iii. BzCl pyr	1:0	239	d	—	239 485
												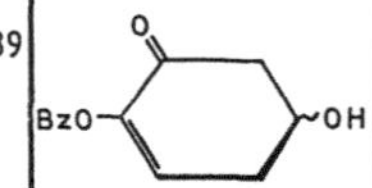 582	Section 4.7 Scheme 199	447 448

(Continued)

Table 11 (contd.)

Entry	Starting sugar (Section)	Starting 6-Halogeno sugar	Hex-5-enopyranoside	Method	Yield (%)	Ref.	Reduction product D (%)	Reduction product L (%)	Method	Ratio D:L	Ref.	Target compound	Method	Ref.
8	559 (2.7.2) Table 7 entry 9	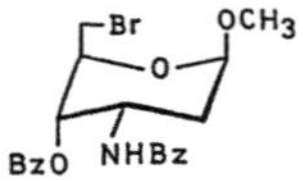 569	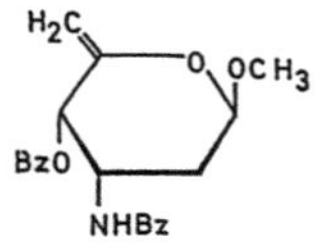 619	AgF/pyr	72	485	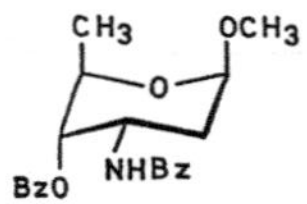 575 (97)	—	Pd/ BaSO4 H2	1:0	485	d	—	485
												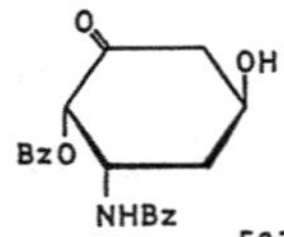 583	Section 4.7 Scheme 200	448
9	560 (2.7.2) Table 7 entry 10	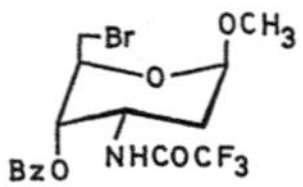 570	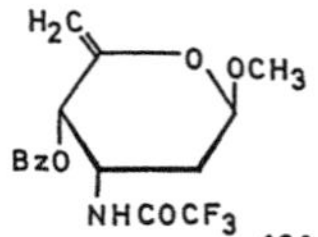 620	AgF/pyr	81	448	—	—	—	—	—	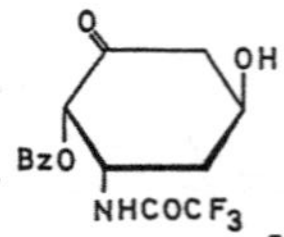 584	Section 4.7 Scheme 200	448

[a] Not isolated
[b] Converted directly into the target sugar
[c] The authors' unpublished results
[d] Attempted preparation of L-daunosamine

640 — prepared from the 4,6-O-benzylidene-3-trifluoroacetamido-glycoside **589** (Table 9 entry 2) — led (380) to N-trifluoroacetyl-O-benzoyl-L-vancosamine **(301)**, a precursor of the branched-chain aminoglycal **305** (Sect. 2.2.6, Fig. 7) that was used for the glycosylation of daunomycinone.

The analogous reaction sequence for the generation of the 6-deoxy function with the simultaneous C-5 inversion of di- (238) and trisaccharides (351), carrying a 3-azido-6-iodo- (or bromo) -2,3,6-trideoxyhexose building block at the reducing end, was also reported. An alternative method for the C-6 deoxygenation of carbohydrates with a concomitant change of the chirality of carbon C-5 involves the development and reductive ring opening of 5,6-anhydro-hexofuranose derivatives. An example of such a reaction (495) is shown in Sect. 2.1.2 (Scheme 13) as part of the discussion of the transformation of **84** into **85** that was used by the Kuszmann group (135) to synthesize L-daunosamine **(12)**.

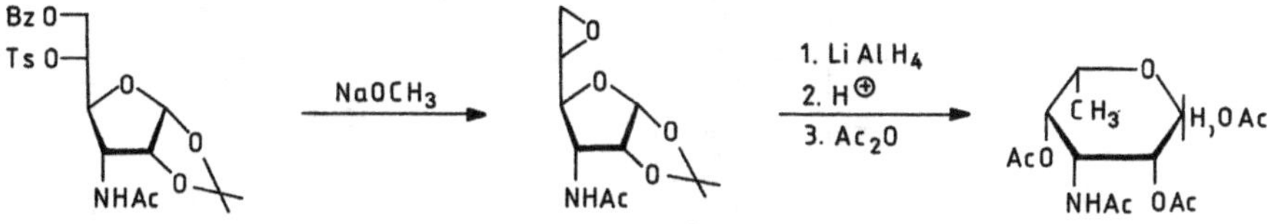

643 644 520

Scheme 115

Analogous strategies with protected C-3 azido- or aminodeoxy-D-hexofuranoses have been developed by Japanese researchers (248, 373) for the synthesis of L-daunosamine. Tanaka (373) converted 3-acetamido-6-O-benzoyl-1,2-O-isopropylidene-4-O-*p*-toluenesulfonyl-3-deoxy-α-D-allofuranose **(643)** into the epoxide **644** (Scheme 115). The reductive epoxide ring cleavage of **644** using lithium aluminum hydride followed by mild acid hydrolysis and acetylation gave the peracetylated 3,6-dideoxy-L-aminosugar **520** with the talo configuration. The deoxygenation of this latter substance at C-2 to form daunosamine by means of the Fischer's glycal procedure is discussed in Sect. 2.6.1 Scheme 94). In the daunosamine synthesis of Yamaguchi and Kojima (248) (Scheme 116), methyl 5,6-anhydro-3-azido-2-S-benzyl-3-deoxy-2-thio-α,β-L-galactofuranoside **(646)** was derived from the D-altrofuranoside **645**. Lithium aluminum hydride reduction of the epoxide function in **646** gave the 3-amino-3,6-dideoxy-α,β-L-galactofuranoside **532**. L-Daunosamine was then prepared from **532** by sequential desulfurization at C-2 and acid hydrolysis, as shown in Sect. 2.6.2 (Scheme 99). An example of the related configurational inversion of a γ-lactone derivative of D-ristosamine at C-5 is discussed in Sect. 3.2.4 (Scheme 152).

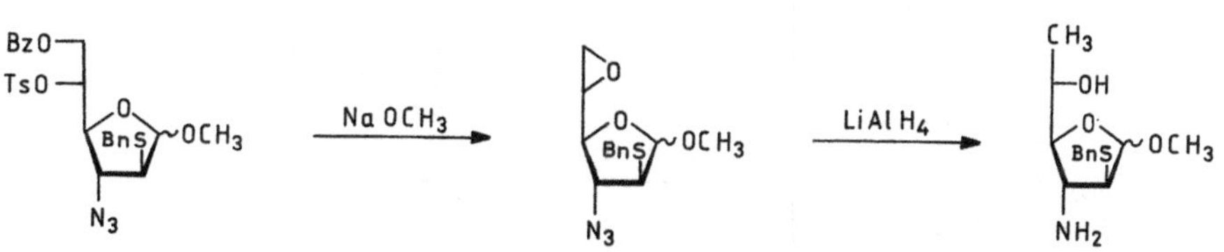

645 646 532

Scheme 116

Table 12. Synthesis and reduction of C-3 nitrogen-substituted 2-fluorodeoxy-hex-5-enopyranosides

Entry	Starting sugar (Section)	6-Halogeno sugar	Hex-5-enopyranoside	Method	Yield (%)	Ref.	Reduction product D (%)	Reduction product L (%)	Method	Ratio D:L	Ref.	Target compound	Method	Ref.
1	351 (2.4.2) Scheme 70	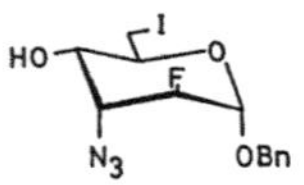 626	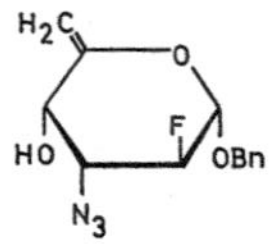 627	AgF/pyr	51	400	—	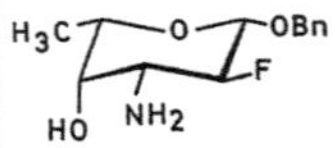 361 (95)	PdC/H_2 Et_3N	0:1	400	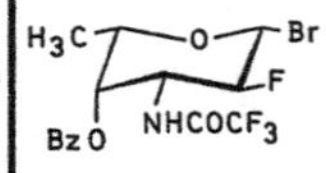 637	i. TFAA ii. BzCl iii. PdC/H_2 iv. Ac_2O/NaOAc v. HBr/HOAc	400
2	343 (2.7.2) Table 8 entry 1	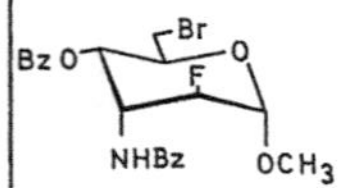 585	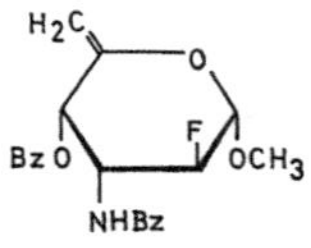 628	AgF/pyr	92	396	—	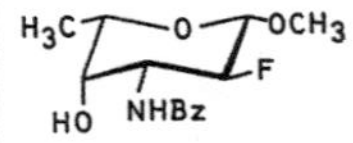 633 (67)	PdC/H_2 Et_3N	a	396	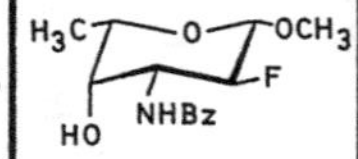 633	—	396
				DBU-HMPT	73	397	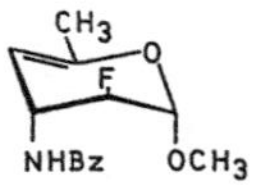 632 (39)	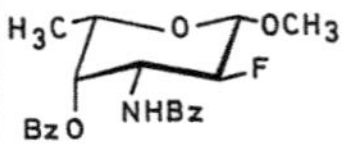 634 (41)	PdC/H_2	1:1	397	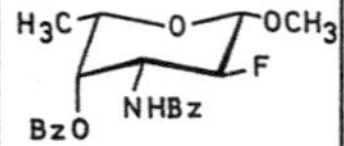 634	—	397
				AgF/pyr	93	398	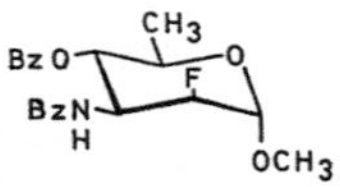 588 (traces)	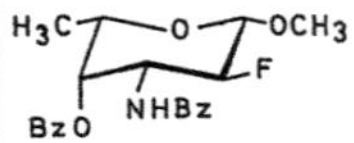 634 (62)	Pd/$BaSO_4$ H_2	~0:1	398	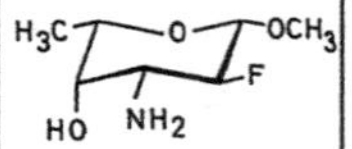 638	aq. NaOH	398

3	347 (2.7.2) Table 8 entry 2	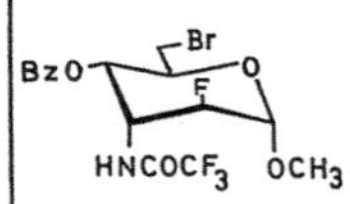 586	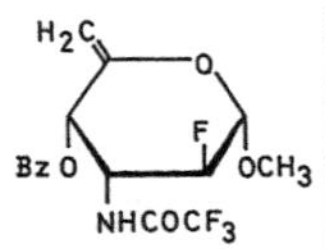 629	AgF/pyr	96	399	—	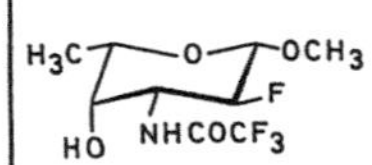 635 (100)	i. $OCH_3^{\ominus}$ ii. PdC H_2	0:1	399	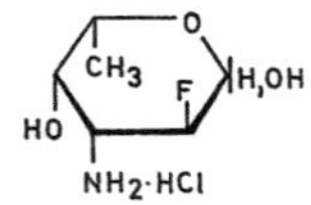 348	HCl	399
4	340 (2.7.1) Scheme 110	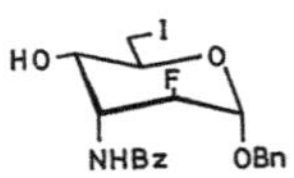 555	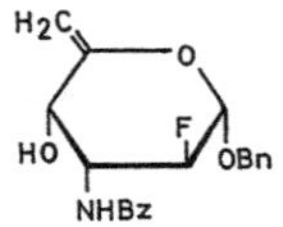 630	AgF/pyr	53	396	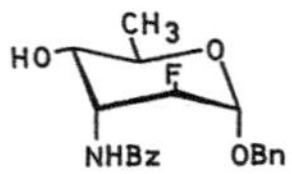 556 Pd/C (8)	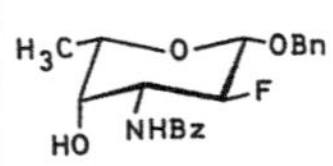 636 Pd/C (58) Rh/Al_2O_3(80)	PdC/H_2 Et_3N Rh/Al_2O_3 H_2	8:58 ~0:1	396	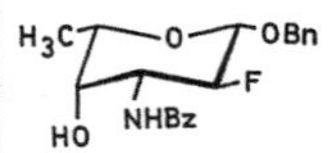 636	—	396
5	360 (2.7.2) Table 8 entry 3	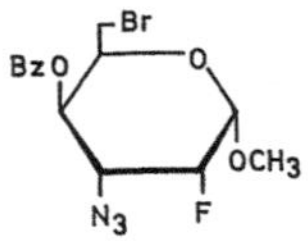 587[b]	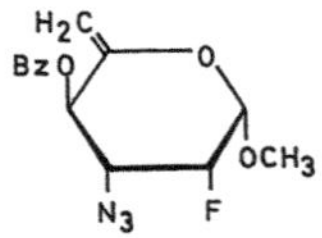 631[b]	AgF/pyr	72	401	—	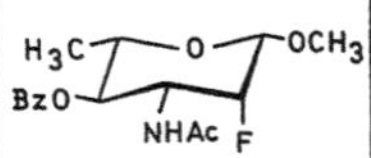 362 (60)	PdC/H_2 MeOH-Ac_2O	a	401	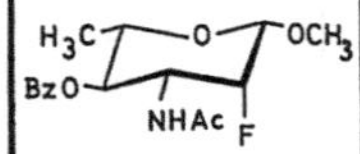 362	—	401

[a] Not reported
[b] Exists in a $^4C_1 \rightleftharpoons {}^1C_4$ conformational equilibrium

Table 13. Synthesis and reduction of C-3 nitrogen-substituted 3-C-methyl-branched deoxy-hex-5-enopyranosides

Entry	Starting sugar	6-Halogeno sugar	Hex-5-enopyranoside	Method	Yield (%)	Ref.	Reduction product D (%)	Reduction product L (%)	Method	Ratio D:L	Ref.	Target compound	Method	Ref.
1	410 Table 9 entry 1	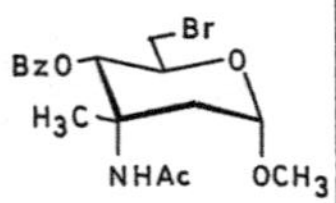 590	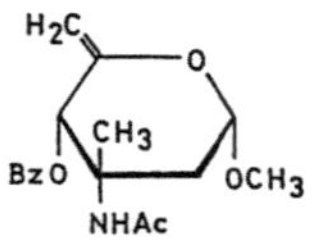 639	AgF/pyr	95	177	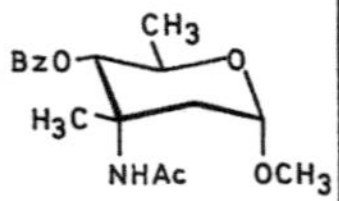 593 (7.8)	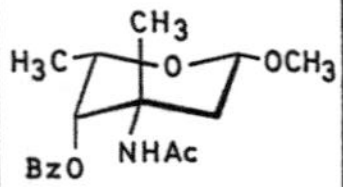 641 (78)	Ni/H_2	1:10	177	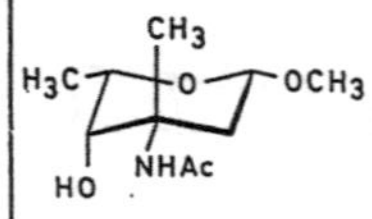 411	$OCH_3^{\ominus}$	177
2	589 Table 9 entry 2	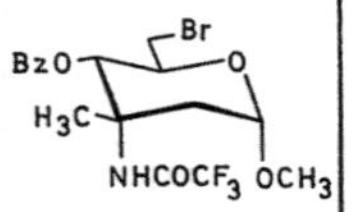 591	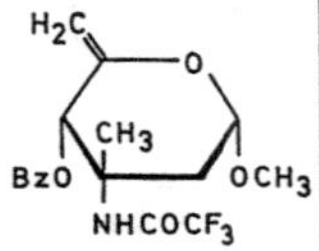 640	AgF/pyr	88	380	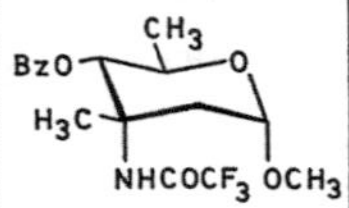 594 (4)	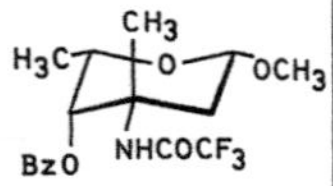 642 (78)	Ni/H_2	1:19.5	380	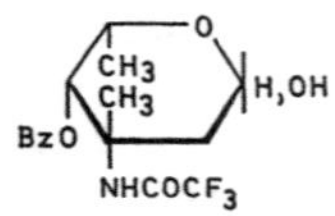 301	i. dil. H_2SO_4 dioxane-H_2O (cf. Section 2.6 Fig. 7)	380

2.8 Interconversion of 3-amino-2,3-dideoxy- and 3-amino-2,3,6-trideoxyhexose derivatives by means of the inversion of the configuration at carbon C-4

Throughout the preceding sections, details of the well-estabilshed procedures that make each of the stereoisomeric 3-amino-2,3,6-trideoxyhexoses and their 3-C-methyl analogs available have been discussed. The value of an efficient method introduced for the synthesis of one of these aminosugars can in practice be significantly enhanced by also extending it to the preparation of the respective C-4 diastereoisomer, which is occasionally obtainable in more complicated way by independent syntheses. The two principal types of procedures for achieving the inversion of the configuration of the C-4 hydroxyl group of the aforementioned aminodeoxy sugars are: (a) the bimolecular nucleophilic substitution of a suitable 4-O-sulfonate ester followed by hydrolysis (or solvolysis in one step) and (b) the reduction of a hexopyranos-4-ulose derived from the starting sugar. The following subchapters summarize such reactions that allow *arabino* → *lyxo*, *lyxo* → *arabino*, *ribo* → *xylo*, and *xylo* → *ribo* interconversions of the stereoisomeric L- and D-3-amino-2,3-di- and 2,3,6-trideoxyhexoses.

2.8.1 The *arabino* → *lyxo* conversion

Because the *arabino* representatives of the foregoing type of aminodeoxy hexoses are readily available from commercial natural sugars, it is quite obvious that the challenge of a step over into the *lyxo* series to obtain the C-4 epimeric daunosamine and vancosamine derivatives has stimulated considerable interest. Almost two-thirds of the related studies were concentrated on the development and improvement of the *arabino* → *lyxo* transformation procedures.

Pioneering efforts in the 1960s with a goal of synthesizing L-daunosamine and derivatives of its enantiomer, first used the conversions of the α-L-*arabino* azide **180** (123) and the methyl α- **(544)** (229, 230) and -β-D-acetamido glycosides **650** (318) into L-daunosamine **(12)**, N-benzoyl-D-daunosamine **(671)** and N-acetyl-D-daunosamine **(545)**, respectively (Table 14 entries 1, 6, and 7), via nucleophilic substitutions of the corresponding 4-O-mesylates **652**, **658**, and **659**. Since then this method and its modified versions, proceeding with moderate to good yield, have been widely applied in aminodeoxy sugar chemistry. Methyl α-D-rhodosaminide **(699)** was prepared (55) from the 3-azidoglycoside **135** (Table 14 entry 2), and the Brimacombe group (187) converted methyl N-acetyl-α-L-acosaminide **(648)** into L-daunosamine **(12)** via the 4-O-mesylate **656** (Table 14 entry 4). The N-methoxycarbonyl derivative **647** of **648** and the N-benzoyl-L-*arabino*-δ-lactone **649**, both synthesized from non-sugar precursors (Sects. 3.2.4 and 3.2.2, Schemes 154 and 133, respectively), were transformed by similar procedures into methyl α-L-daunosaminide **(260)** (496, 497) and N-trifluoroacetyl-L-daunosamine **(580)** (498), respectively (Table 14 entries 3 and 5).

Notice that the mesylation-substitution procedure with the 3-N-benzoyl-3-C-methyl-α,β-*arabino* glycoside **651** proceeded (499, 500) with only a 5% overall yield to give N-benzoyl-L-vancosamine (**688**, Table 14 entry 8), and the best results with the

Table 14. Inversion of the configuration at C-4. *Arabino* → *lyxo* conversion via 4-O-sulfonates

Entry	Starting sugar (Section)	Arabino		Ref.	Lyxo product	Method	Yield (%)	Target aminosugar	Method	Ref.
		4-Hydroxy compound	4-O-Sulphonate							
1	174 (2.2.2) Fig. 5	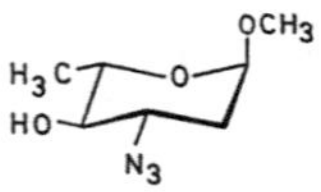 180	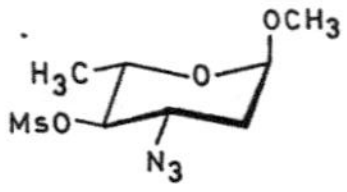 652	123 155	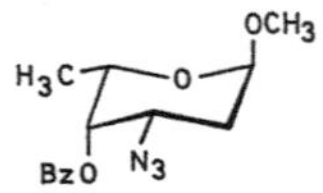 661	NaOBz DMF	a	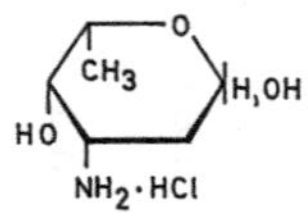 12	i. aq. NaOH ii. H_2/PdC iii. HCl	123
					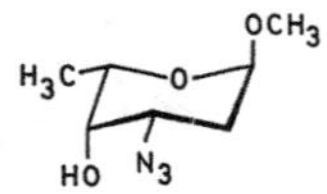 662	i. NaOBz DMF ii. aq. NaOH	25		i. H_2/PtO_2 ii. HCl	155
2	121 (2.2.1) Fig. 3 or the enantiomer of 174 (2.2.2)	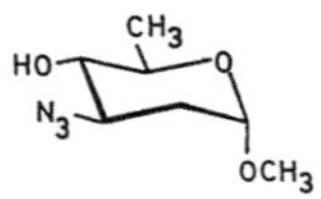 135	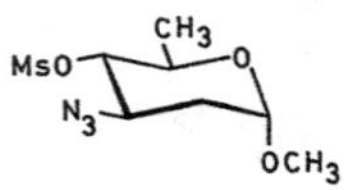 653	55	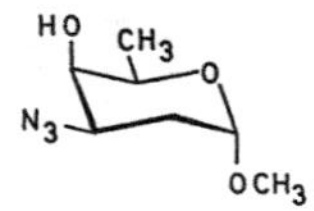 663	i. NaOBz DMF ii. $OCH_3^{\ominus}$	~80 92	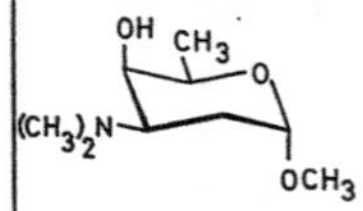 669	H_2/PdC HCHO-MeOH	55
			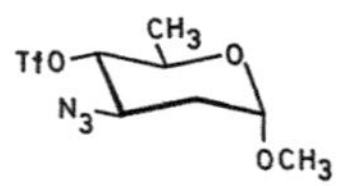 654[b]	269		i. Tf_2O ii. Bu_4NOAc CH_3CN iii. $OCH_3^{\ominus}$	88 from 135	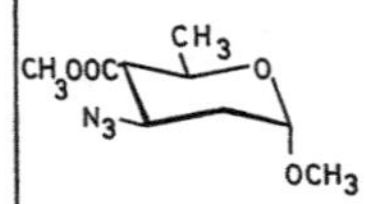 670	Section (4.3.3) Scheme 185	269
3	940 (3.2.4) Scheme 154	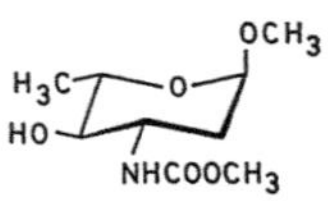 647	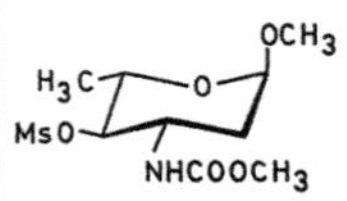 655	496 497	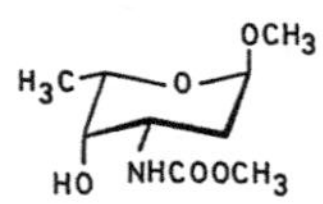 664	DMF-H_2O 65:35 105°C	50[c]	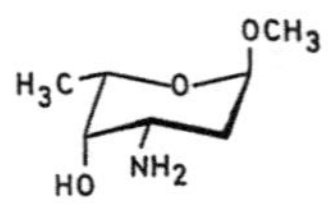 260	aq. $Ba(OH)_2$	496 497

4	123 via 137 (2.2.1) Fig. 3	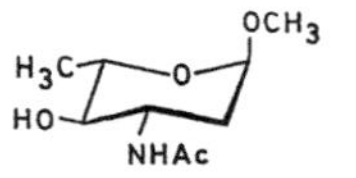 648	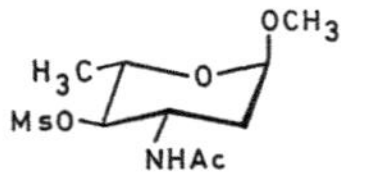 656[b]	187	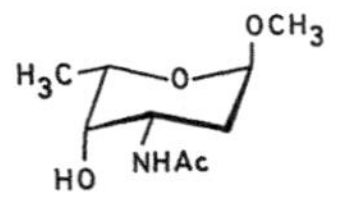 261	i. MsCl / pyr ii. aq. $NaHCO_3$ refl.	50 from 648	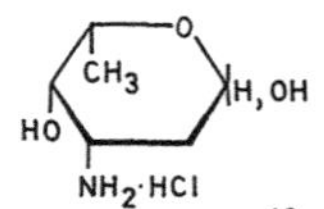 12	i. $Ba(OH)_2$ ii. HCl	187
5	800 (3.2.2) Scheme 133	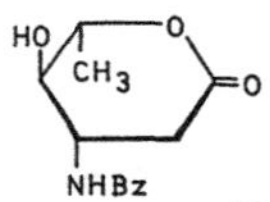 649	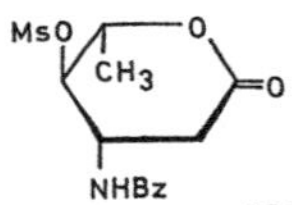 657[b]	498	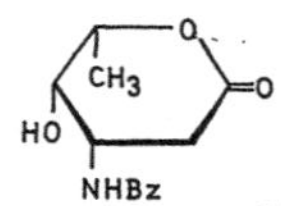 665	i. MsCl / pyr ii. aq. NaOAc	~65 from 649	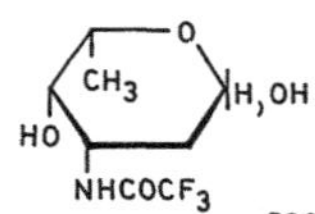 580	i. $(CF_3CO)_2O$ ii. DIBAL	498
6	146 (2.2.1) via 197 (2.7.1) Scheme 107	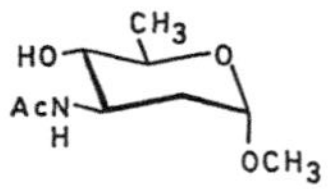 544	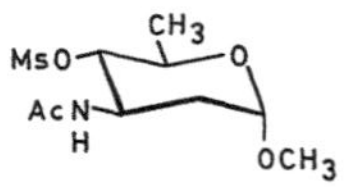 658[b]	229 230 231	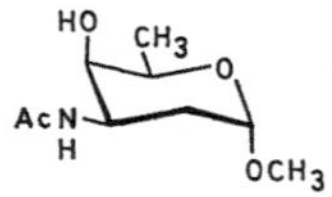 666	i. MsCl / pyr ii. aq. $NaHCO_3$ refl.	42	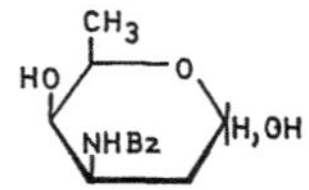 671	i. aq. NaOH ii. Bz_2O / EtOH iii. $H^{\oplus}$	229, 230
							53	666	—	231
7	229 (2.2.4) via 551 (2.7.1) Scheme 108	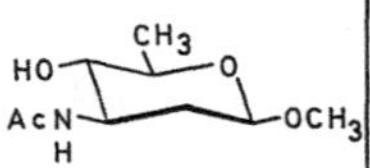 650	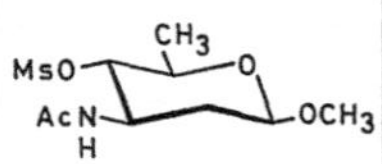 659	318	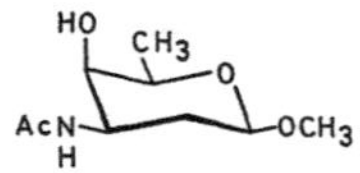 667	$NaOAc \cdot 3H_2O$ $CH_3O(CH_2)_2OH$	70	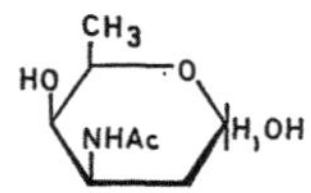 545	i. $H^{\oplus}$	318
8	895 (3.2.3) Scheme 148	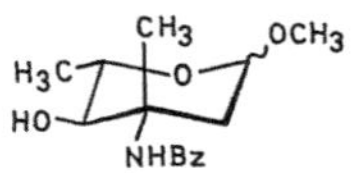 651	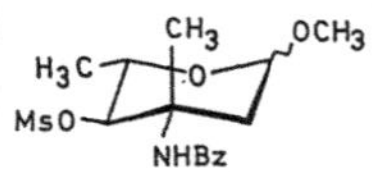 660[b]	499 500	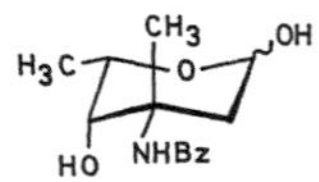 668	i. MsCl / pyr ii. NaOAc H_2O iii. $H^{\oplus}$	5	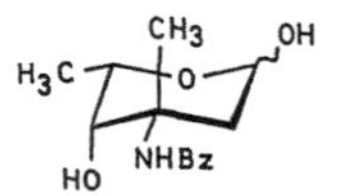 668	—	499 500

[a] No yield is reported
[b] Not isolated
[c] Not optimized yield

Table 15. Inversion of the configuration at C-4. *Arabino* → *lyxo* conversion via C-4 ulosides

Entry	Starting sugar (Section)	Arabino 4-Hydroxy compound	C-4-Ulose	Method of oxidation	Yield (%)	Ref.	Lyxo product	Reducing agent	Yield (%)	Target sugar	Method	Ref
1	147 (2.2.1) Scheme 27	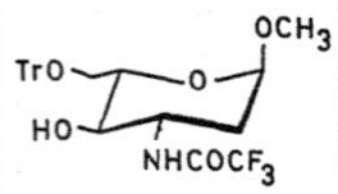 672	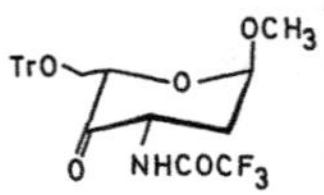 674	RuO_4	56	236	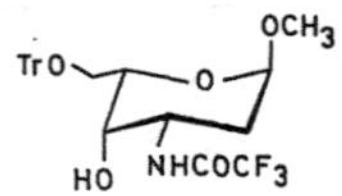 679	L-Selectride THF	83	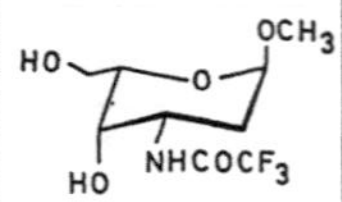 683	60% AcOH heat 60°C	236
2	137 (2.2.1) Fig 3.	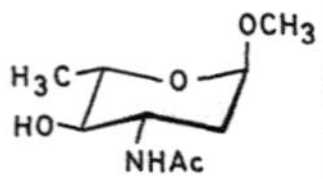 648	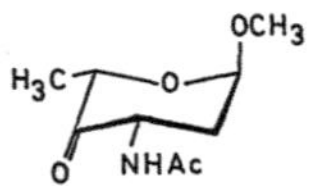 675	PCC $(CF_3CO)_2O$-DMSO	60 63	187	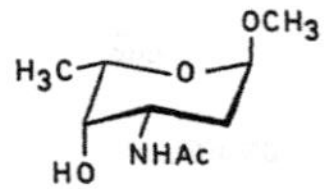 261	L-Selectride THF	90	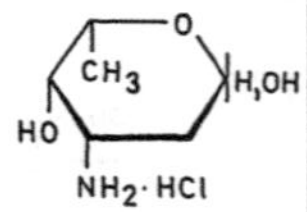 12	i. $Ba(OH)_2$ ii. HCl	187
3	208 (2.2.3) Scheme 39	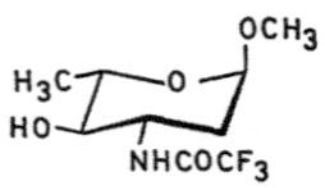 211	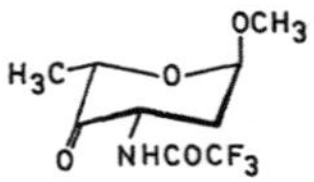 676	RuO_4 $CrO_3 \cdot 2\,Pyr$ Ac_2O	70 88	190 191	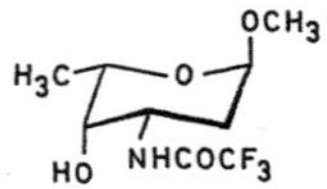 680	L-Selectride THF	93	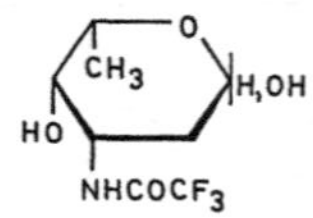 580	$H^{\oplus}$	190 191

4	371, 372 (2.5.1) Scheme 76 via 420 (2.6.1) Scheme 92 or (2.6.2) Schemes 95-96	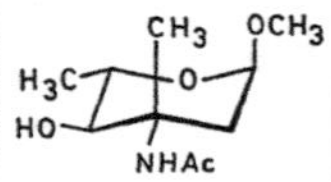 673	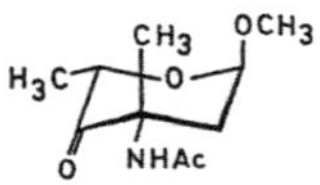 677	PCC	66	406	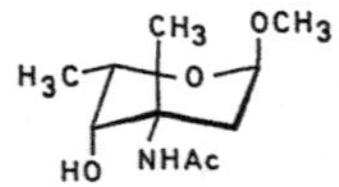 681	L-Selectride THF	80[a]	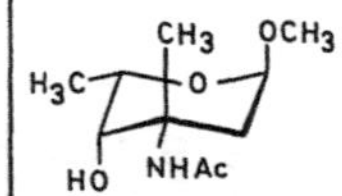 681	—	406
				RuO_4	45	405	Complete recovery of 673	$NaBH_4$ MeOH	100	Attempted inversion of configuration	—	405
5	481 Section (2.5.4) Scheme 85 via 592 (2.7.2) Table 9 entry 3	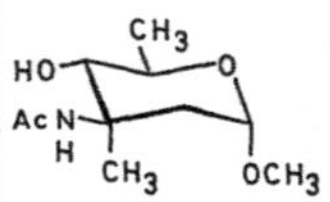 482	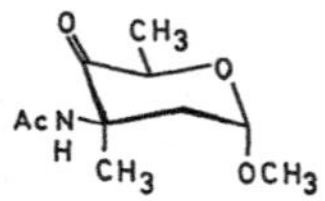 678[b]	PCC	58	183	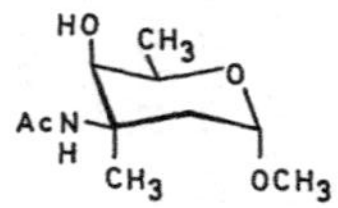 682	L-Selectride THF	64[c]	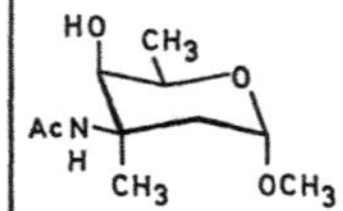 682	—	183

[a] The starting *equatorial* alcohol **673** was obtained in 12% yield
[b] Used directly for reduction
[c] Isolated yield

arabino → *lyxo* conversion in the "4-O-sulfonate route" were obtained when a triflate leaving group was employed (296) (Table 14 entry 2). The L-*lyxo*-α-D-azidoglycoside **663** so prepared (Table 14 entry 2) was converted (Sect. 4.3.3, Scheme 185) into **670**, a precursor of beta-lactam antibiotics (269).

The alternative route for the *arabino* → *lyxo* configurational conversion involves the oxidation of the C-4 hydroxyl group of protected methyl 3-amino-2,3,6-trideoxy-*arabino*-hexopyranosides and results in *threo*-hexopyranosid-4-uloses (such as compounds **674**–**678**, Table 15). It has become more popular than the previous method leading through 4-O-sulfonates. This is clearly because of the exceedingly high stereoselectivity observed upon reduction of cyclic ketones using lithium tri-*sec*-butylborohydride (L-Selectride) to produce products with *axially* disposed hydroxyl group (501), almost exclusively. Oxidations of the α-L- and D-*arabino* glycosides **672**, **648**, **211**, **673**, and **482** were carried out using various oxidizing agents (Table 15 entries 1–5). The highest yield (88%) was obtained (190, 191) with the chromium trioxide-dipyridyl complex in the presence of acetic anhydride (Table 15 entry 3), introduced to carbohydrate chemistry by Garegg and Samuelsson (285).

The L-Selectride reductions of methyl 3-acetamido-2,3,6-trideoxy-α-L-*threo*-hexopyranosid-4-ulose **(675)** and its N-trifluoroacetyl analog **676**, prepared from **648** and **211**, respectively, gave the corresponding L-daunosamine glycosides **261** (187) and **680** (190, 191) in over 90% isolated yields (Table 15 entries 2 and 3). Upon reduction of the 3-C-methyl-branched ketone **677** using the same reagent, a 20:3 mixture of methyl N-acetyl-α-L-vancosaminide **(681)** and the starting *arabino* compound **673** (Table 15 entry 4) was isolated (406). In contrast, when the reduction of **677** was performed with sodium borohydride, the *arabino* glycoside was completely recovered (405). Methyl N-acetyl-α-D-vancosaminide **(682)** was synthesized (183) by an essentially similar method via the C-4 uloside **678** derived from **482** (Table 15 entry 5).

2.8.2 The *lyxo* → *arabino* configurational transformation

The only method (502) reported for the *lyxo* → *arabino* configurational transformation (Scheme 117) led to the preparation of N-trifluoroacetyl-L-acosamine **(684)** that

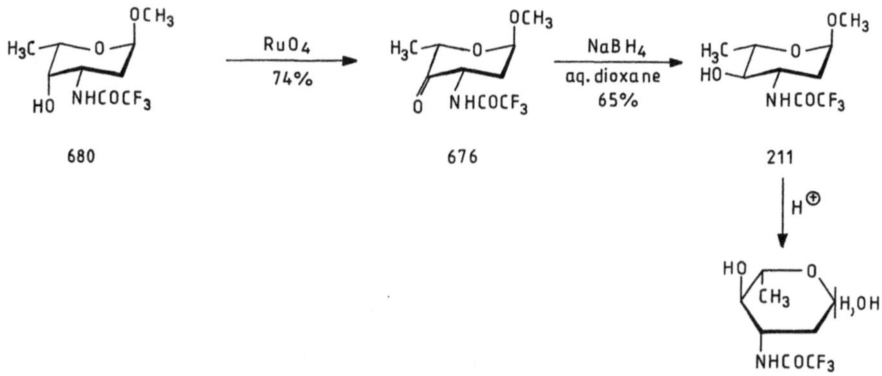

680 676 211 684

Scheme 117

was used, after conversion into the 1-chloro derivative, to glycosylate anthracyclinones. This study led in 1975 to the development of 4′-epidaunomycin **(4)** and 4′-epi-adriamycin **(5)** possessing as high antitumor activity but significantly less cardiotoxicity than the parent antibiotics carrying L-daunosamine moiety. Methyl α-L-daunosaminide **(260)** was trifluoroacetylated to give methyl N-trifluoroacetyl-α-L-daunosaminide **(680)**. Oxidation of this latter product using ruthenium tetroxide yielded the α-L-*threo*-uloside **676** in 74% yield. The reduction of **676** using sodium borohydride in aqueous dioxane gave a 65% yield of methyl N-trifluoroacetyl-2,3,6-trideoxy-α-L-*arabino*-hexopyranoside (**211**, methyl N-trifluoroacetyl-α-L-acosaminide). Hydrolysis of the glycosidic bond of **211** under mild acidic conditions then gave N-trifluoro-acetyl-L-acosamine **(684)**.

The Arcamone group also elaborated the direct conversion of daunomycin into 4′-epidaunomycin by a similar strategy which is discussed in Sect. 2.8.5, (Scheme 119).

2.8.3 The *ribo* → *xylo* C-4 configurational inversion

Early investigations of the possibilities of converting methyl 3-acylamino-2,3,6-trideoxy-L-*ribo*-hexopyranoside derivatives into the corresponding *xylo* C-4 epimers were promoted by structure–activity relationship studies on semisynthetic anthracycline glycoside antibiotics. In contrast to the 4′-epi-derivatives, the 3′-epi-analogs of such antibiotics with the L-*xylo* aminodeoxyhexose moiety practically do not possess antitumor activity.

Following the 4-O-mesylation-substitution methodology, the Arcamone group transformed the L-*ribo* methyl glycosides **203** and **597** (both prepared from the benzylidene acetal L-**157**) into the 6-hydroxy- **(689)** (236) and 6-deoxyl-L-*xylo* **(688)** (503) glycosides via the 4-O-mesylates **685** and **686**, respectively (Table 16 entries 1 and 2). Inversion of the configuration of the C-4 hydroxyl group of the ristosaminide **597** was achieved (122) via the α-L-*erythro*-hexopyranosid-4-ulose **691**, the reduction of which using sodium borohydride resulted in a readily separable 1:1 mixture of the *xylo* glycoside **688** and the starting *ribo* compound (Table 17 entry 1).

The N-acetyl methyl glycoside **424** of the aminodeoxy hexose (3-epi-L-vancosamine, **21**) occurring in antibiotic A 35512 B was synthesized (185) by the reaction sequence **378** → **422** → **690** → **692** → **424** (Table 17 entry 2). In this case reduction of the 4-uloside **692** with L-Selectride proceeded stereoselectively to produce the target *xylo* compound **424** in 82% yield.

When the enantiomer **693** (prepared from **410** via **461**) of the L-*erythro* ketone **692** was subjected to treatment with lithium tri-*sec*-butylborohydride (Table 17 entry 3), the stereoselective formation of the *axial* aminoalcohol **(457)** was again observed. This substance served as the key intermediate (182, 414) for both D-rubranitrose **(241)** and methyl α-D-kijanoside **(437)**, as discussed in Sect. 2.5.3 (Fig. 8 and Schemes 81–83). In contrast, the reduction of the N-trifluoroacetyl analog **(694)** of the 3-acetylamino ketone **693** was found (182, 414) to be extremely sluggish, producing, under various conditions, only a 12% yield of the desired product **464** (Table 17 entry 4) and a large amount of the unchanged ketone **694**.

Table 16. Inversion of the configuration at C-4. *Ribo* → *xylo* conversion via 4-O-sulfonates

Entry	Starting sugar (Section)	Ribo		Ref.	Xylo product	Method	Yield (%)	Target aminosugar	Method	Ref.
		4-Hydroxy compound	4-O-Mesylate							
1	L-157 (2.2.3) Scheme 37	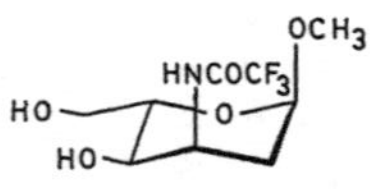 203	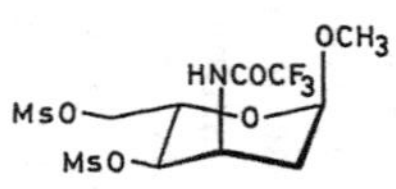 685	236	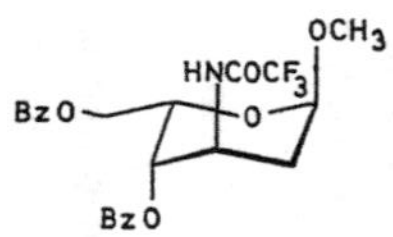 687	NaOBz DMF	80	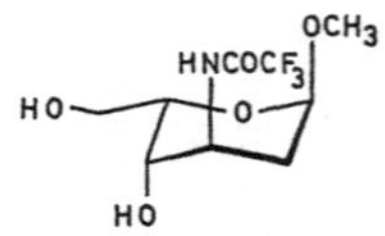 689	$OCH_3^{\ominus}$	236
2	203 (2.7.3) Scheme 112	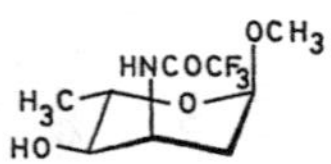 597	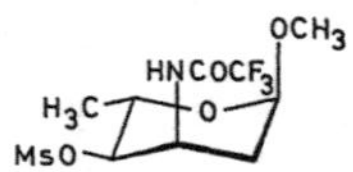 686	503	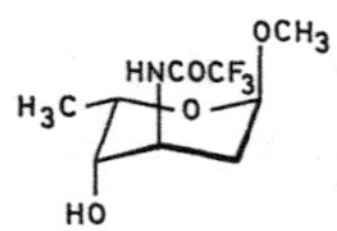 688	i. NaOBz DMF ii. $OCH_3^{\ominus}$	70	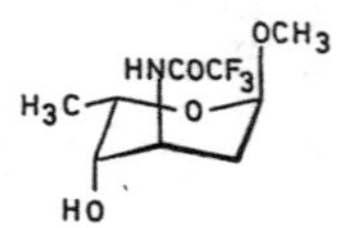 688	—	503

Table 17. Inversion of the configuration at C-4. *Ribo* → *xylo* conversion via C-4 ulosides

Entry	Starting sugar (Section)	*Ribo* 4-Hydroxy compound	C-4-Ulose	Method of oxidation	Yield (%)	*Xylo* product	Reducing agent	Yield (%)	Target aminosugar	Method	Ref.
1	203 (2.7.3) Scheme 112	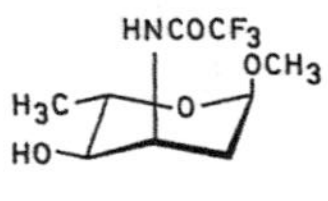 597	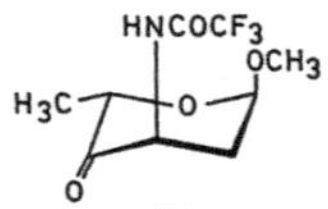 691	RuO4	40	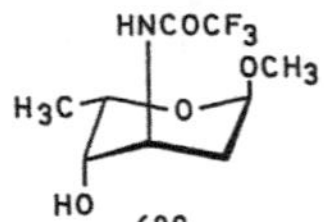 688	NaBH4 dioxane	45[a]	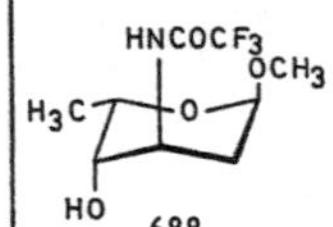 688	—	122
2	378 via 422 (2.5.2) Table 6 entry 4	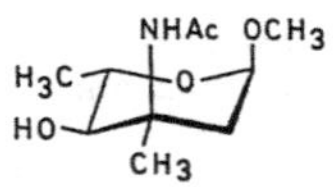 690	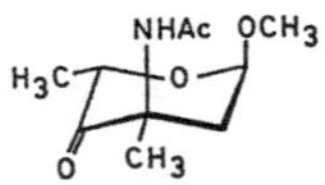 692	PCC CH2Cl2	87	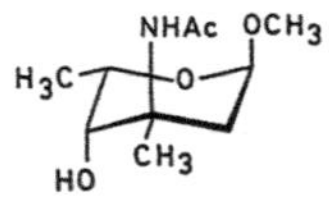 424	L-Selectride THF	82	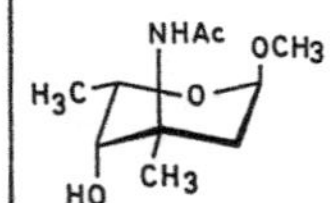 424	—	185
3	410 via 593 (2.7.2) Table 9 entry 1	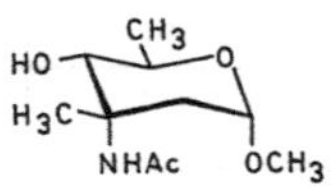 461	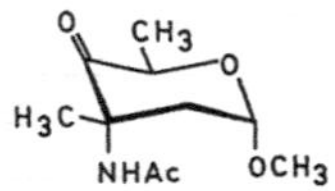 693	PCC CH2Cl2	74	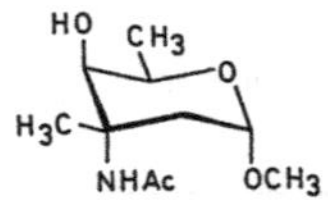 457	L-Selectride THF	92	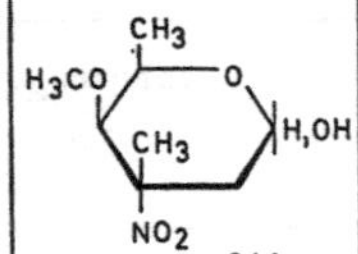 241	via 462 and 463 see Section 2.5.3 Fig. 8	182 414
4	589 via 591 (2.7.2) Table 9 entry 2	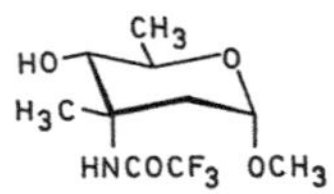 595	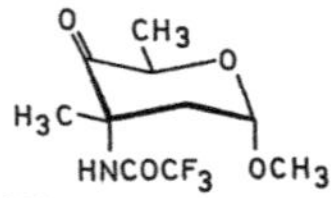 694	PCC CH2Cl2	93	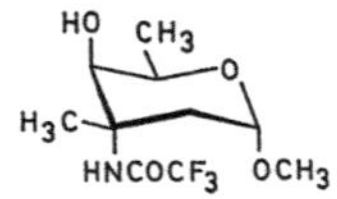 464	L-Selectride THF	12[b]	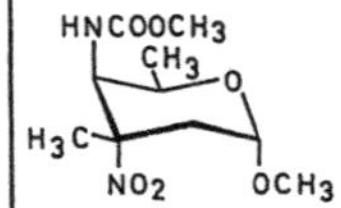 437	c	182 414

[a] A 1:1 mixture of the *xylo* compound **688** and the *ribo* glycoside **597** was obtained

[b] Unchanged ulose **694** was recovered in 81% and 47% yields when the reduction was conducted at —40 °C and 0 °C, respectively.

[c] For a more efficient route via **466** and **471** see Sect. 2.5.3, Fig. 8 and Schemes 81–83

Table 18. Inversion of the configuration at C-4. *Xylo* → *ribo* conversion via 4-O-sulfonates

Entry	Starting sugar (Section)	Xylo		Ribo	Method	Yield (%)	Target aminosugar	Method	Ref.
		4-Hydroxy compound	4-O-Sulphonate	product					
1	610 via 132 (2.7.4) Table 11 entry 1	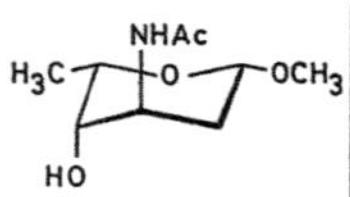 625	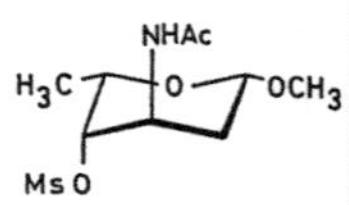 696 [a]	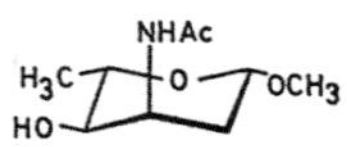 698	i. MsCl pyr ii. NaOAc AcOH $NaHCO_3$	50 from 625	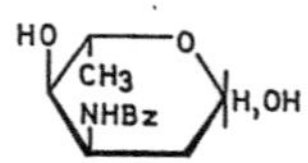 700	i. $Ba(OH)_2 + H_2O$ ii. $BzCl/(CH_3)_2CO$ iii. $H^{\oplus}$	231
2	805 (3.2.2) Scheme 134	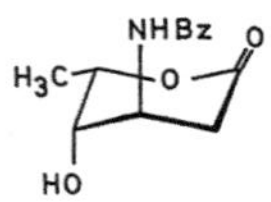 695	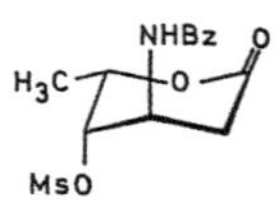 697	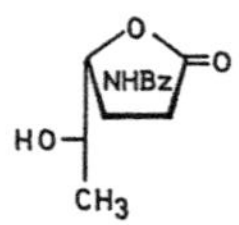 699	aq. NaOAc heat	75	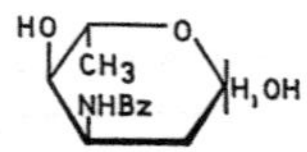 700	DIBAL	504

[a] Not isolated

2.8.4 The *xylo* → *ribo* configurational step-over

The practical value of methods enabling the configurational change of a 3-amino-2,3,6-trideoxy-L-*xylo*-hexose derivative into the *ribo* (ristosamine) C-4 epimer was shown by the studies of the groups of Horton (232), Monneret (231), and Arcamone (500, 504). These methods make N-acylated derivatives (such as **625** and **695**, Table 18 entries 1 and 2) of 3-amino-2,3,6-trideoxy-L-*xylo*-hexopyranose (**21**, R=H) readily available from either D-glucose (see Sects. 2.2.1, 2.7.2, and 2.7.4) or (2R,3R)-tartaric acid (Sect. 3.2.2).

Methyl 3-acetamido-2,3,6-trideoxy-α-L-*xylo*-hexopyranoside **(625)** and the N-benzoyl-L-*xylo*-δ-lactone **695** were converted through mesylates **696** and **697** into the L-*ribo*-aminosugars **698** (231) and **699** (504), respectively, using the 4-O-mesylate → epimeric 4-O-acetate way of inverting the configuration at C-4. Subsequent operations (Table 18 entries 1 and 2) then allowed the preparation of N-benzoyl-L-ristosamine **(700)**, a stable crystalline derivative of the parent aminodeoxy sugar **18** isolated from the vancomycin-type antibiotics (58) ristomycin (ristocetin) and avoparcins.

In contrast to previously discussed results (cf. Tables 15 and 17), the reduction of the C-4 ulose **702**, derived from the α-methyl glycoside **(701)** of the L-*xylo* sugar **896** (Sect. 3.2.3, Scheme 148) using lithium tri-*sec*-butylborohydride, gave (500) a significant amount (42%) of the target 3-benzamido-3-C-methyl-α-L-*ribo*-glycoside **703** together with a 35% yield of the starting *xylo* compound **701** (Scheme 118). The unusually high proportion of the *equatorial* alcohol produced by the L-Selectride reduction might be attributed to the presence of the bulky, *axial* 3-N-benzoyl function adjacent to the reaction center.

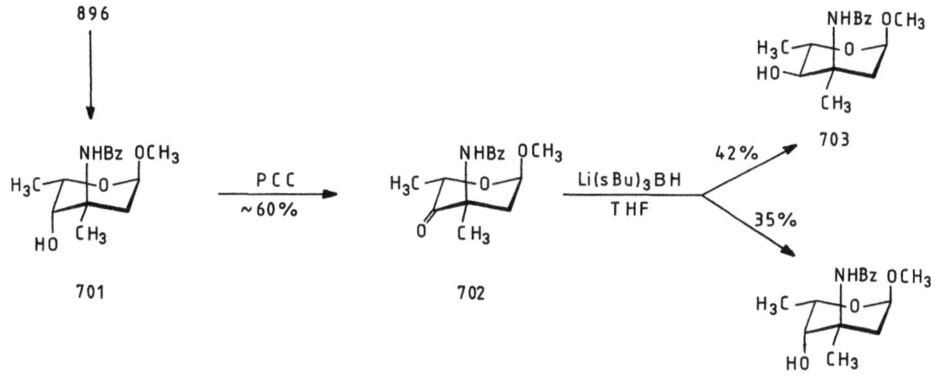

Scheme 118

2.8.5 Methods for the configurational interchange at the carbohydrate portion of the intact anthracycline antibiotics

It has been mentioned previously that many of the aminosugars prepared by means of the inversion of the configuration of a given steroisomer at C-4 (Sects. 2.8.1–2.8.4) have been used to glycosylate anthracyclinones to obtain semisynthetic antibiotic analogs for structure–activity relationship studies. To avoid multistep syntheses of the aminodeoxy hexose portion of such antibiotic derivatives, the Arcamone group has introduced (505) an ingenious straightforward strategy for the direct transformation of the N-trifluoroacetylated intact molecule of daunomycin **(704)** into the corresponding L-*arabino*- **(705)**, L-*ribo*- **(706)**, and L-*xylo*- **(707)** analogs (Scheme 119). Swern-oxidation of **704**, prepared by N-trifluoroacetylation of **1**, gave the L-*threo* C-4 ulose **708** in 80% yield. Reduction of this substance using sodium borohydride at −70 °C gave exclusively the *arabino* isomer: N-trifluoroacetyl-4′-epidaunomycin **(705)**. At higher temperatures a small quantity of the original *lyxo* alcohol **704** was also produced.

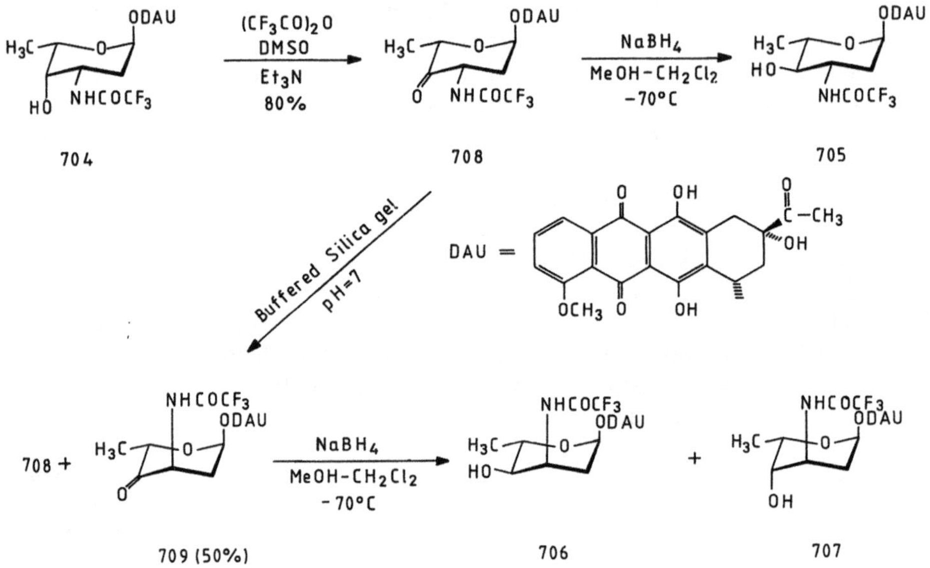

Scheme 119

When the *threo* ketone **708** was treated with buffered silica gel (phosphate buffer, pH = 7), a mixture of the *erythro*-ulose **709** and the unreacted **708** was isolated, from which the C-3-isomerized product **(709)** crystallized in 50% yield. Reduction of **709** using sodium borohydride at −70 °C produced a separable 1:1 mixture of the antibiotic analogs carrying L-*ribo*- **(706)** and L-*xylo* **(707)** 3-amino-2,3,6-trideoxyhexose moieties. The use of a related technique for the direct production of the 4′-deoxy analogs of anthracycline glycoside antibiotics is discussed in Sect. 4.3.1 (Scheme 173).

3. Synthesis of 3-Amino-2,3,6-Trideoxyhexoses by Using Non-carbohydrate Precursors

The common basic principle of the syntheses discussed throughout Chap. 2 is that the target aminodeoxy sugars are prepared by means of manipulations (inversion, elimination, etc.) on the chiral centers of multichiral carbohydrate-type substrates. In the majority of such conversions the already existing asymmetric centers and the more or less stable conformation of the cyclic sugar molecules ensure the generation of the new chiral centers in a highly diastereoselective manner. However, the situation is completely different when the target carbohydrates are synthesized starting with non-sugar precursors (61, 506). The reparation of a desired stereoisomer from among the four enantiomeric pairs of 3-amino-2,3,6-trideoxyhexoses (cf. Chap. 1, Table 1) is rather complicated, but at the same time, it is one of the most challenging problems in the field of carbohydrate chemistry. As demonstrated by the increasing number of publications in the past few years, the goal of synthesizing daunosamine-type aminodeoxy sugars from achiral or small chiral molecules, has encouraged organic chemists to develop diverse strategies — sometimes in a close competition.

Because of the wide-spread scale of the reported methodologies to be reviewed in this chapter, the classification of the synthetic procedures is a rather difficult problem. From the numerous possibilities, a system that is based, principally, on the construction of the carbon chain of the target aminodeoxy sugar was chosen. According to this classification, syntheses applying six-carbon precursors, bearing the entire carbon framework of the desired aminosugar essentially assembled as targets for the further development of the chiral centers possessing the disired substitution and stereochemical patterns are discussed first (methods type "6C").

In the second large group of related strategies, those aimed at the construction of the six-carbon chain by forming one- or more C—C bonds are collected (methods type "nC + mC"). It is quite apparent that in either of the above two principal groups of syntheses, operations directed to configurational inversion (universally involved as crucial steps in the procedures described in Chap. 2) are of secondary importance. Instead, a more or less selective saturation (amination or oxygenation) of a diastereotopic molecular fragment (sp^2 plane) is emphasized. Moreover, such manipulations often emerge as crucial steps in synthetic strategies employing carbon-carbon bond formation and represented here by the symbol "nC + mC". In the outlines of the individual procedures, the place of attachment of the carbon atoms resulting in the complete six-carbon framework of the aminosugars to by synthesized is indicated, as is whether the synthesis is chiral (*) or achiral (±). Finally, a sub-classification of the methods according to the reaction-type of the key step is also attempted. Because of the diverse methodologies, this kind of systematization proved to be extremely difficult — and sometimes impossible.

3.1 Synthesis of 3-amino-2,3,6-trideoxyhexoses from six-carbon substrates

Many of the approaches to the aminodeoxy sugars are based on unsaturated six-carbon starting materials and involve their derivatization through saturation of the sp^2 systems. In the classical acosamine synthesis of Dyong and Bendlin (507) (Scheme 120), sorbic acid (*trans-trans*-hexadienecarboxylic acid) was used as the starting material. Regiospecific epoxidation of the 4,5-double bond of methyl sorbate **(710)** with peracetic acid gave the racemic epoxide **711**, the Lewis acid-catalyzed ring cleavage in acetone of which led to the formation of the 4,5-O-isopropylidene derivative **712** with the *erythro* configuration. (Due to the *trans* arrangement of

6 C (±)

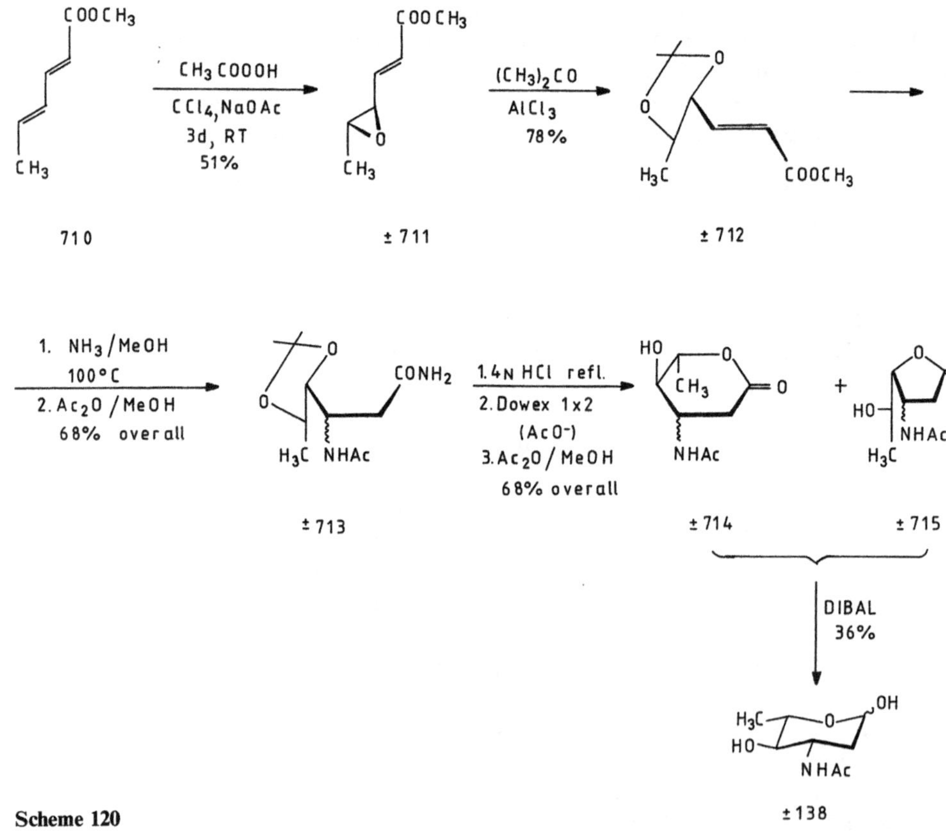

Scheme 120

the double bonds in **710**, the epoxide is of the *threo* configuration and the usual *trans* opening of the anhydro ring results in the formation of the *erythro*-diol **712**). Introduction of the amino function to C-3 was achieved by the addition of ammonia across the double bond of **712** to obtain, after acetylation, a racemic mixture **(713)** of the *arabino* and *ribo* amides. Hydrolysis of this latter mixture with hydrochloric acid, followed by N-acetylation gave a complex unseparable mixture consisting of the antipodes of the C-3 epimeric δ- **(±714)** and γ- **(±715)** lactones. Reduction of the lactone function with di-isobutylaluminum hydride (DIBAL) produced a crude product from which N-acetyl-D,L-acosamine **(±138)** could be isolated in moderate yield by fractional crystallization.

6C (*)

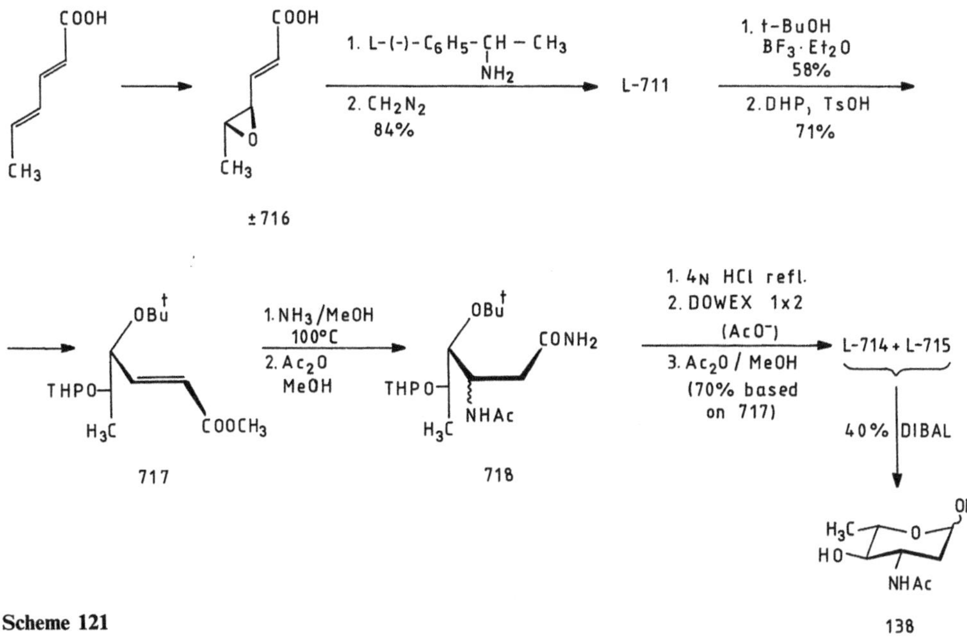

Scheme 121

In a modified version (507) of this procedure (Scheme 121), the racemic 4,5-anhydro-carboxylic acid **716**, prepared from sorbic acid, was resolved with L-(—)-α-phenylethylamine, and the L-enantiomer was converted into the methyl ester L-**711**. Boron trifluoride in *tert*-butanol catalyzed the opening of the anhydro ring. Reaction with dihydropyrane then gave the enantiomerically pure mixed diether **717**, which was first transformed into the C-3 epimeric acid amide **718** and then into the L-*arabino* + *ribo* lactones (L-**714** and L-**715**), as shown for the racemic mixtures. Reduction of this latter mixture with DIBAL furnished N-acetyl-L-acosamine **(138)**. Despite the formation of rather complex mixtures in these two pioneering syntheses, the simple isolation of the desired aminosugar by crystallization makes them quite attractive.

A simple route (508) (Scheme 122) that leads to the γ-lactone of D,L-megosaminic acid (**±719**, a derivative of megosamin **19**) can be regarded as an incompletely ela-

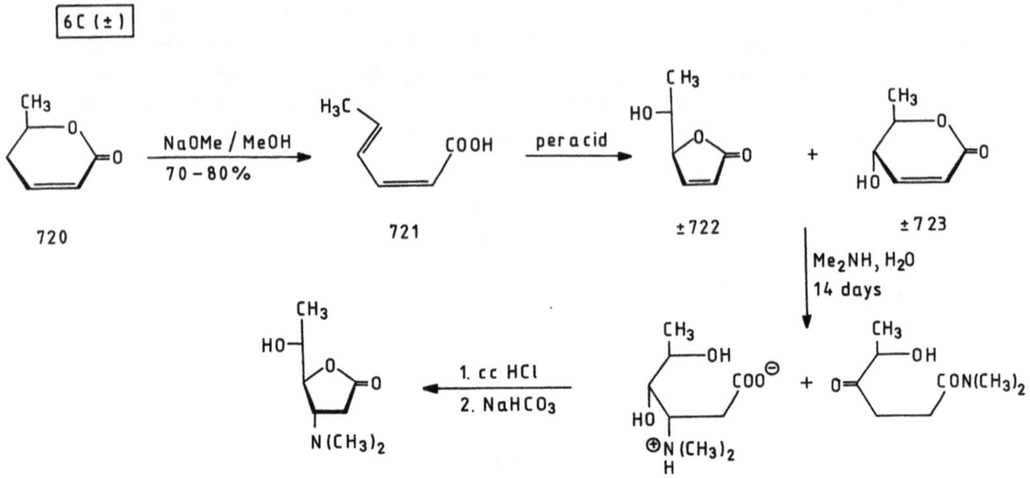

Scheme 122

borated alternative to the Dyong method (507). Peracid oxidation of *Z,E*-sorbic acid **(721)**, prepared from parasorbic acid **(720)**, gave a mixture of D,L-osmunda-γ- **(±722)** and δ- **(±723)** lactones. Addition of dimethyl amine across the double bond of these lactones furnished compounds **±724** (41%) and **±725** (36%), from which the 4,5-dihydroxy-3-dimethylamino-D,L-hexanoic acid **(±724)** was isolated. It was then transformed into D,L-megosaminic acid-γ-lactone **(±719)** by treatment with hydrochloric acid. However, DIBAL reduction did not convert **±719** into D,L-megosamine **(±19)**.

A synthesis of N-benzoyl-D,L-daunosamine **(±671)** reported by Hirama et al. (509) is also based on *E,E*-sorbic acid (Scheme 123). Hydroxylation of the 4,5-double bond of ethyl sorbate **(726)** followed by partial silylation of the resulting diol gave the 5-O- **(727)** and 4-O- **(728)** mono silyl ethers in a ratio of 2:1. The major component **727**, isolated by chromatography, was subjected to protecting group manipulations involving tetrahydropyranylation, removal of the silyl ether, teatment with chlorosulfonyl isocyanate to produce the 5-O-carbamate, hydrolysis of the 4-O-tetrahydropyranyl ether, and trimethylsilylation. The intramolecular Michael addition of the resulting racemic 2,3-unsaturated ester **730** — the key step of the reaction sequence — proceeded with high 1,3-*anti* selection and produced a separable 5:1 mixture of the diastereoisomeric cyclic carbamates **731** and **732** in 75% yield.

Note that the analogous Michael addition with the 5-O-carbamoyl-4-O-*tert*-butyldimethylsilyl ether **729**, prepared from the minor component **728** of the hydroxylation reaction, provided only a 2:1 *anti* stereoselection **(733** and **734)**. Sequential alkaline hydrolysis and N-benzoylation of **731** gave the D,L-*lyxo*-γ-lactone **±735**, a precursor of N-benzoyl-D,L-daunosamine **(±671)**. The synthetic utility of this procedure was also demonstrated by the conversion (509) of the 5-O-monosilyl ether **727** into the D,L-*xylo*-lactone **±736**, the C-3 epimer of **±735** (Scheme 124). Thus, **727** was transformed into the 4-O-carbamate **737** or the 4,5-di-O-carbamate **738**, and the Michael addition of both compounds readily produced the oxazolidinones **739** and **740** with excellent regioselectivity (27:1 and 23:1, respectively).

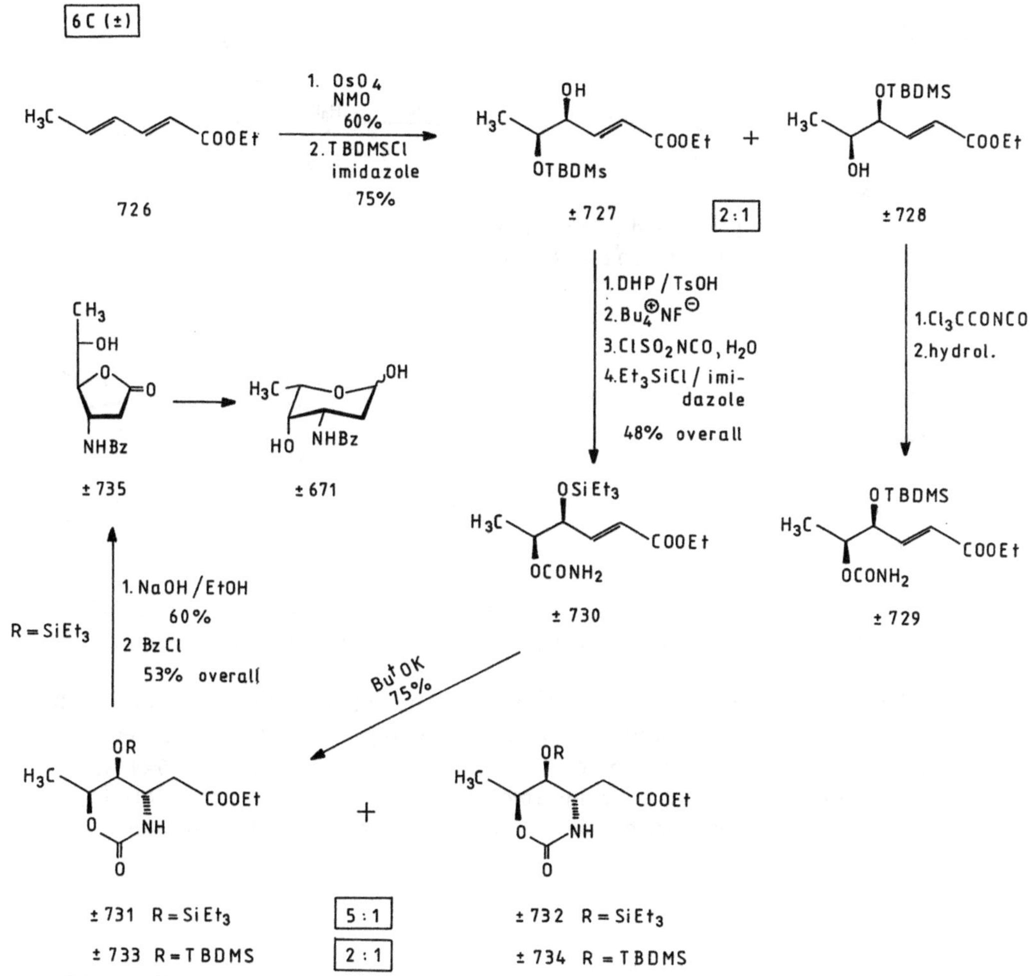

Scheme 123

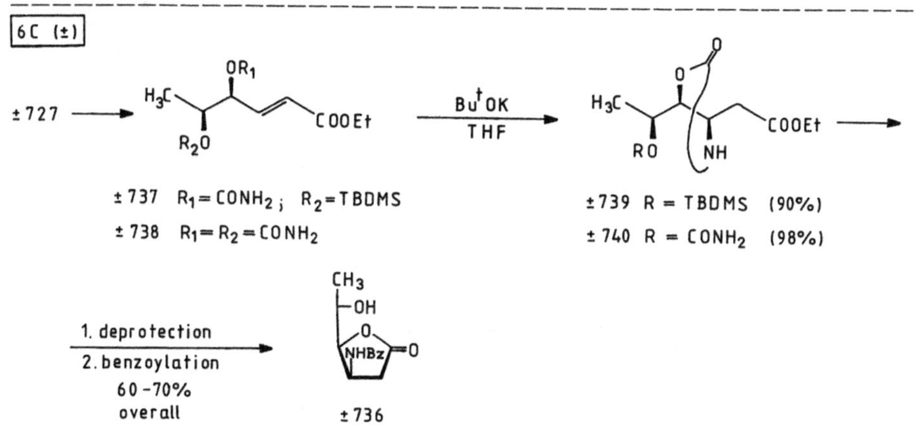

Scheme 124

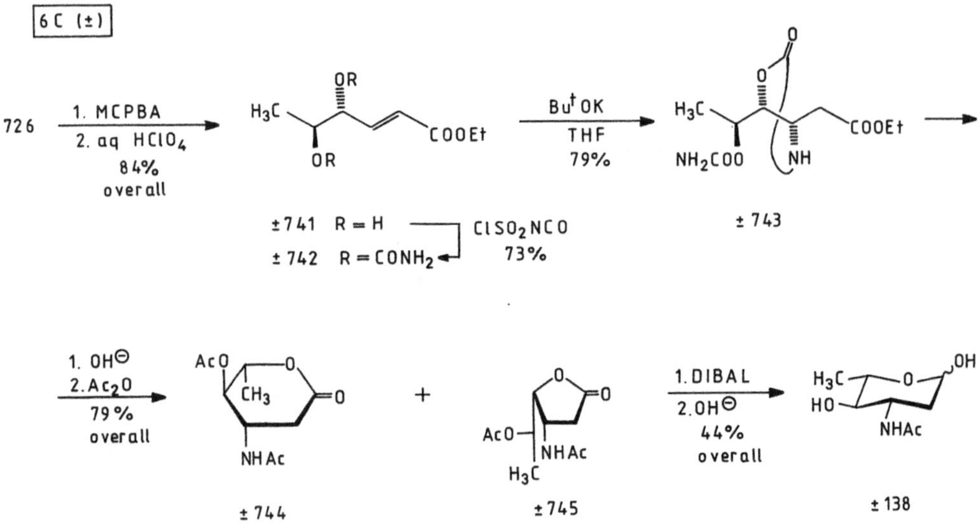

Scheme 125

Alkaline hydrolysis of **739** and **740**, followed by N-benzoylation and removal of the silyl protecting group then gave the *xylo*-γ-lactone ±**736**, a precursor of 3-amino-2,3,6-trideoxy-D,L-*xylo*-hexopyranose.

The Hirama group extended the Michael addition method to the preparation (510) of derivatives of D,L-acosamine (±**8**) and D,L-ristosamine (±**18**). The epoxide prepared from ethyl sorbate (**726**) was hydrolyzed into the *erythro*-diol **741** whose carbamoylation yielded **742**. The intramolecular Michael addition of this latter substance gave almost exclusively the oxazolidinone **743** with the D,L-*arabino* con-

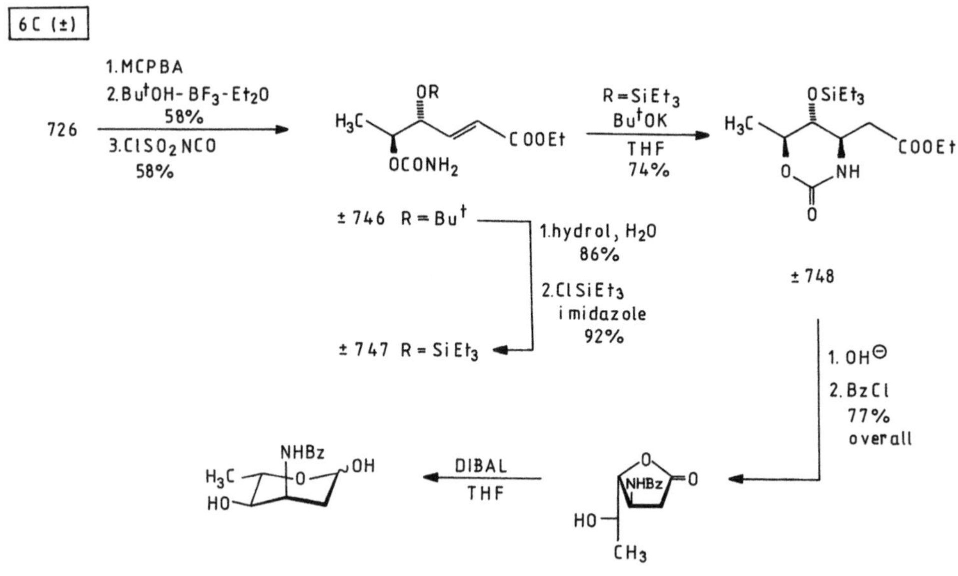

Scheme 126

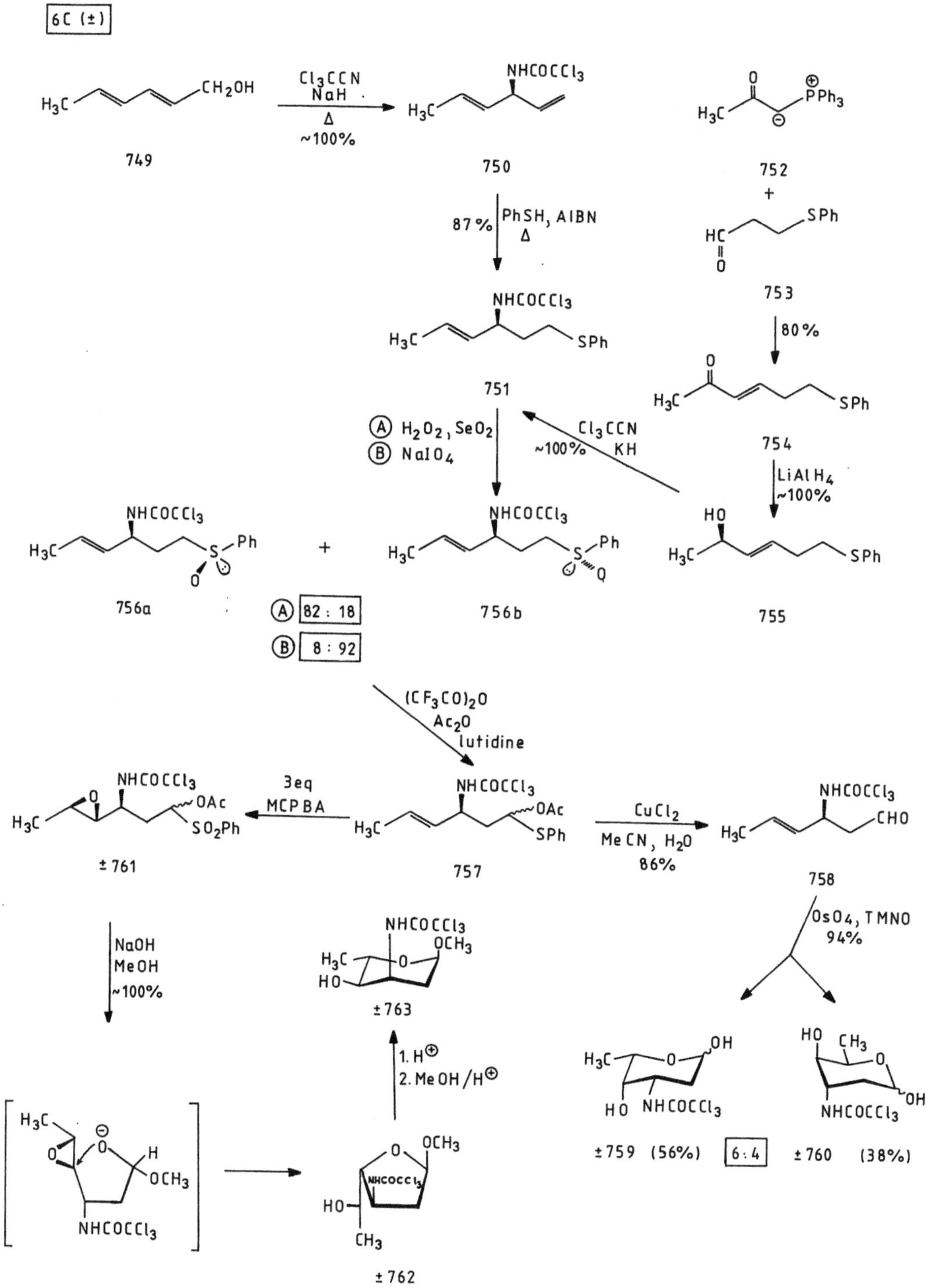
6C (±)
Cl_3CCN
NaH
Δ
~100%
749
750
752
753
754
755
751
87% PhSH, AIBN Δ
80%
(A) H_2O_2, SeO_2
(B) $NaIO_4$
Cl_3CCN KH ~100%
$LiAlH_4$ ~100%
756a
756b
(A) 82 : 18
(B) 8 : 92
$(CF_3CO)_2O$
Ac_2O
lutidine
3eq MCPBA
$CuCl_2$
MeCN, H_2O
86%
± 761
757
758
OsO_4, TMNO 94%
NaOH
MeOH
~100%
± 763
1. $H^⊕$
2. MeOH/$H^⊕$
± 762
± 759 (56%)
6 : 4
± 760 (38%)

Scheme 127

figuration (Scheme 125). Alkaline hydrolysis of the racemic **743**, followed by acetylation led to the mixture of the D,L-*arabino*-N,O-diacetyl-δ- and γ-lactones (**744** and **745**, respectively). Reduction of this mixture with DIBAL and subsequent O-deacetylation produced N-acetyl-D,L-acosamine (**±138**).

In order to synthesize (510) ristosamine (the *ribo* stereoisomer) with this method, the C-5 homoallyl carbamate derivative (**747**) of **741** was necessary. To obtain this compound, a sequence of reactions was carried out starting with ethyl sorbate (**726**) and involving epoxidation, oxirane-ring opening by boron trifluoride-catalyzed solvolysis in *tert*-butanol, and carbamoylation with chlorosulfonyl isocyanate at O-5 to give the 4-O-*tert*-butyl-5-O-carbamoyl derivative **746** (Scheme 126). Aqueous hydrolysis of the 4-O-*tert*-butyl group of **746**, followed by silylation with triethylchlorosilane and imidazole gave the racemic homoallyl carbamate ±**747**, the key intermediate of this procedure. Cyclization of this latter substance gave exclusively the cyclic carbamate ±**748**, from which the racemic γ-lactone of N-benzoyl-D,L-ristosaminic acid (**±699**) was obtained upon alkaline hydrolysis and Schotten-Baumann benzoylation. This latter compound was then transformed into N-benzoyl-D,L-ristosamine (**±700**) by DIBAL reduction.

In a recent diastereoselective synthesis of derivatives of D,L-daunosamine and D,L-ristosamine, Hauser and associates (511) employed sorbyl alcohol (**749**) as the six-carbon substrate to be derivatized (Scheme 127). The Overman reaction (512) of **749** yielded the aminodiene derivative **750**, which was regioselectively converted into the (*E*)-1-phenylthio compound **751** (87% yield) through free radical addition of benzenethiol. As an alternative route, the sulfide **751** was also prepared by the Wittig reaction of the phosphorane **752** with 3-(phenylthio)-propanal (**753**), followed by hydride reduction of the resulting enone **754** into the allylic alcohol **755** and Overman reaction of this latter substance with trichloroacetonitrile, as was used to obtain **750**.

By the proper choice of the oxidizing agent, the sulfide **751** could be selectively oxidized to form a mixture of the diastereoisomeric sulfoxides **756a** and **756b**.

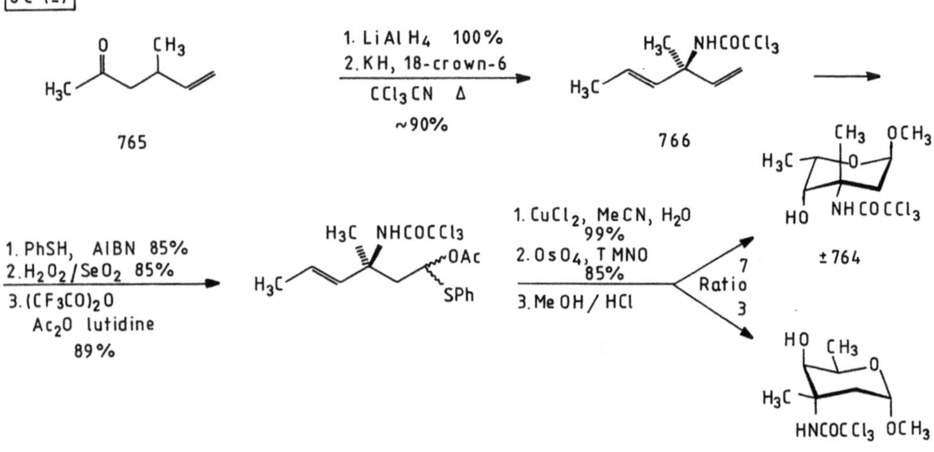

Scheme 128

The Pummerer reaction of these S-oxides readily afforded the mixed O,S-acetal **757** in 92–95% yield. This then served for the preparation of the derivatives of both D,L-daunosamine and D,L-ristosamine. Hydroxylation of the *trans* double bond of the aldehyde **758** (obtained upon deacetalation of **757** with cupric chloride) with OsO_4-trimethylamine oxide gave a readily separable 6:4 mixture of N-trichloroacetyl-D,L-daunosamine (±**759**) and its D,L-*xylo* isomer (±**760**) in 56% and 38% yields, respectively.

For the synthesis of the corresponding ristosamine derivative (±**763**), *cis*-hydroxylation of the 4,5-double bond of **757** was effected. Epoxidation with three equivalents of *m*-chloroperbenzoic acid resulted in the *trans*-epoxide ±**761**, the treatment of which with sodium hydroxide in methanol produced the furanoside ±**762** through *trans*-opening of the oxirane function by the intramolecular attack of the methoxide ion.

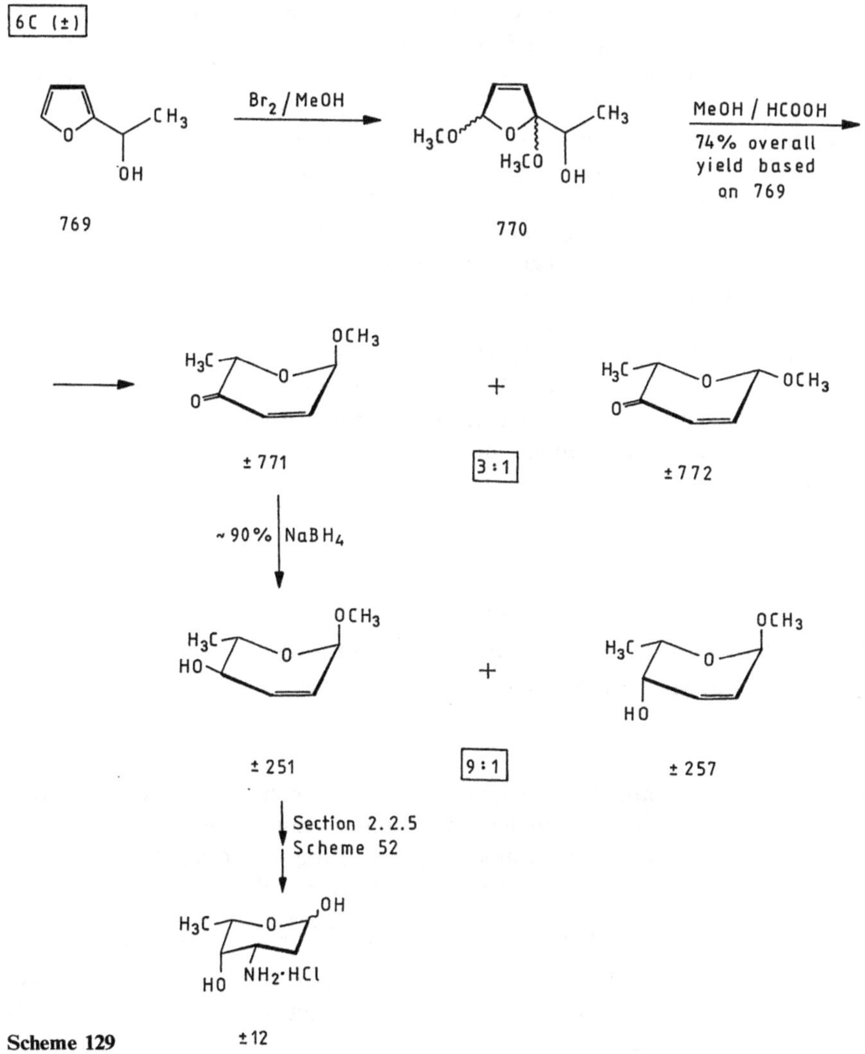

Scheme 129

Acid hydrolysis of the furanoside ±**762** and methyl glycosylation led to methyl N-trichloroacetyl-α-D,L-ristosaminide (±**763**). Note that the O,S-acetal **757** was also utilized in a synthesis of phenyl 1-thio-N-trichloroacetyl-α,β-D,L-daunosaminide (see Sect. 4.1.1, Scheme 159).

By using an essentially similar strategy, Hauser et al. (511) synthesized methyl N-trichloroacetyl-α-D,L-vancosaminide (±**764**) starting with the branched-chain enone **765** (Scheme 128). The application of the Overman-reaction, followed by the development of the latent aldehyde function by means of the Pummerer rearrangement and hydroxylation of the *trans* double bond according to the sequence **765** → **766** → **767** gave 7:3 mixture of the methyl α-D,L-vancosaminide derivative ±**764** and its D,L-*xylo* diastereoisomer (±**768**).

Furan derivatives often serve as starting materials for the synthesis of monosaccharide derivatives (513). Correspondingly, in one of the procedures worked out for the preparation of racemic daunosamine (±**12**) by Samuels and Thetford (347), 1-(2-furyl)-ethanol (**769**) was employed as the substrate of the target sugar (Scheme 129). Methoxylation of **769** was carried out with bromine in methanol to give the dimethoxy derivative **770**, the acid-catalyzed methanolysis of which furnished the racemic α- and β-methyl enulosides **771** and **772**, respectively, in a 3:1 ratio. Reduction of the keto function of the separated α-anomer ±**771** using sodium borohydride gave a 9:1 mixture of methyl 2,3,6-trideoxy-α-D,L-*erythro*-hex-2-enopyranoside (±**251**) and the corresponding *threo*-glycoside (±**257**) that was separable by column chromatography. The *erythro* glycoside ±**251** was then transformed into D,L-daunosamine (±**12**) by a modified Mitsunobu reaction via the sequence ±**251** → ±**262** → ±**263** → ±**264** → ±**12** (cf. Scheme 52), as discussed in Sect. 2.2.5. Because the operations involved in this procedure are simple and the starting **769** can be obtained (347, 514) in enantiomerically pure form by the enantioselective reduction of furyl methyl ketone, this method appears to also be one of the most effective and economical procedures for the production of the natural L-daunosamine (**12**).

3.2 Synthesis of 3-amino-2,3,6-trideoxyhexoses from chiral and achiral precursors by means of carbon-carbon bond formation reactions

In this large group of synthetic strategies, the six-carbon framework of the target aminodeoxy sugars is built from smaller chiral units, achiral units or both. In this section the related methodologies are subdivided according to the chain-length of the substrate to be elongated, and the place of attachment of the added molecular fragment is indicated. For example, the designation $\boxed{4\,C + 2\,C(\pm)C_2{-}C_3}$ is used for an achiral synthesis in which a four-carbon precursor is elongated with a two-carbon fragment and the carbon-carbon bond forms between carbons C_2 and C_3 of the target aminodeoxy sugar.

3.2.1 Construction of the six-carbon framework with a 5C + 1C ascent

A five-carbon substrate can be lengthened into a hexose derivative by the addition of either the C-1 or C-6 carbon atom of the desired sugar derivative. As C-2 in the resulting 2-deoxyhexose is achiral, carbon-carbon bond formation in the first case does not involve the problems of diastereoselectivity that occur in the alternative strategy.

Anticipating that the Morgan-Elson reaction of aminodeoxy hexoses — resulting in oxazolines — can be reversed, Wong et al. (515) used the oxazole **773** for a synthesis (Scheme 130) of racemic daunosamine. Allylic bromination of the methyl group of **773** gave the bromomethyl derivative **774**, which was treated with cyanide representing the 5C + 1C elongation step, to obtain the nitrile **775**. Upon methanolysis and acid hydrolysis, **775** was converted into the carboxylic acid **776**

5C + 1C (±) C_1–C_2

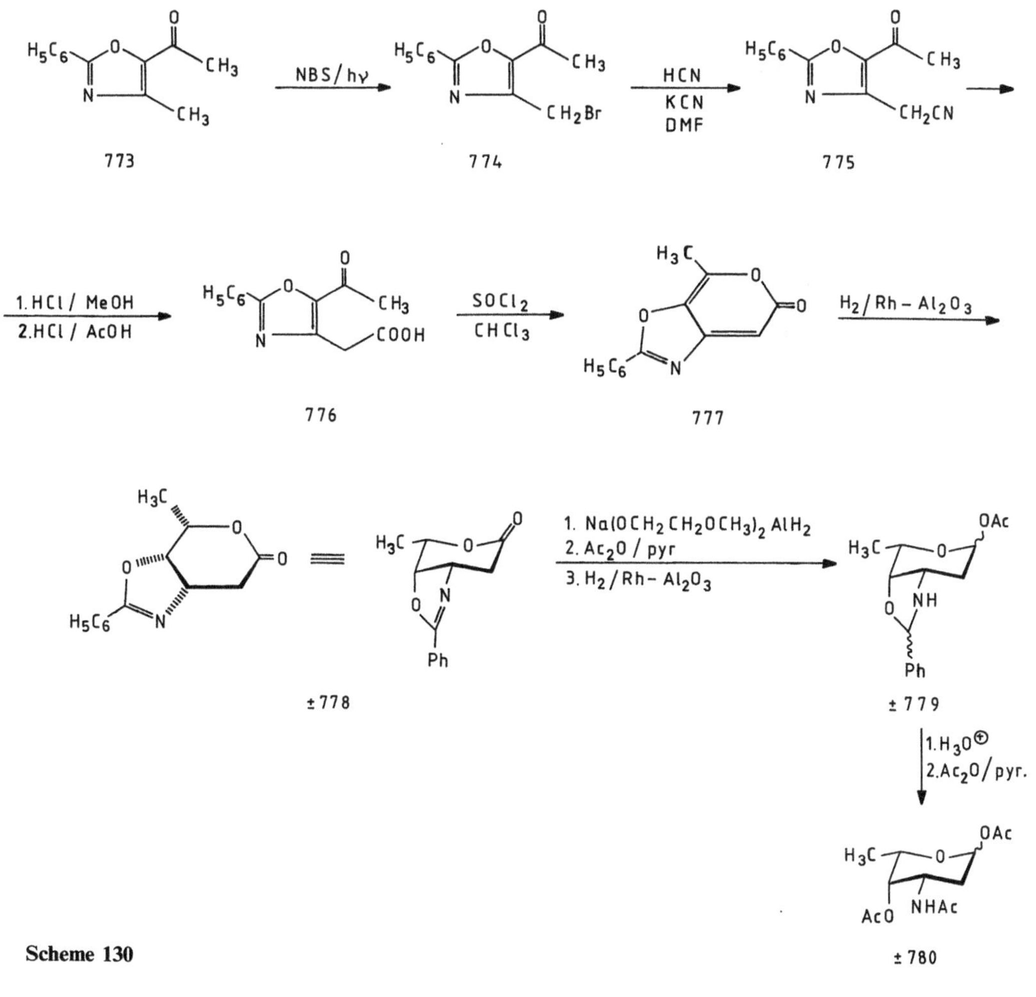

Scheme 130

from which the oxazolo-α-pyrone **777** was prepared by the reaction with thionyl chloride. Hydrogenation of **777** over rhodium-on-alumina (Rh/Al_2O_3) catalyst produced the *all-cis* diastereoisomeric oxazolino-lactone ±**778** exclusively. When the reduction was conducted over Adams catalyst, the δ-lactone of 4-deoxy-D,L-daunosamine (±**1029**) was produced (515, 516), as discussed in Sect. 4.3.1 (Scheme 176). Reduction of the lactone carbonyl group, followed by acetylation and saturation of the double bond of the oxazoline ring by means of rhodium-catalyzed hydrogeno-

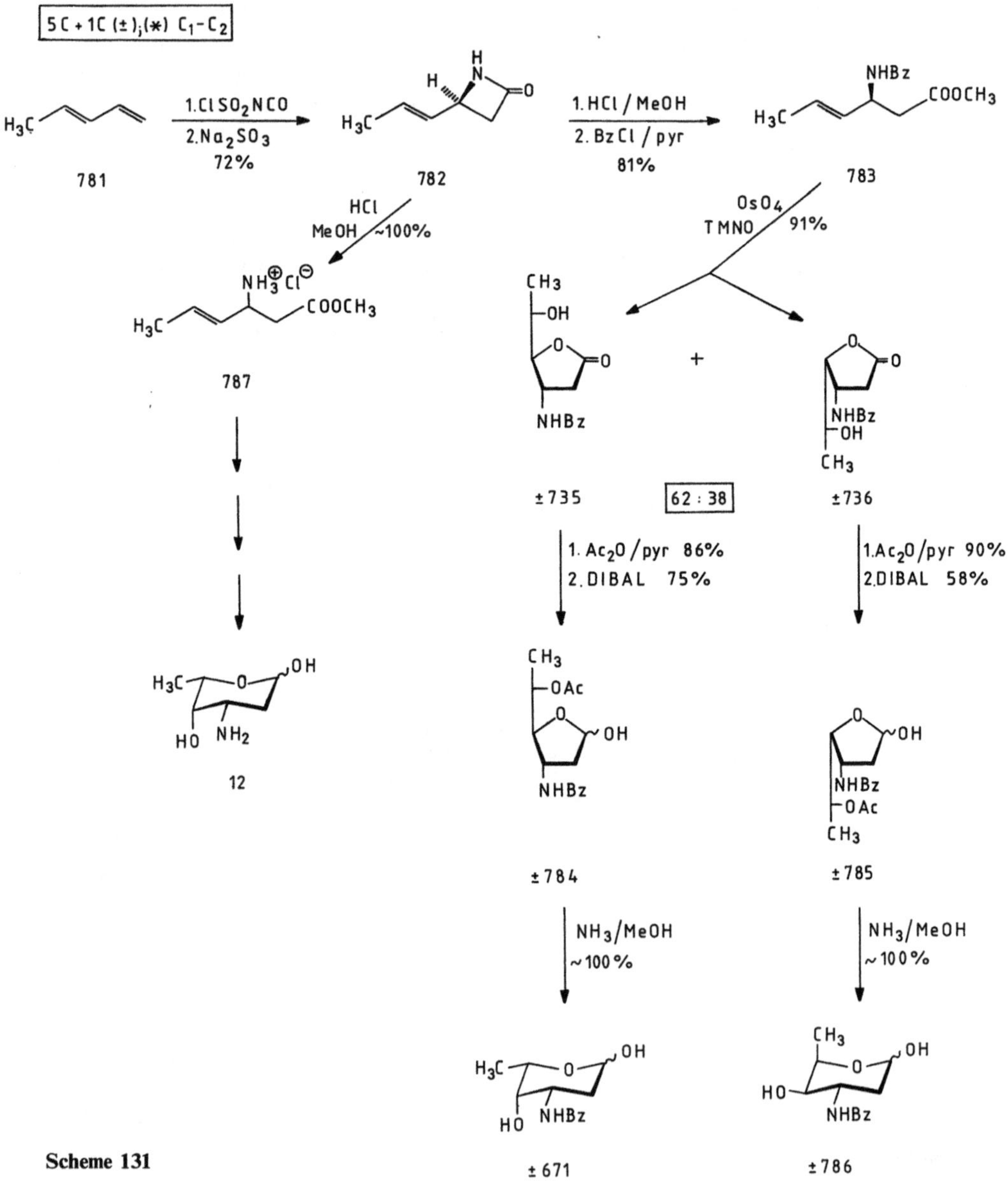

Scheme 131

lysis yielded the oxazolidine ±**779**, a target for conversion into the peracetyl derivative (±**780**) of D,L-daunosamine (±**12**) by acid hydrolysis and subsequent acetylation.

A total synthesis of the optically active N-benzoyl-L-daunosamine, as well as of the corresponding racemic analog (Scheme 131), elaborated by Hauser et al. (517, 518), is also based on the 5C + 1C construction of the hexose framework. The key step of the procedure, a $[{}_2\pi_s + {}_2\pi_a]$ cycloaddition of chlorosulfonyl isocyanate across the double bond of (*E*)-1,3-pentadiene (**781**), is followed by reductive cleavage of the N-chlorosulfonyl moiety by sodium sulfide to form the azetidinone **782** in 72% yield. This reaction ensures the simultaneous 5C + 1C chain lengthening and introduction of the C-3 amino group.

Methanolysis of the propenyl-azetidinone **782** and benzoylation gave methyl (±)-3-benzamido-4-(*E*)-hexenoate (**783**), the *cis*-hydroxylation of which gave a separable 62:38 mixture of the *lyxo-* (±**735**) and *xylo-* (±**736**) -N-benzoyl-γ-lactones in 91% yield. Acetylation of the separated diastereoisomers, followed by DIBAL reduction gave the furanose aminosugars ±**784** and ±**785**, respectively. Upon treatment with ammonia in methanol, these were converted into 3-benzamido-2,3,6-trideoxy-D,L-*lyxo-* (±**671**) and -D,L-*xylo*-hexopyranose (±**786**).

Since methyl (±)-3-amino-4-(*E*)-hexenoate (**787**), produced from the azetidinone **782** by methanolysis, could be readily resolved through the diastereoisomeric methyl-phenyl urea derivatives, this simple procedure made L-daunosamine (**12**) also available according to the reaction sequence just described. The propenyl-azetidinone **782** was also prepared by a 4C + 2C approach (519) by Manhas et al. (Sect. 3.2.2, Scheme 139).

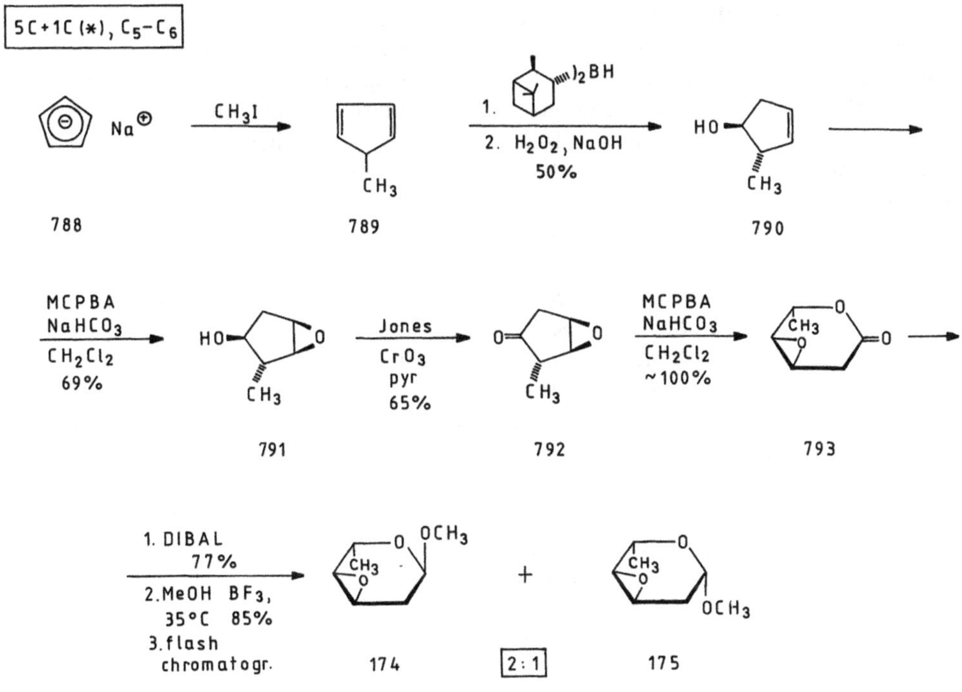

Scheme 132

In the ingenious asymmetric synthesis of Grethe et al. (520, 155), aimed at the preparation of L-daunosamine (Scheme 132), the starting material cyclopentadiene was methylated in the form of the sodium salt **788** to generate the C_5—C_6 bond in the target aminosugar. As the crucial step of the procedure, asymmetric hydroboration of the product **789** with (—)-di-3-pinanylborane gave a 50% yield of 2(s)-methyl-3-cyclopenten-1(s)-ol **790** with an optical purity greater than 95%. Stereoselective epoxidation of this latter substance produced a 69% yield of **791**, which was oxidized to the ketone **792**. This was subjected to a Bayer-Villiger ring enlargement that furnished the epoxylactone **793** in quantitative yield and in stereoselective manner. DIBAL reduction of **793** followed by careful a boron trifluoride-catalyzed methyl-glycosylation furnished a separable 2:1 mixture of methyl 3,4-anhydro-2,6-dideoxy-α-L-*ribo*-hexopyranoside **(174)** and its β-anomer **(175)**. The major product **174** represented the starting material (155) of the syntheses of L-daunosamine **(12)** and L-acosamine **(8)** that were elaborated by Goodman et al. (123) and Gupta (266), respectively, in the pioneering period of the synthetic research on these two important aminodeoxy hexoses.

3.2.2 Synthesis of 3-amino-2,3,6-trideoxyhexoses according to the 4C + 2C chain-lengthening principle

Related synthetic procedures can be based on the formation of either the C_2—C_3 or C_4—C_5 bond of the aminodeoxy hexoses that are to be synthesized by chain extension. Since in the target sugars the functional groups at positions 3, 4, and 5 are attached to chiral carbons, methods allowing the preferential diastereoselective formation of the coupled products carrying the respective substituents with the required stereochemistry have emerged as particularly useful. Involved in related strategies carbanion additions, Diels-Alder reactions, and nitrone-additions are the most frequently used key steps. In the other group of syntheses, the newly generated carbon-carbon bond does not yet carry the required substituent. The subsequent introduction of functional groups is achieved by manipulations involving the saturation of the olefinic linkages.

The Fuganti group has prepared numerous aminodeoxy sugars starting with either natural C_4-carbon compounds or the products of the baker's yeast-mediated coupling reaction (521, 522) of acetaldehyde with *trans*-cinnamaldehyde. Microbial reduction of the intermediary ketol **794** produced by acyloin condensation gives the (2s,3R)-diol **795** in 25–30% yield (Scheme 133). This versatile chiral compound already carries the C-4 and C-5 asymmetric centers of the danosamine-type aminodeoxy sugars.

Cleavage (498) of the double bond of the isopropylidenated diol **796** by ozonolysis, followed by the Wittig reaction of the resulting aldehyde **797** with ethoxycarbonylmethylenetriphenylphosphorane, produced the unsaturated ester **798** in 65% overall yield based on **796**. The Michael addition of ammonia to **798** gave stereoselectively the β-amino acid ester **799**. Acid hydrolysis of this latter substance led to the L-*arabino*-γ-lactone **800**, from which N-trifluoroacetyl-L-acosamine **(684)** was prepared by trifluoroacetylation and subsequent DIBAL reduction. Alternatively, the γ-lactone **800** was converted into the δ-lactone **649**, the inversion of the configuration

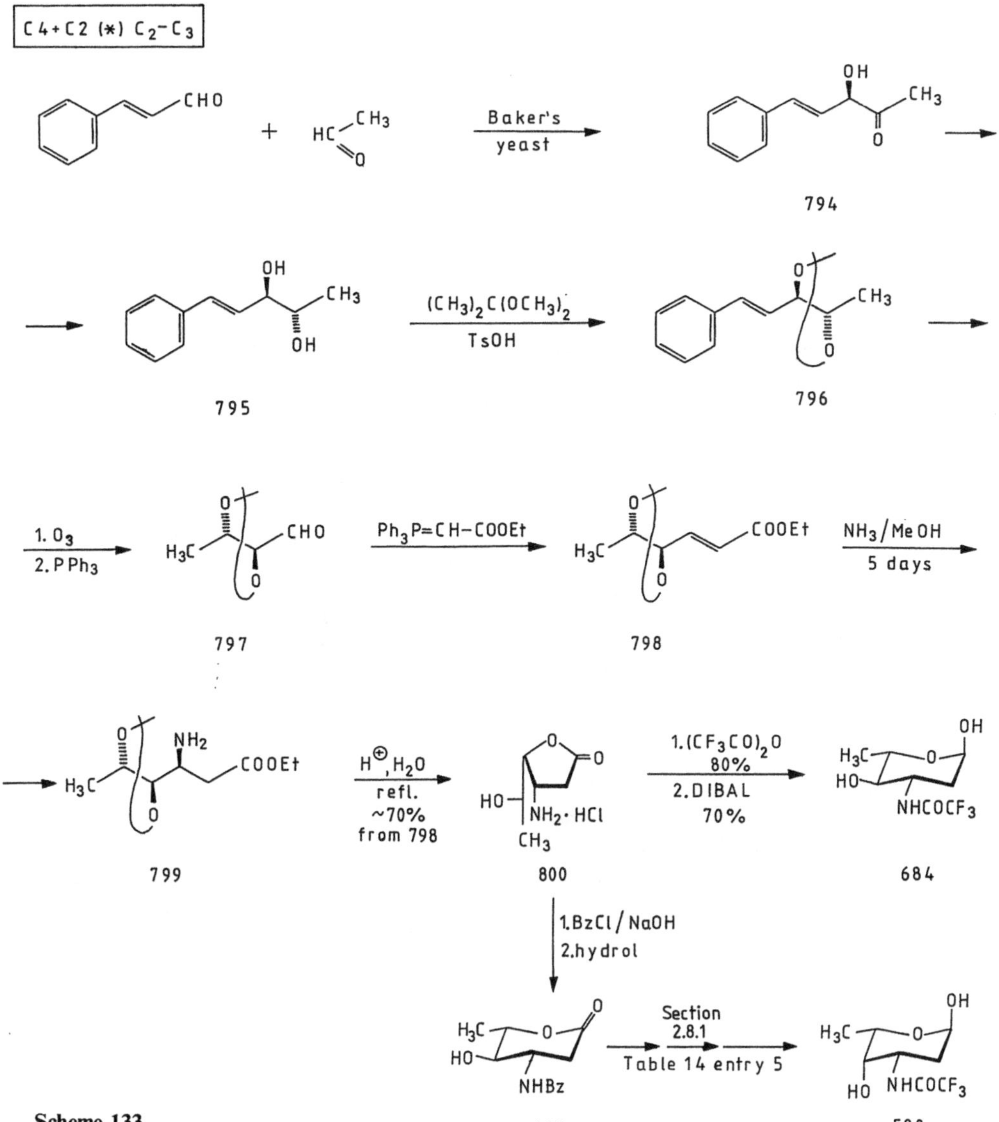

Scheme 133

of which at C-4 via the 4-O-mesylate **657** produced N-trifluoroacetyl-L-daunosamine **(580)** as shown in Sect. 2.8.1. (Table 14 entry 5).

By the application of an essentially similar methodology, starting with L-tartaric acid **(801)** and L-threonine **(806)**, the Fuganti group (523) developed chiral total syntheses for the derivatives of 3-amino-2,3,6-trideoxy-L-*xylo*-hexopyranose and its enantiomer (Schemes 134, 135). Thus, (2R,3R)-tartaric acid **(801)** was converted into the mono-tosylate **802** according to known procedures (524, 525) and deoxygenation was achieved by subsequent hydride reduction. Oxidation of the resulting (2R,3S)-triol **803** gave the aldehyde **804** which was subjected to a Wittig reaction that yielded a 13:7 *E*,*Z*-mixture of the unsaturated ester **805**. Addition of ammonia in methanol

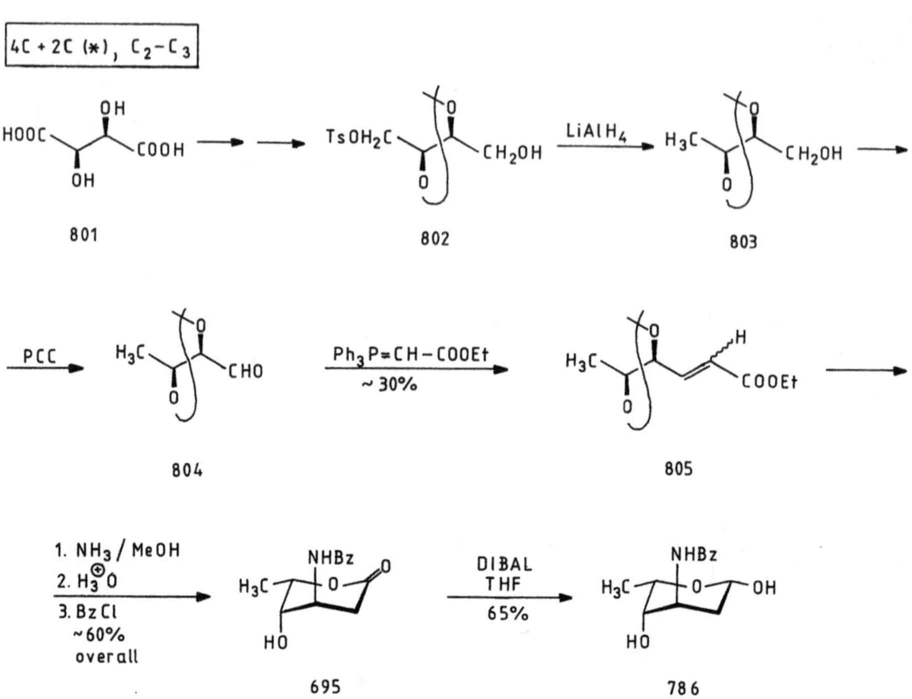

Scheme 134

and then hydrolysis of the protecting groups led to the δ-lactone **695** with L-*xylo* configuration in ca. 60% yield based on **805**. DIBAL reduction of this latter then yielded 3-benzamido-2,3,6-trideoxy-L-*xylo*-hexopyranose (**786**, 3-epi-L-daunosamine, Scheme 134). The N-benzoyl-L-*xylo*-δ-lactone **695** served (504) also for the synthesis of N-benzoyl-L-ristosamine (**700**) through inversion of the configuration at C-4 of the 4-O-mesylate **697** as discussed in Sect. 2.8.4 (Table 18 entry 2).

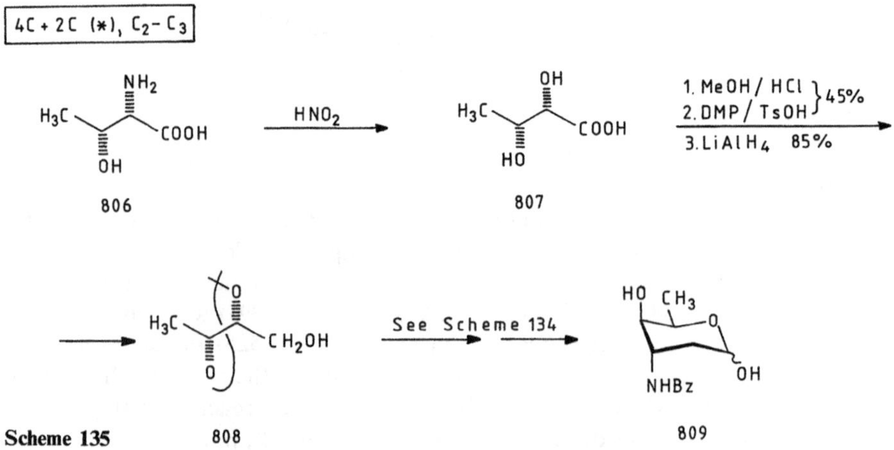

Scheme 135

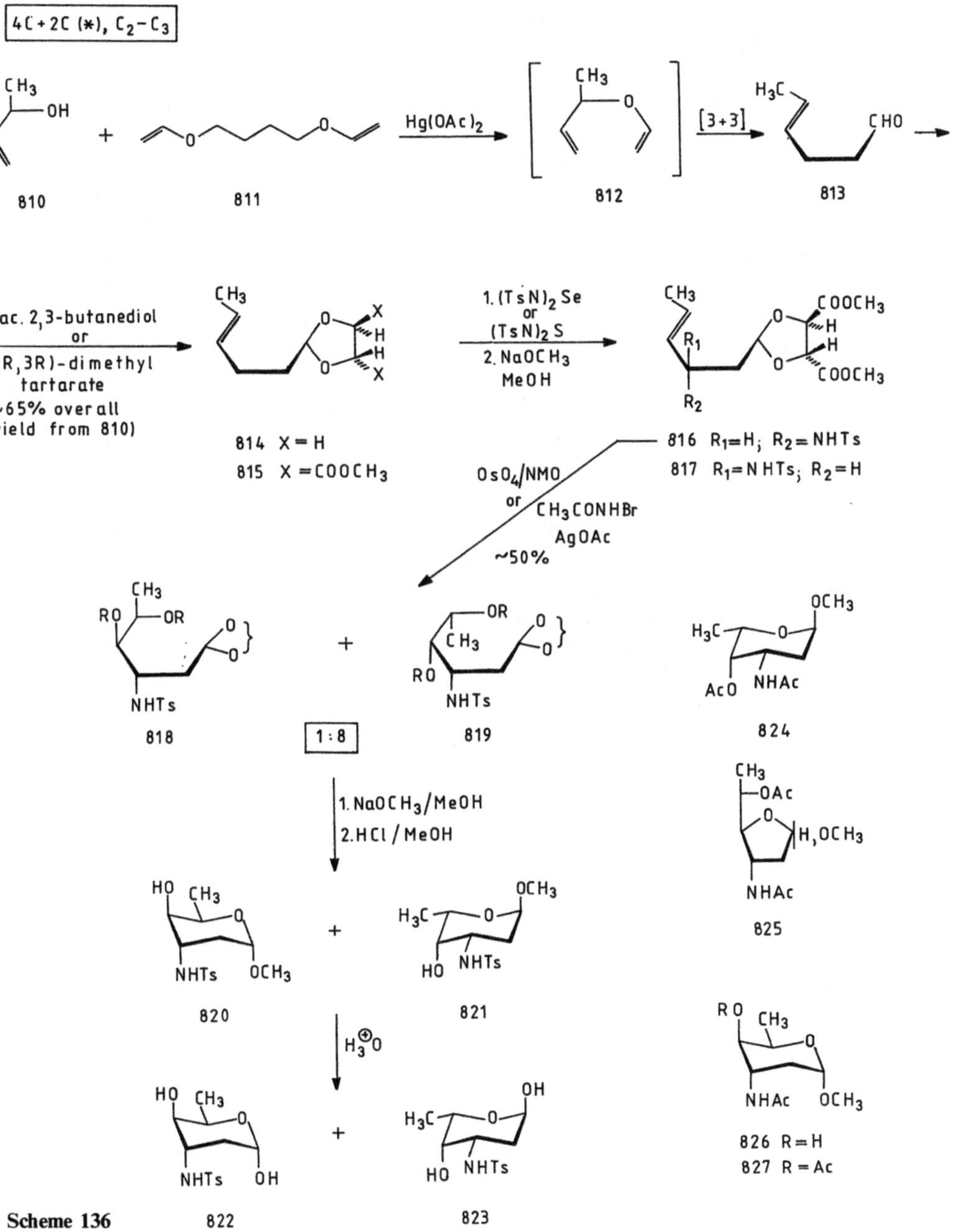

Scheme 136

The D-enantiomer **(809)** of the L-*xylo* sugar **786** was prepared (523) from L-threonine **(806)** that had been deaminated into (2S,3R)-dihydroxybutyric acid **(807)**. Transformation of this substance into the protected triol **808** was carried out in three steps involving acid-catalyzed esterification, isopropylidenation, and hydride reduction, as shown on Scheme 135. The (2S,3R)-alcohol **808** was converted into

3-benzamido-2,3,6-trideoxy-D-*xylo*-hexopyranose **(809)** as shown for the enantiomeric (2R,3S)-triol **803** (cf. Scheme 134).

In another total synthesis based on the 4C + 2C chain construction principle and aimed at the preparation of L-daunosamine and the C-3 epimeric L-*xylo* isomer, Dyong et al. (526–529) generated the C_2—C_3 carbon bond of the desired aminosugar by the [3.3]-sigmatropic rearrangement of the vinyl-allyl ether **812**, obtained upon vinylation of 3-buten-2-ol **(810)** with 1,4-bis(vinyloxy)butane **811** (Scheme 136). The aldehyde function of the resulting *trans*-4-hexenal **813** was protected through formation of the achiral (526) or chiral (527–529) acetals **814** and **815**, respectively, with racemic 2,3-butanediol and (2R,3S)-dimethyl tartarate, allowing in the latter case the asymmetric synthesis of the target aminodeoxy hexose.

The introduction of the C-3 amino function was based on the Sharpless method (530) and its improved version (531), employing N,N′-ditosylselenodi-imide or N,N′-ditosylthiodi-imide. The *trans*-double bond of the separated **816** component of the resulting diastereoisomeric mixture was *cis*-hydroxylated or *cis*-acyloxylated upon treatment with osmium tetroxide-dimethylamine oxide or N-bromoacetamide-silver acetate, respectively, to obtain a ca. 1:8 mixture of the D-*xylo* **(818)** and L-*lyxo* **(819)** amines. O-Deacylation and removal of the acetal function by methanolysis yielded the methyl glycosides **820** and **821**, from which N-tosyl-L-daunosamine **(823)** and its D-*xylo* isomer **(822)** were isolated by chromatography.

Methyl N,O-diacetyl-α-L-daunosaminide **(824)** (528), its furanoside ring-isomer **825** (528), methyl 3-acetamido-2,3,6-trideoxy-α-D-*xylo*-hexopyranoside **(826)** (529), and the corresponding N,O-diacetyl derivative **827** (529) were also prepared by the exchange of the N-tosyl group for acetyl in the methyl glycosides **821** and **820**. Dyong et al. (532, 533) worked out syntheses of methyl N-tosyl-α-L-vancosaminide and its C-3 epimeric L-*xylo* stereoisomer by an essentially similar methodology.

In a synthesis of racemic daunosamine derivatives, described by DeShong et al. (534, 535), the formation of the C_2—C_3 connection was achieved by a dia-

4C + 2C (±), C_2-C_3

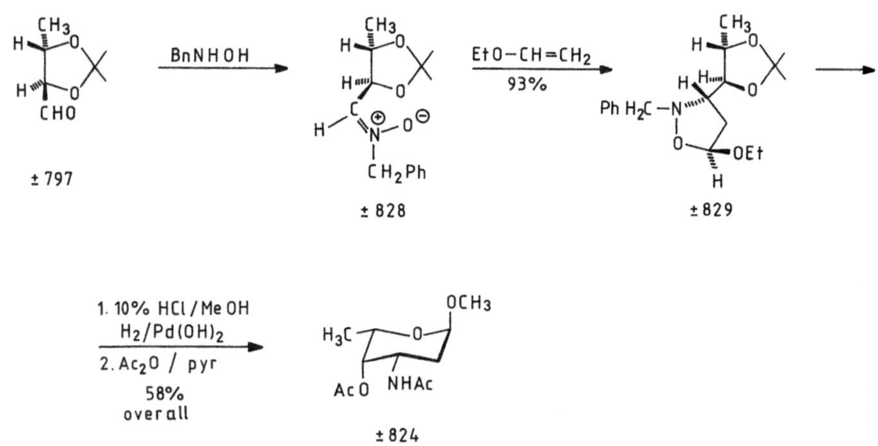

Scheme 137

stereoselective nitrone cycloaddition reaction (Scheme 137). Thus, cycloaddition of ethyl vinyl ether with the Z-nitrone **828** derived from the racemic aldehyde ±**797**, gave a single diastereoisomer (±**829**) in 93% yield. Hydrogenolysis of this compound over Pearlman's catalyst and subsequent acetylation led to methyl N,O-diacetyl-α-D,L-daunosaminide (±**824**) in 58% overall yield. The simultaneous development of the C_2—C_3 bond of the six-carbon framework and diastereoselective introduction of the amino functionality in the key step make this simple procedure quite attractive.

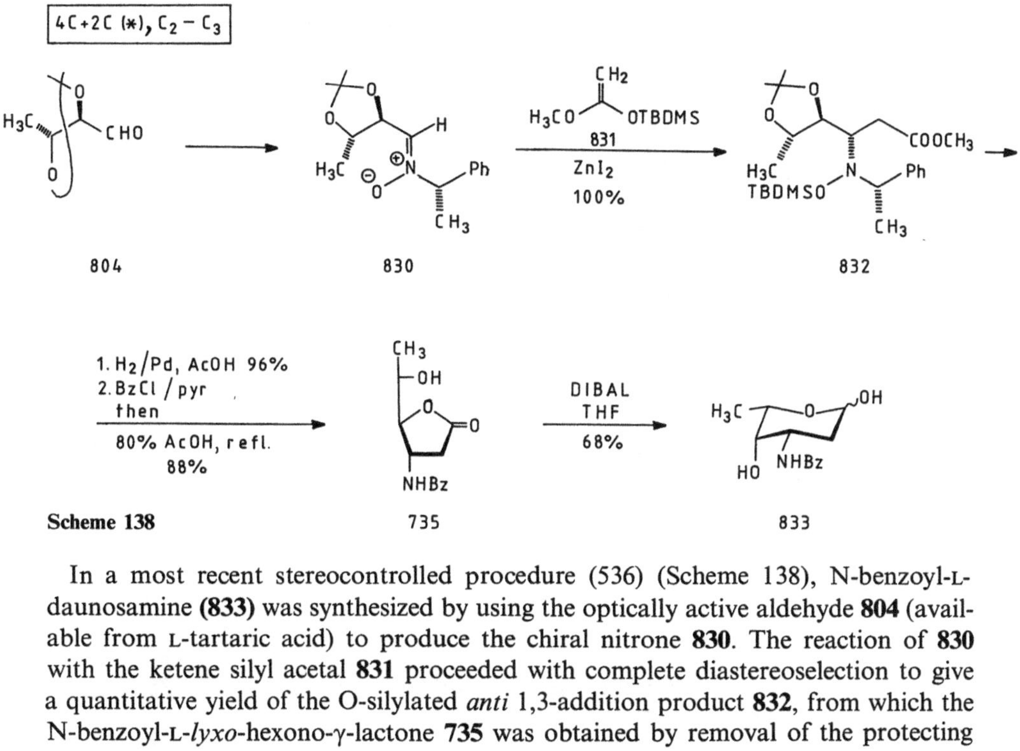

Scheme 138

In a most recent stereocontrolled procedure (536) (Scheme 138), N-benzoyl-L-daunosamine (**833**) was synthesized by using the optically active aldehyde **804** (available from L-tartaric acid) to produce the chiral nitrone **830**. The reaction of **830** with the ketene silyl acetal **831** proceeded with complete diastereoselection to give a quantitative yield of the O-silylated *anti* 1,3-addition product **832**, from which the N-benzoyl-L-*lyxo*-hexono-γ-lactone **735** was obtained by removal of the protecting

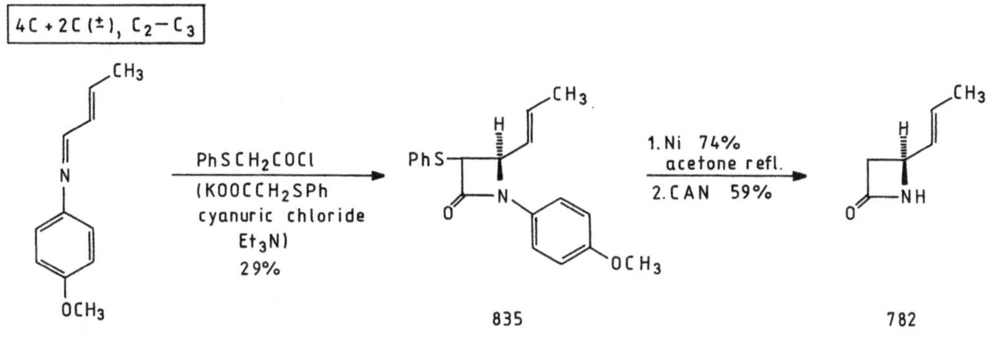

Scheme 139

groups. DIBAL reduction of this latter substance then resulted in N-benzoyl-L-daunosamine **833**.

In an approach to the total synthesis of aminodeoxy sugars from β-lactams, the Manhas group (519) introduced the C-3 substituent and the 2C unit, simultaneously (Scheme 139). Thus, *trans*-2-butenal-4-methoxyphenylhydrazone **(834)** was cyclized into *trans*-1-(4-methoxyphenyl)-3-phenylthio-4-(prop-1-enyl)-azetidin-2-one **(835)** using cyanuric chloride and potassium (phenylthio)acetate. Removal of the protecting groups of **835** by successive reductive desulfurization and oxidation with cerium ammonium nitrate gave 4-(prop-1-enyl)-azetidin-2-one **(782)**. This was also prepared by Hauser et al. (517, 518) by a 5C + 1C ascent. It is a key intermediate in the total synthesis of the racemic daunosamine (cf. Sect. 3.2.1, Scheme 131).

An additional achiral synthesis of daunosamine, described by Iwataki and associates (537), is based on the *cis*-hydroxylation of 1,1-ethylenedioxy-4-hexen-3-one **(838)**, prepared by means of 4C + 2C(C_2—C_3) chain-lenghtening of *trans*-crotonyl chloride **(836)** with vinyl chloride, followed by the formation of the ethylene acetal of the resulting hexadienone **837** (Scheme 140). Treatment of **838** with potassium permanganate gave a 35% yield of the *threo*-diol **839**, which was oximated and then hydrogenated over Adams catalyst to obtain a diastereoisomeric mixture of the amines **840** and **841**. Methanolysis of the amino-acetal **840** produced methyl α-D,L-daunosaminide **(±260)** in rather low overall yield.

4C + 2C (±), C_2−C_3

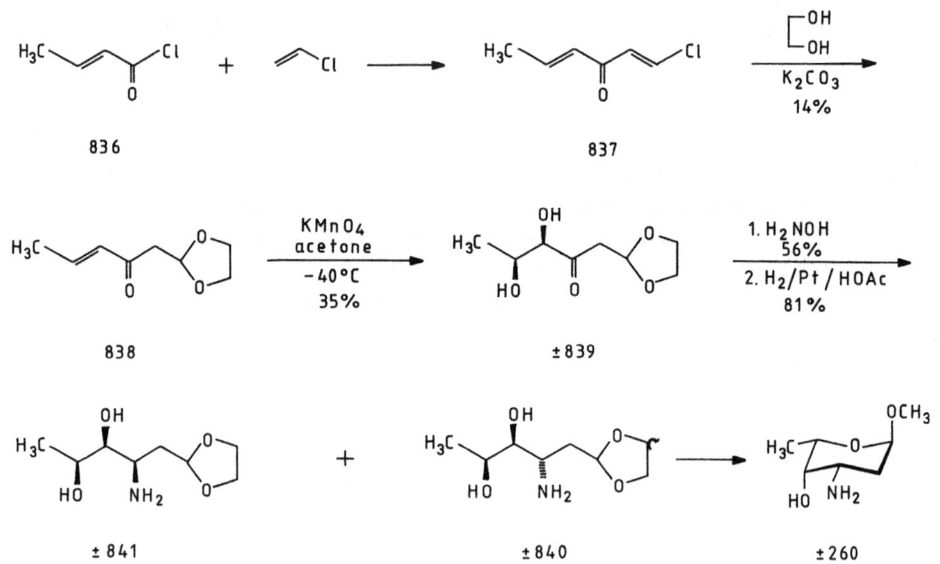

Scheme 140

The key-step of a chiral route (538) to L-daunosamine derivatives involves the zinc salt-catalyzed diastereoselective carbanion addition to the azomethine **843**, derived from 2,3-O-cyclohexylidene-4-deoxy-L-threose **842**. This latter sugar was readily obtained from (2R,3R)-diethyl tartarate via the reaction sequence shown on

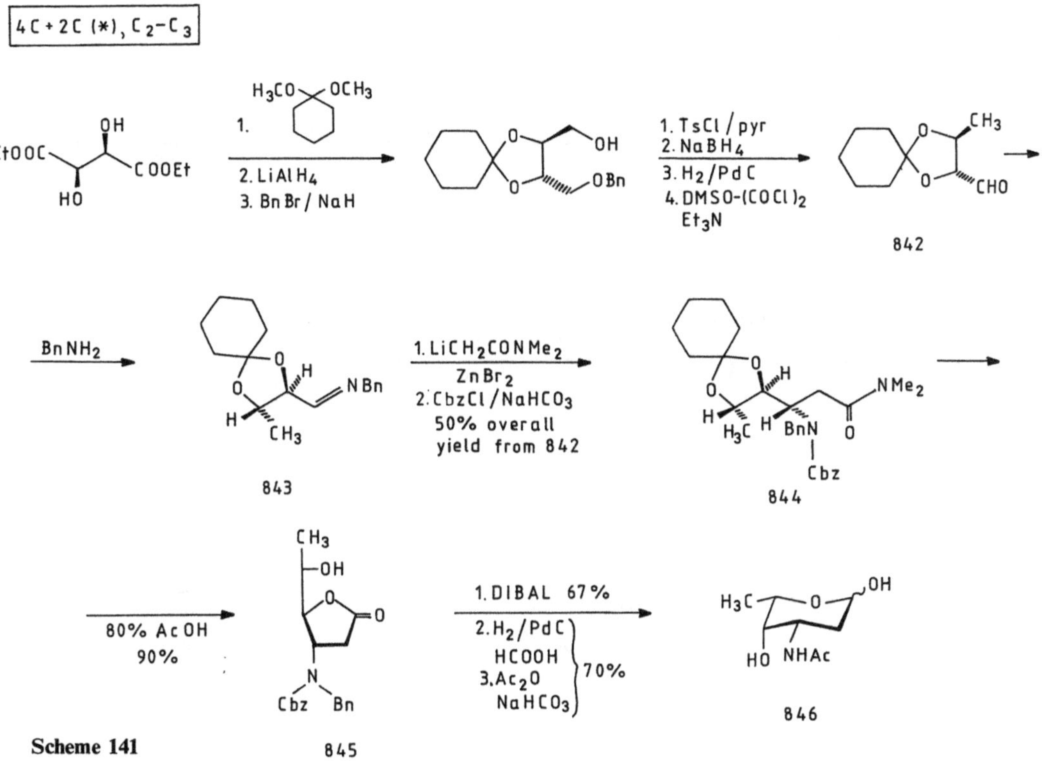

Scheme 141

Scheme 141. Treatment of the imine **843** with α-lithio-N,N-dimethylacetamide in the presence of zinc bromide gave almost exclusively the hexono-amide **844** with the L-*lyxo* configuration upon benzyloxycarbonylation. Acid hydrolysis transformed **844** into the L-*lyxo*-hexono-γ-lactone **845**, from which N-acetyl-L-daunosamine **(846)** was obtained by the reduction of the lactone function followed by N-deprotection and amine-selective acetylation.

Another asymmetric synthesis (539) of daunosamine (Scheme 142), designed according to the 4C + 2C(C_2—C_3) principle, uses (2R,3R)-tartaric acid as the starting material. The crucial step of this effective (although somewhat laborious) procedure is also a diastereoselective carbanion addition. The Grignard reaction of the aldehyde **847**, obtained in four steps, with [(1,3-dioxolan-2-yl)-methyl]magnesium bromide **(848)** provides exclusively the masked six-carbon aldehyde **849**. In this method introduction of the amino function at C-3 was achieved by means of the Mitsunobu reaction. It proceeded with a complete inversion of the configuration to produce the N-phthalimido-aminosugar **850** in 94% yield. Following deoxygenation at C-6 (four steps) and removal of the N-phthalyl group, the resulting amine was benzoylated **(851)** and converted into N-benzoyl-L-daunosamine **(833)** by deprotection. This ten-step procedure could synthesize N-benzoyl-L-daunosamine in 31% overall yield — based on the protected threose derivative **847**.

In two syntheses of the Hiyama group aimed at the preparation of N-benzoyl-L-daunosamine **(833)** (540) and N-benzoyl-L-acosamine **(863)** (541), the key-step is the addition of the magnesium enolate of *tert*-butyl acetate **(857)** to (2S,3S)-2,3-

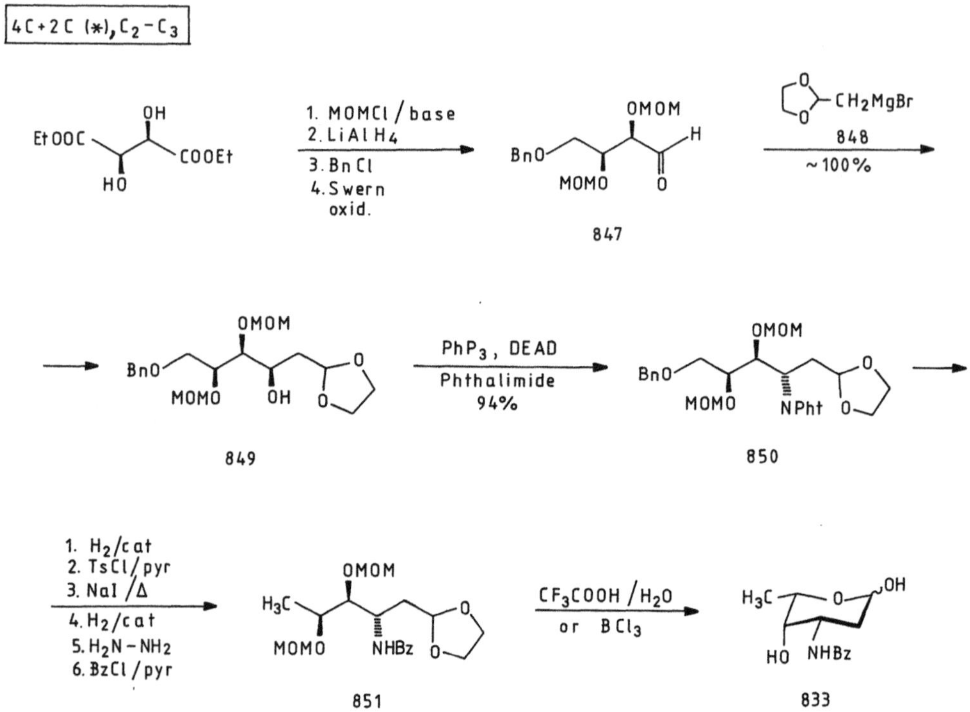

Scheme 142

(cyclohexylidenedioxy)-butanenitrile **(855)** and to the corresponding (2R,3S) diastereoisomer **856**, respectively (Scheme 143). These two intermediates were produced from ethyl (S)-lactate **(852)** through conversion into ethoxyethyl (S)-lactaldehyde **(853)** and its condensation with acetone cyanohydrin to obtain a mixture of the diastereoisomeric cyanohydrins **854**, which was cyclohexylidenated to produce a separable 45:55 mixture of **855** and **856**. The (2S,3S) isomer **855** could also be prepared from the L-threose derivative **842**, available from diethyl L-(+)-tartarate (cf. Scheme 141). Treatment of **855** with the enolate **857** gave a 99% yield of the β-aminoacrylate **858**, which was acetylated and hydrogenated to obtain a single diastereoisomeric amine **859**. Removal of the protecting group of this latter, followed by benzoylation, gave the L-*lyxo*-hexono-δ-lactone **860**, which was converted into N-benzoyl-L-daunosamine **(833)** by DIBAL reduction. Essentially the same procedure using the (2R,3S) nitrile **856** proceeded through the sequence: **(856 + 857) → 861 → 862 → 649** to produce N-benzoyl-L-acosamine **(863)**.

In connection with studies on the *hetero*-Diels-Alder reactions of enaminones Tietze et al. (542) treated **866** (prepared by the formylation of phenylthioacetone **(864)**, followed by ammonolysis and benzoylation of the resulting dioxo compound **865**) with ethyl vinyl ether (Scheme 144). The Diels-Alder cyclization produced a 1:2 mixture of the diastereoisomeric lactols ±**867** and ±**868**, which were separated by column chromatography. The α-anomer ±**868** was used to prepare a derivative of 4-deoxy-D,L-daunosamine **(±1038)**, as shown in Sect. 4.3.1 (Scheme 178).

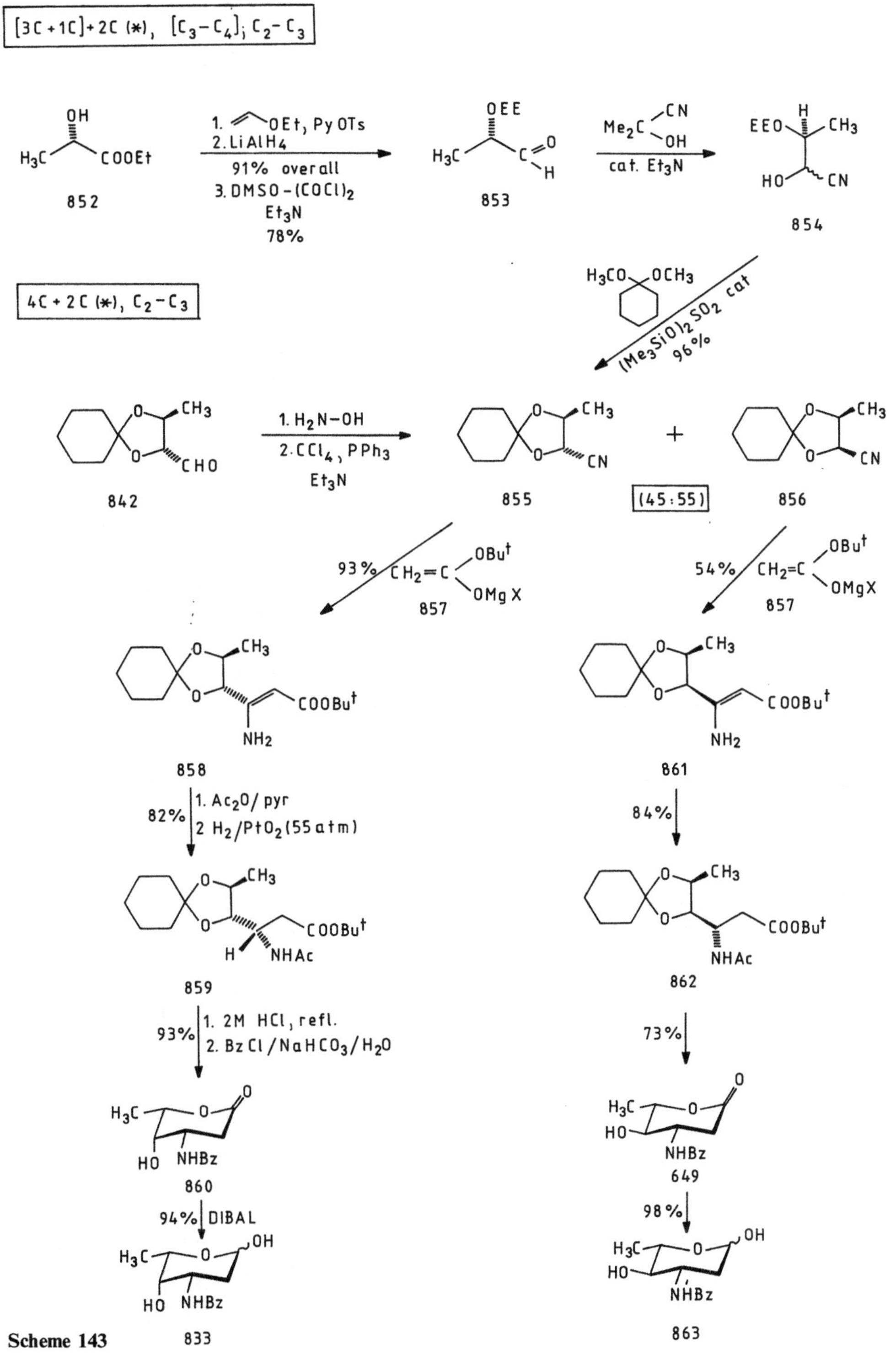

Scheme 143

4C + 2C (±), C_2–C_3

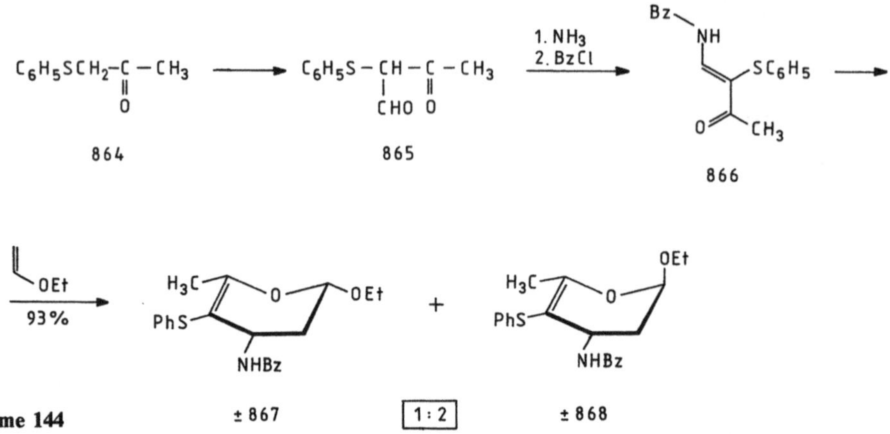

Scheme 144

The Diels-Alder reactions developed in the research group of Danishefsky have found wide utility in the total syntheses of carbohydrates. One of these procedures offered the possibility of preparing (543) racemic daunosamine derivatives. The, transformation of the Danishefsky diene **(869)** into (*E*)-1-(benzoyloxy)-4-methoxy-3-buten-2-one **(871)** was carried out in three steps as shown on Scheme 145. The reaction of **871** with tert-butyldimethylsilyl triflate gave a three-component mixture of the dienes **872a, 872b,** and **872c** in a ratio of 4:1:2. The zinc chloride-catalyzed Diels-Alder cycloaddition of this mixture with acetaldehyde produced a 90% yield of a 3.3:1 mixture of 1,5-anhydro-4-O-benzoyl-2,6-dideoxy-D,L-*threo*-hex-1-en-3-ulose **(±873)** and the corresponding D,L-*erythro* isomer **(±874)**. In this case the 4C + 2C chain ascent was used to generate the C_4—C_5 carbon bond of the target amino-deoxy sugar. After separation the *threo* compound **±873** was methoxymercurated and demercurated (cf. Chapter 2.1.1) to obtain a 57% yield of methyl 4-O-benzoyl-2,6-dideoxy-β-D,L-*threo*-hexopyranosid-3-ulose **(±400)**. Oximation of the latter, followed by acetylation and reduction with the borane-tetrahydrofuran complex was carried out as described in Sect. 2.2.3 (cf. Schemes 39 and 40). However, contrary to earlier findings (189–191) with the oxime-acetates **208** and **212** (carrying the C-1 and C-4 substituents with opposite stereochemistry) the borane reduction did not proceed stereoselectively with the oxime of **±400** and produced a 53% yield of methyl N,O-diacetyl-β-D,L-daunosaminide **(±875)** and a 25% yield of its C-3 epimeric D,L-*xylo* analog **±622**.

3.2.3 Construction of the carbon framework of 3-amino-2,3,6-trideoxyhexoses by means of the [4C + 3C]-1C methodology

The [4C + 3C]-1C strategy of building-up the carbon chain of the foregoing sugars can be regarded as an alternative to the 4C + 2C construction system. With one exception (vide infra) the new carbon-carbon bond is always generated between C_2 and C_3 of the target amino sugar in diastereoselective coupling reactions of four-

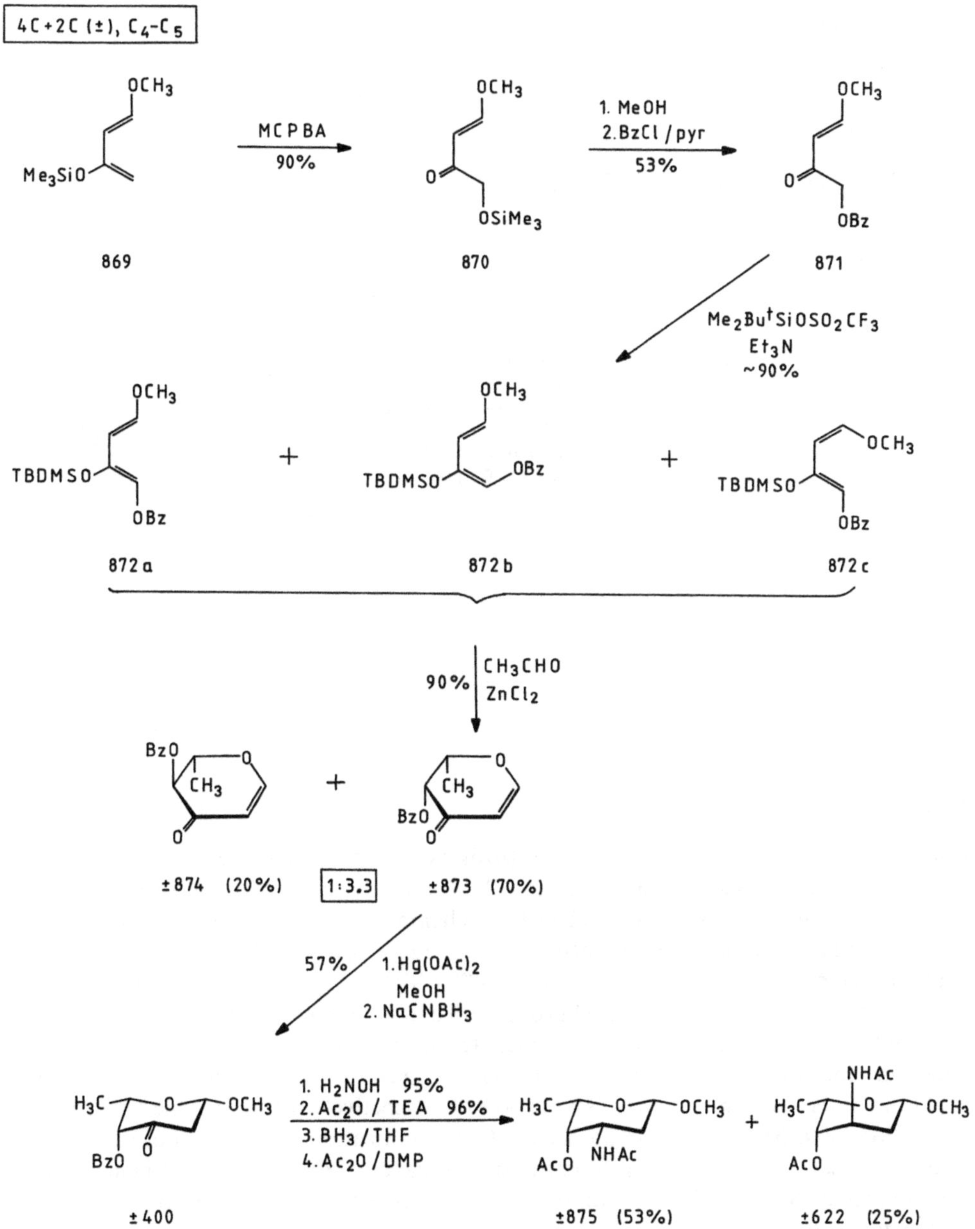

Scheme 145

carbon dichiral aldehydes or their azomethine derivatives with allyl-metal compounds, which as C-3 precursors ensure the required stereocontrol. The final step of each of these procedures involves the oxidative destruction of the excess one-carbon unit.

In one of the chiral synthetic routes based on related methodology, Fuganti et al. (544) prepared 2,3-O-cyclohexylidene-4-deoxy-L-threose **(842)**, the starting

[4C + 3C]-1C(*), C_2-C_3

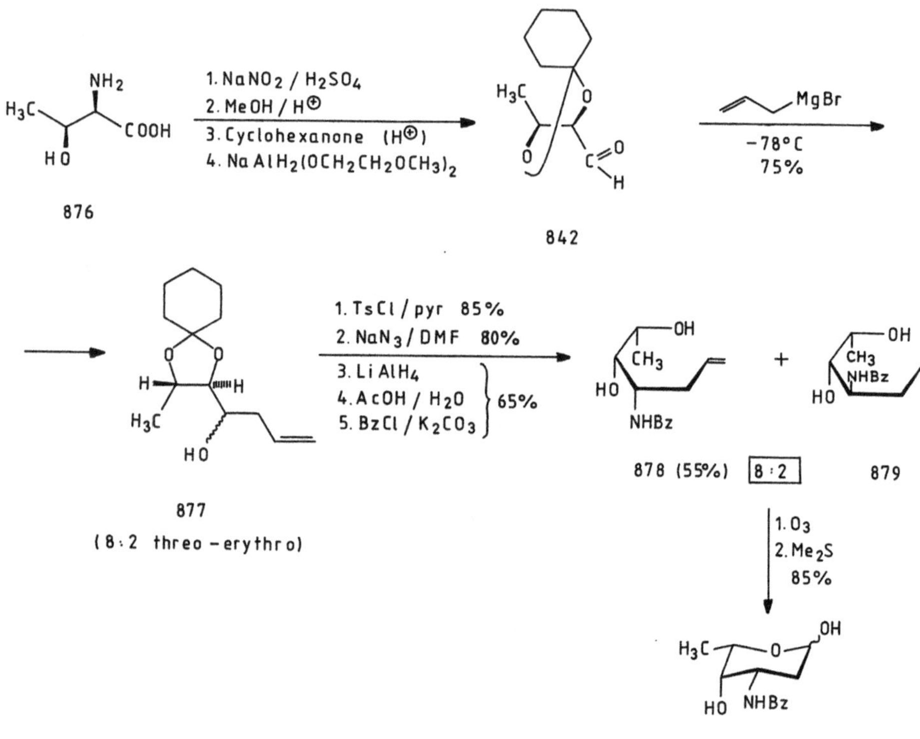

Scheme 146

four-carbon substrate from D-threonine **(876)** as shown on Scheme 146. It is to be noted that **842** is also available from (2R,3R)-diethyl tartarate, according to a reaction sequence discussed previously (cf. Scheme 141). The coupling of **842** with allylmagnesium bromide (representing the 3C unit) proceeded with a 8:2 *threo-erythro* stereocontrol, giving the mixture **877** which could not be separated. The replacement of the C-4 free hydroxyl group with azide (inversion of the configuration) followed by the reduction of the azides to amines, the removal of the acetal function, and a Schotten-Baumann benzoylation gave a 8:2 mixture of the *erythro* and *threo* seven-carbon compounds **878** and **879**, respectively. From this the major isomer **(878)** could be separated in 55% yield by crystallization. Ozonolysis of the latter and subsequent treatment with dimethyl sulfide gave an 85% yield of N-benzoyl-L-daunosamine **(833)**. The *threo* amine **879** was converted into 3-benzamido-2,3,6-trideoxy-L-*xylo*-hexopyranose **(786)** in a similar manner.

In an additional process elaborated by the Fuganti group (545), the crucial step is the addition of diallylzinc onto the phenylsulfenimines **880** and **882**, leading to N-benzoyl-L-daunosamine **(833)** and N-benzoyl-L-ristosamine **(700)**, respectively. In both cases the introduction of the C-3 amino group is carried out simultaneously with the construction of the C_2—C_3 bonding of the target aminosugar (Scheme 147). For the preparation of N-benzoyl-L-daunosamine **(833)**, the (2R,3S)-aldehyde **804** — available either from L-threonine (**876**, Scheme 146) or from (2R,3R)-tartaric

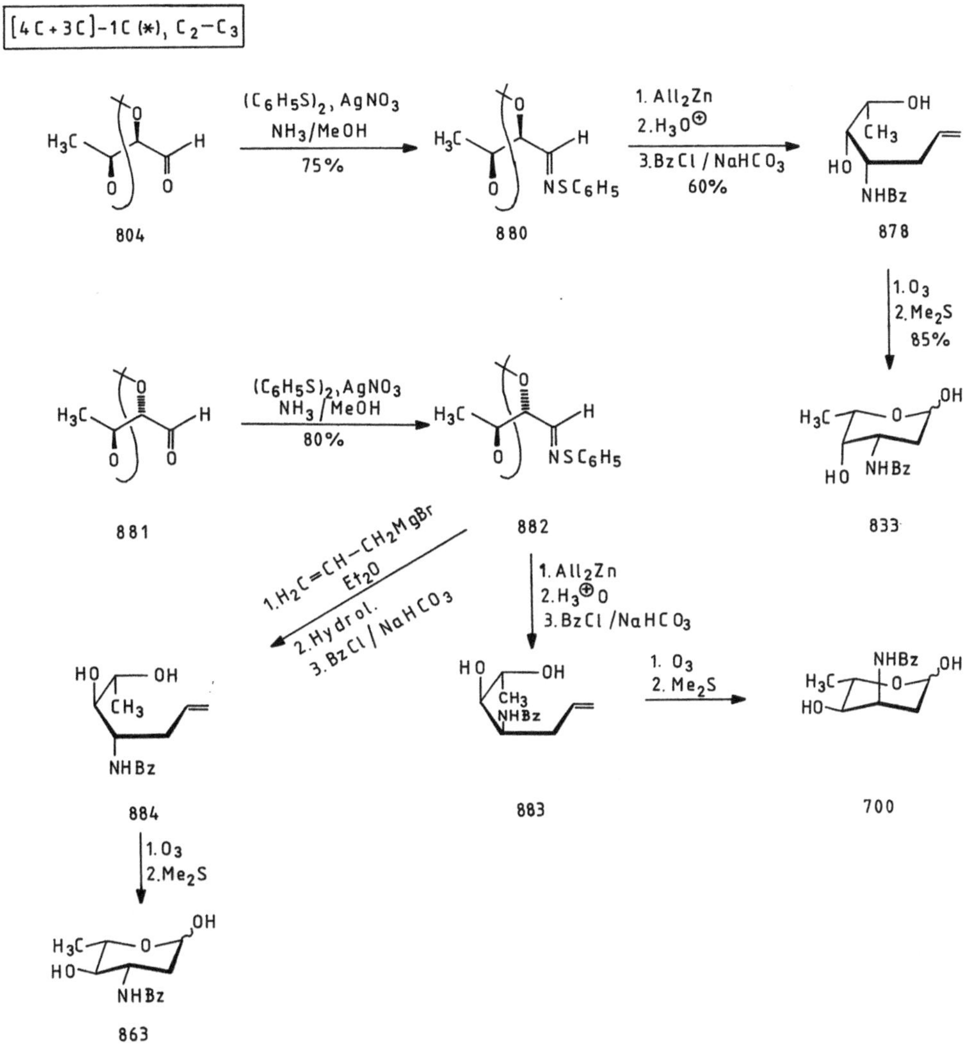

Scheme 147

acid (**801**, Sect. 3.2.2, Scheme 134) — was converted into the phenylsulfenimine **880**. Diastereoselective addition of diallylzinc onto the C=N bond of **880**, followed by hydrolysis and N-benzoylation gave the *erythro* C_7 adduct **878**, which upon ozonolysis afforded N-benzoyl-L-daunosamine **(833)** and the C_3-epimeric L-*xylo* isomer **(786)** in a 75:1 ratio.

Similar high *erythro*-selectivity was observed (545) when the phenylsulfenimine **882**, derived from the (2s,3s)-aldehyde **881**, was allowed to react with diallylzinc, yielding the 4,5-*erythro* adduct **883** (the 4,5-diepimer of **878**). This product is an appropriate stereoisomer to be transformed into N-benzoyl-L-ristosamine **(700)** in an essentially similar way.

3. Synthesis of 3-amino-2,3,6-trideoxyhexoses

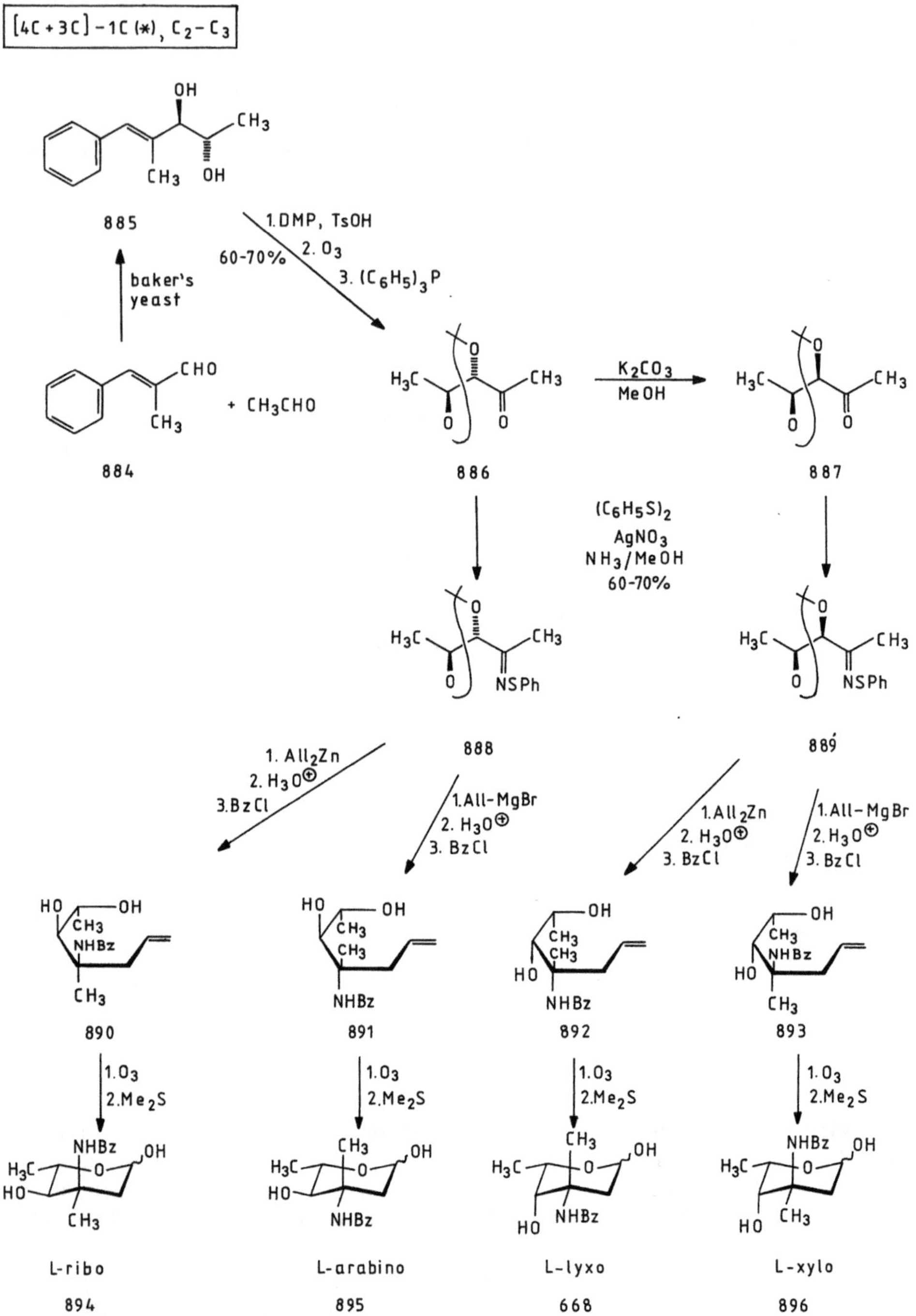

Scheme 148

Based on the recognition that the addition of methylmagnesium bromide onto the sulfenimine **882** results in a ca. 3:7 mixture of the 4,5-*erythro*- **(883)** and *threo*- **(884)** isomers (the opposite diastereoselectivity as compared with the additions with diallylzinc), N-benzoyl-L-acosamine **(863)** (546) was also synthesized using this procedure.

This relatively simple methodology of the Fuganti group gave an excellent opportunity for the preparation (499, 500) of each of the four stereoisomeric 3-amino-2,3,6-trideoxy-3-C-methyl-hexoses as well (Scheme 148). The starting material for these syntheses was the (2S,3R)-methyl-diol **885**, obtained by the fermentation of α-methyl-cinnamaldehyde **(884)** with acetaldehyde in the presence of baker's yeast. Upon isopropylidenation and ozonolytic splitting of the double bond of **885**, the (3S,4S)-methyl-ketone **886** was obtained. This was isomerized into the C-3 epimer **887** via the respective enolate under alkaline conditions. The two ketones were converted into the phenylsulfenimines **888** and **889**, respectively.

As expected on the basis of previous findings (544, 545), the reaction of **888** and **889** with diallylzinc proceeded with high 4,5-*erythro*-selectivity to give the branched C_7 adducts **890** and **892**, respectively. In contrast, the addition of allylmagnesium bromide onto the C=N bond of the sulfenimines **888** and **889** was carried out with the preferential production of the 4,5-*threo* compounds **891** and **893**. Ozonolytic destruction of the excess 1C units of **890–893** resulted in the N-benzoyl-3-C-methyl-branched aminodeoxy hexoses **894, 895, 668,** and **896,** respectively. The L-*arabino* sugar **895** was converted into the methyl glycoside **651**, which was configurationally inverted at C-4, through the 4-O-mesylate **660**, into the *lyxo* analog **668** only in rather low yield (see Sect. 2.8.1, Table 14 entry 8). In order to perform a similar C-4 inversion, the L-*xylo* sugar **896** was methyl-glycosylated, and the resulting **701** was subjected to the oxidation-reduction sequence shown in Sect. 2.8.4 (Scheme 118).

Fuganti et al. (547) also prepared the 1-C- and 5-C-methyl derivatives of L-acosamine using a methodology essentially similar to that summarized in Scheme 148.

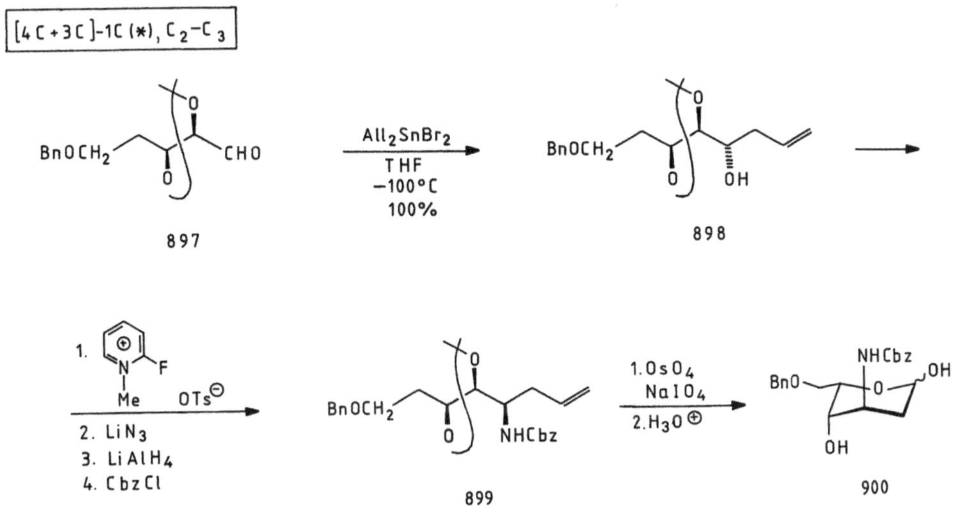

Scheme 149

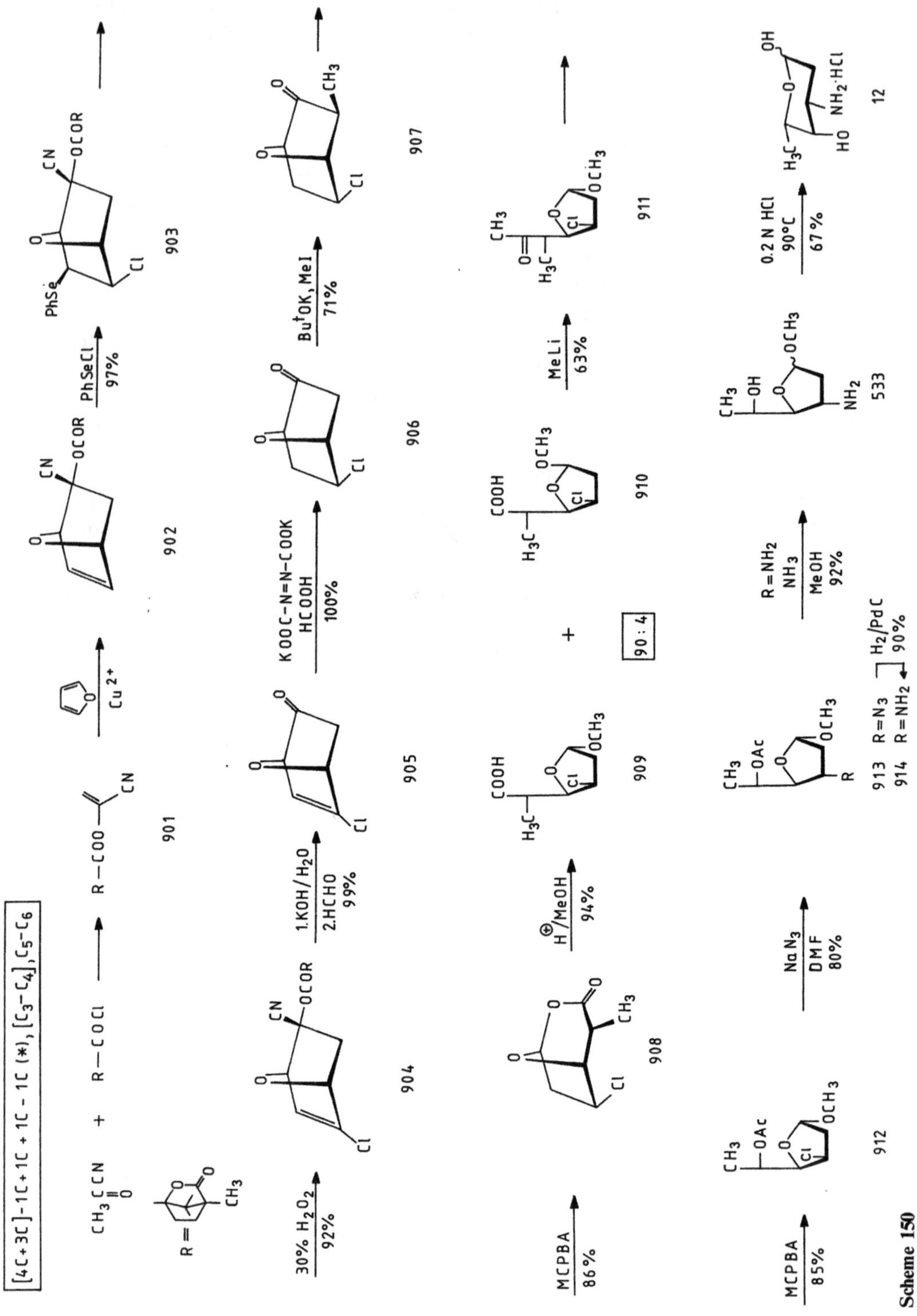

Scheme 150

In closely related work Mukaiyama et al. (548) treated 4-O-benzyl-2,3-O-isopropylidene-L-threose **(897)** with diallyltin(IV)dibromide to obtain — with 90% stereoselectivity — the *anti*-homoallyl alcohol **898** (Scheme 149). Azidolysis of this latter with lithium azide and 1-methyl-2-fluoropyridinium tosylate, followed by hydride reduction and benzyloxycarbonylation of the resulting amine, furnished the C_7 aminotriol **899**. Oxidative splitting off of the C_1-portion and subsequent acid hydrolysis then produced 6-O-benzyl-3-benzyloxycarbonylamino-2,3-dideoxy-L-*xylo*-hexopyranose **(900)** in 53% yield.

The starting material in the total synthesis of L-daunosamine reported (549, 550) by Warm and Vogel (Scheme 150) is the chiral 7-oxabicyclo-[2.2.1.]-hept-5-ene derivative **902**, produced in the Diels-Alder cyclization of camphanoyloxy acrylonitrile **(901)** and furane, representing the simultanous generation of the C_4—C_5 bond and development of the C-4 chiral center of the daunosamine molecule. The diastereoselective addition of phenylselenyl chloride — induced by the camphanoyl moiety — onto **902** gave **903**, which was oxidized to the unsaturated nitrile **904**. Upon saponification and treatment with formaldehyde, the β,γ-unsaturated ketone **(905)** was obtained. Stereospecific reduction of this latter substance to **906**, followed by C-methylation, resulted exclusively in the 5-*endo*-chloro-3-*exo*-methyl ketone **907**. Since this last step corresponds to the development of the C_5—C_6 bond of the target sugar, **907** thus carries the completely assembled carbon framework of L-daunosamine.

The ketone **907** was converted into the lactone **908** by Bayer-Villiger oxidation, and methanolysis of this latter gave a 90:4 mixture of the methyl α- **(909)** and β- **(910)**-hexofuranosiduronic acid, that was than treated with methyl lithium to obtain the 5-C-methyl-α-D-*arabino*-hept-6-ulosofuranoside **911**. Peracid oxidation of this compound proceeded with complete retention of the configuration at C-5 to give methyl 5-O-acetyl-3-chloro-2,3,6-trideoxy-β-L-*xylo*-hexofuranoside **(912)**. Azide displacement of the chloro atom in **912** and subsequent reduction of the resulting azide **913** produced the amine **914**, which upon O-deacylation yielded methyl 3-amino-2,3,6-trideoxy-α,β-L-*lyxo*-hexofuranoside **(533)**. Mild acid hydrolysis of **533** then gave L-daunosamine hydrochloride **(12)** in 21.8% overall yield, based on the "naked sugar" **902**.

3.2.4 3C + 3C Type construction of 3-amino-2,3,6-trideoxyhexoses

In 3C + 3C type reactions the connection of two 3C precursors along the C_3—C_4 bond of the required aminodeoxy sugar molecule can be achieved with or without the simultaneous functionalization of these two carbons. Since in the target hexose both C_3 and C_4 are asymmetric and carry different substituents, the C—C coupling reaction combined with the introduction of the substituent(s) must proceed with a satisfactorily high diastereoselectivity. In one of the related procedures, developed by Hirama et al. (551), the starting 3C substrates are the silylated (s)-lactaldehyde **915** and the lithium acetylide of methyl propiolate. Their reaction gave a 5:1 *erythro-threo* mixture **916** (Scheme 151). This was oxidized, and the resulting C-4-ketone was stereoselectively reduced with lithium tri-*sec*-butylborohydride to obtain the 4,5-*threo*-diol **917** almost exclusively. Following protecting group manipulations,

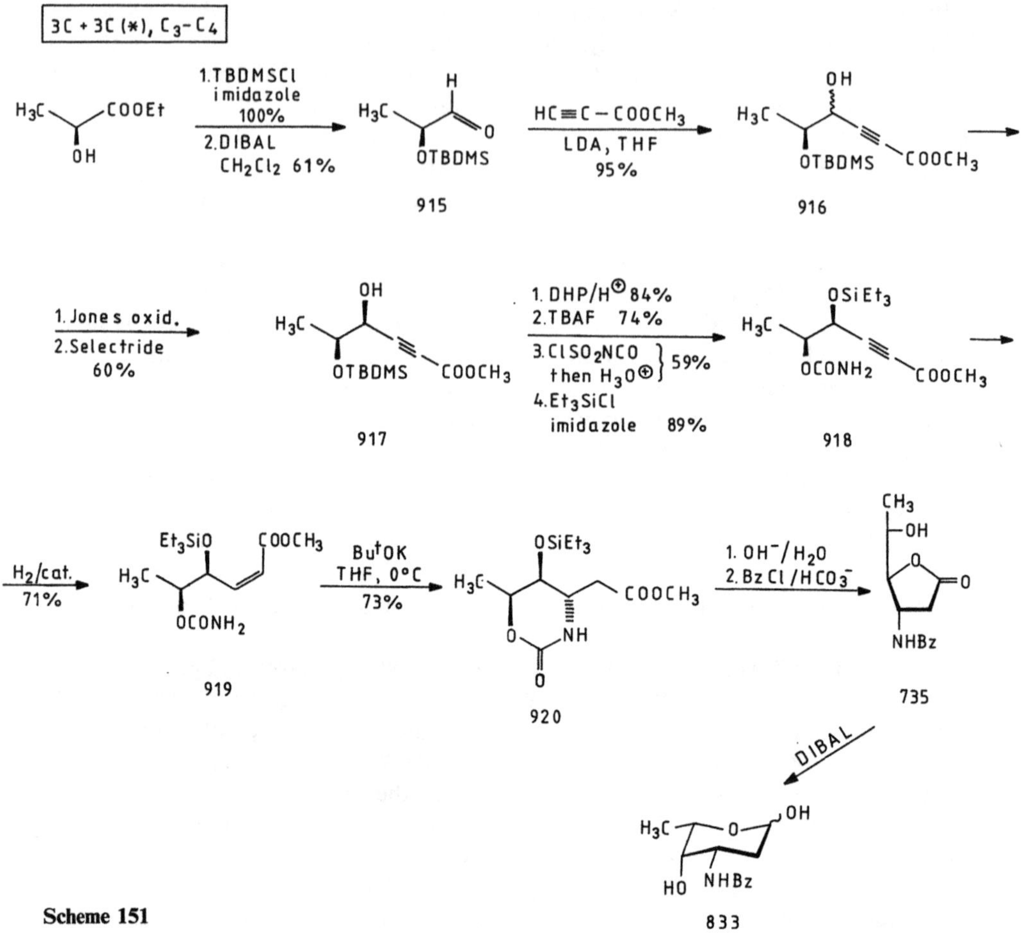

Scheme 151

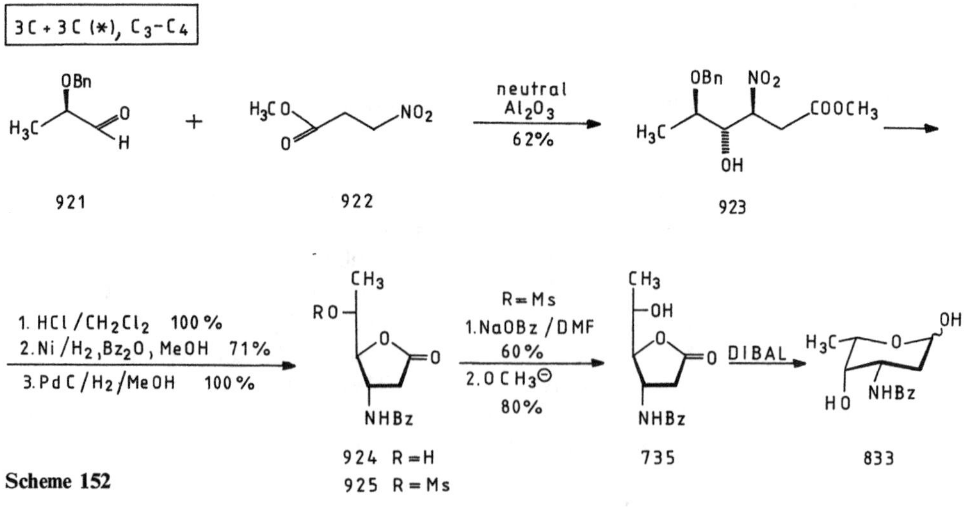

Scheme 152

917 was converted into the 4-O-silyl-5-O-carbamoyl ester **918**, the *Z*-specific reduction of which yielded the olefinic carbamate **919**. The crucial step of the synthesis, intramolecular Michael addition of the amino group of the carbamate onto the double bond, proceeded with exclusive 1,3-*anti*-diastereoselectivity to provide the cyclic carbamate **920**. This was then converted into the L-*lyxo*-hexono-γ-lactone **735** and then into N-benzoyl-L-daunosamine **(833)**.

Hanessian and Kloss (552) elaborated a simple chiral strategy for the synthesis of each of the four stereoisomeric 3-amino-2,3,6-trideoxyhexoses in the L-series based on the 3C + 3C system of building up. (Schemes 152 and 153). The base-catalyzed nitroaldol condensation of O-benzyl-D-lactate **(921)** with methyl 3-nitropropionate **(922)** gave a three-component mixture, from which the major D-*ribo* isomer **923** could be isolated by crystallization (Scheme 152). Lactonization of **923**, followed by the reduction of the nitro group with concomitant N-benzoylation and removal of the O-benzyl function, yielded the D-*ribo*-hexono-γ-lactone **924**. Inversion of the configuration at C-5 of the latter substance was achieved by nucleophilic substitution of the intermediary 5-O-tosylate **925** with sodium benzoate, giving the L-*lyxo*-hexono-γ-lactone **735** after O-debenzoylation. This was then converted into N-benzoyl-L-daunosamine **(833)**.

An analogous nitroaldol condensation of O-benzyl-L-lactaldehyde **(926)** with **922** was performed by applying various catalysts to obtain the additional three stereoisomers of daunosamine. Using aluminum oxide, the debenzylation and isopropylidenation of the crude mixture of nitroaldol products again produced the *ribo* stereoisomer **(927)** as the preponderant component (43%) together with the L-*lyxo*- **(928)**, L-*arabino*- **(929)**, and L-*xylo*- **(930)** acetals (Scheme 153). Upon equilibration of the isolated individual stereoisomers, each of the four 3-nitro compounds could be prepared in the quantities required for further transformations into the N-benzoyl-γ-lactones **699, 735, 931,** and **932** with the L-*ribo*-, L-*lyxo*-, L-*arabino*-, and L-*xylo* configurations, respectively, using the sequence of reactions shown for **923** in Scheme 152. DIBAL reduction of these lactones then allowed the preparation of the N-benzoylamino sugars **700, 833, 863,** and **786.** In parallel work, Brandänge and Lindquist (553) reported an essentially similar method for the synthesis of the N-benzoyl-L-*ribo*-hexono- **(699)** and N-benzoyl-L-*lyxo*-hexonolactone **(735)** starting with the same three-carbon building blocks.

The crucial step in the ingenious chiral acosamine synthesis of Wovkulich and Uskokovic (496, 497) is a diastereoselective and regiospecific [3 + 2]nitrone-olefin cycloaddition reaction (Scheme 154). The *all-trans* azomethine **(935)**, prepared from *trans*-propenyl acetate **(933)** and bis-dimethylamino-*tert*-butoxymethane **(934)**, was treated with the oxalate salt of (s)-(—)-N-hydroxy-α-methyl-benzenemethanamine to give the nitrone **936**. Intramolecular cyclization of this latter substance led to a 82:18 mixture of the isoxazolidines **937** and **938**. Thus this single reaction step generated the C_3—C_4 bond of L-acosamine and simultaneously developed the C_3, C_4, and C_5 chiral centers carrying the required substituents with the proper relative and absolute stereochemistry. Moreover, this cycloaddition exhibits a unique regioselectivity, providing the first example of a preferential reverse addition of the nitrone-oxygen atom to the double bond of an enolether. Conversion of the isoxazolidine **937** into derivatives of L-acosamine can be carried out in two alternative ways.

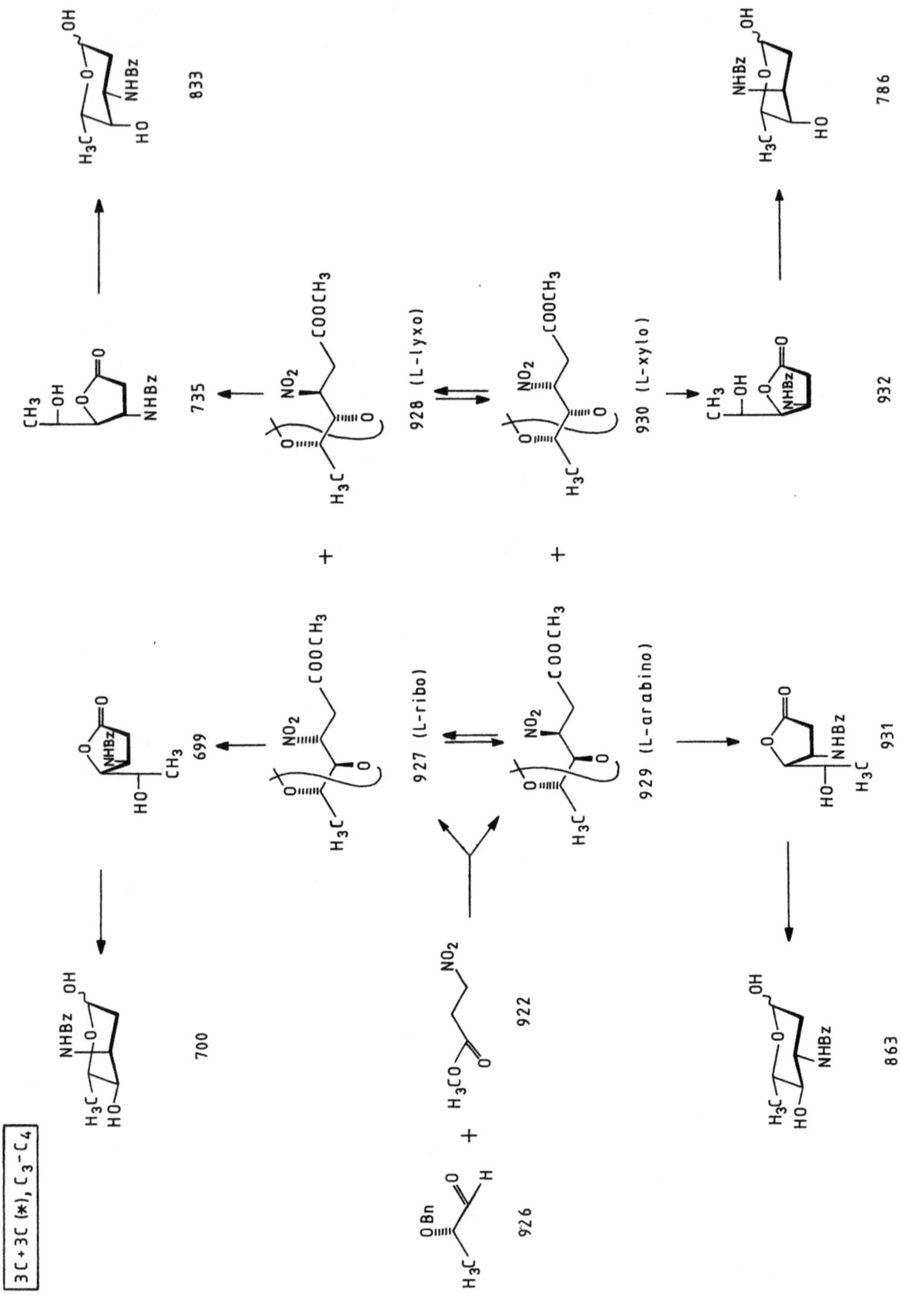

Scheme 153

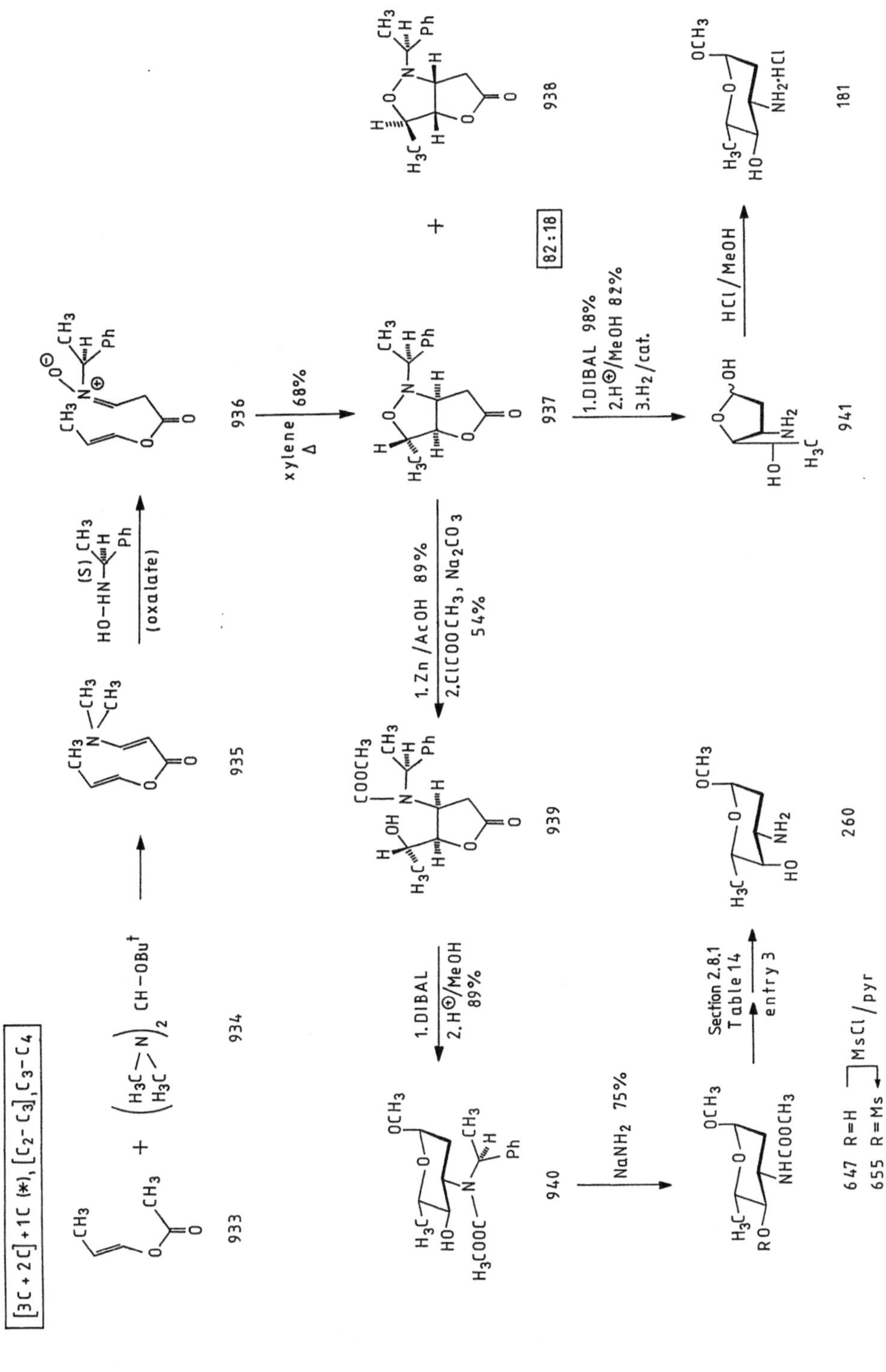
[3C + 2C] + 1C (*), [C2 - C3], C3 - C4
933
934
935
(S)
(oxalate)
936
xylene
Δ
68%
937
938
82 : 18
1. Zn / AcOH 89%
2. ClCOOCH3, Na2CO3
54%
939
1. DIBAL
2. H⊕/MeOH
89%
940
1. DIBAL 98%
2. H⊕/MeOH 82%
3. H2/cat.
941
HCl/MeOH
181
NaNH2 75%
Section 2.8.1
Table 14
entry 3
260
647 R = H
655 R = Ms
MsCl/pyr

Scheme 154

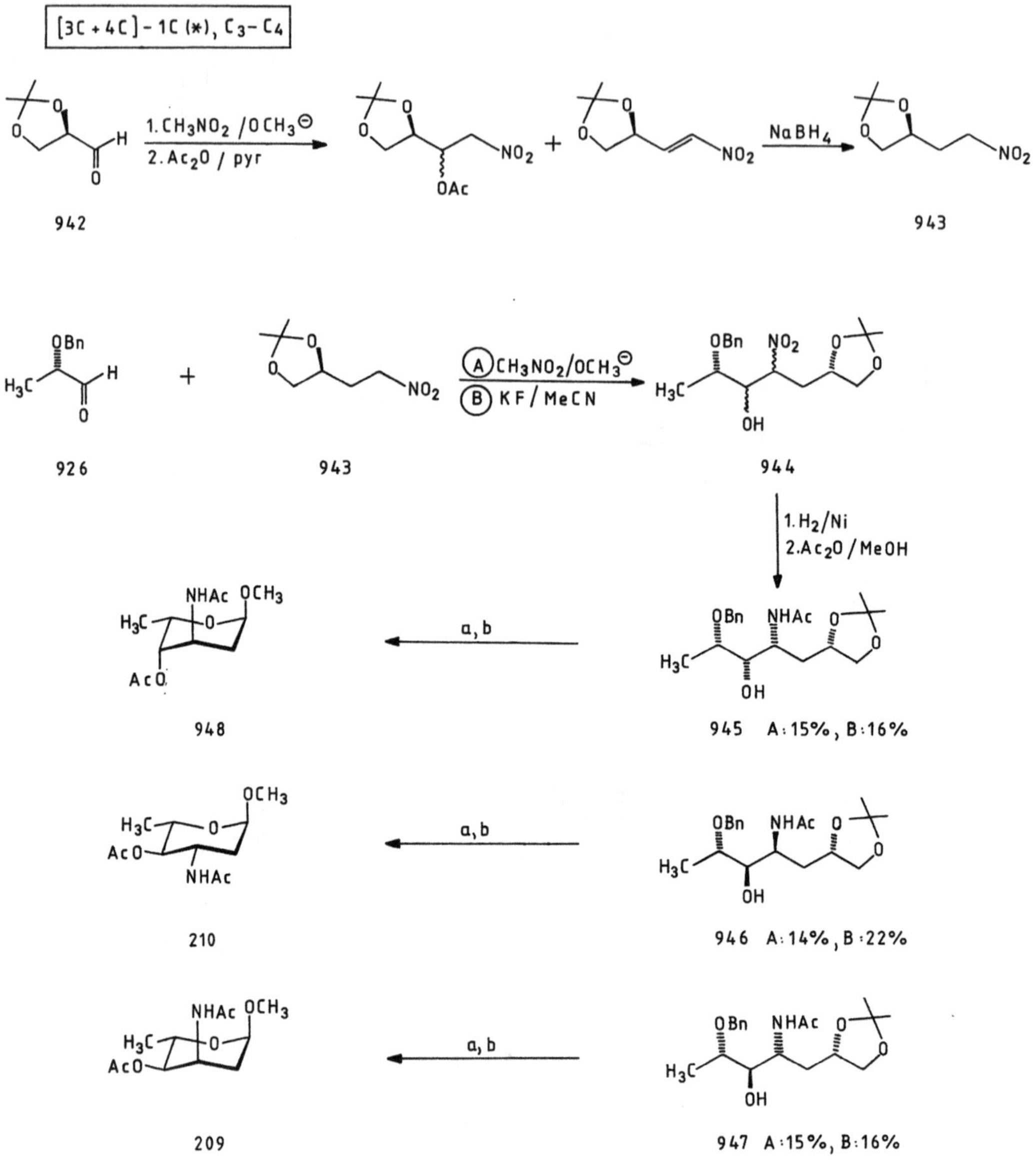

a. i, Ac_2O / pyr; ii, H_2O–AcOH; iii, $Pd(OH)_2$ cyclohexene; iv, MeOH–$NaIO_4$ (~78% overall)
b. HCl / MeOH, ~40%

Scheme 155

Cleavage of the isoxazolidine ring of **937** with zinc in acetic acid, followed by N-methoxycarbonylation, yielded the γ-lactone **939**, which upon DIBAL reduction and methanolysis gave the methyl α-L-acosaminide **940**. Removal of the chiral N-substituent of this latter product led to methyl N-methoxycarbonyl-α-L-acosaminide

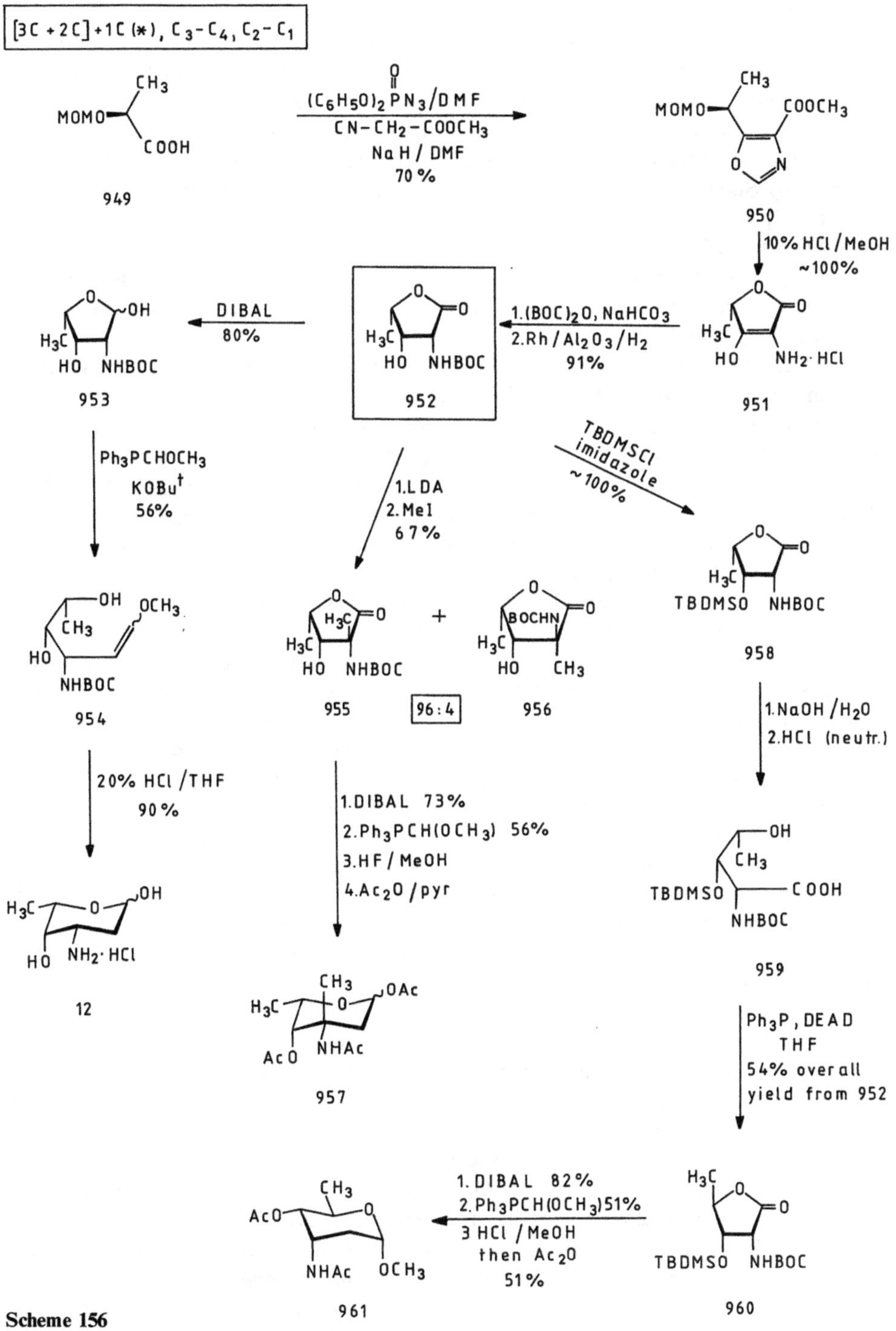
[3C + 2C] + 1C (*), C3-C4, C2-C1
949
950
951
952
953
954
955
956
957
958
959
960
961
12
DIBAL
80%
10% HCl / MeOH
~100%
96:4

Scheme 156

(647). This was transformed via the 4-O-mesylate **655**, into methyl α-L-daunosaminide **(260)** by the inversion of the configuration at C-4, as shown in Sect. 2.8.1 (Table 14 entry 3). The alternative route to methyl α-L-acosaminide **(181)** starting with **937** involved sequential DIBAL reduction of the lactone function and removal of the N-substituent, followed by methanolysis of the intermediary furanose sugar **941**.

A somewhat laborious synthesis of L-acosamine- and L-ristosamine derivatives, reported by Suami et al., (554) is also based on the nitroaldol-dimerization reaction ([3C + 4C]-1C, see Scheme 155). The three-carbon starting precursor, O-benzyl-L-lactaldehyde **(926)** was prepared from D-mannitol in 16% overall yield by an unnecessarily long twelve step sequence. The four-carbon synthon, (2s)-1,2-O-iso-propylidene-4-nitro-1,2-butanediol **(943)**, was synthesized from 2,3-O-isopropylidene-D-glyceraldehyde **(942)**.

The Henry reaction of **926** and **943** gave a mixture **(944)** of diastereoisomeric addition products that was hydrogenated and then N-acetylated to allow the separation of three 4-acetamido-heptanetetrols **945–947**. Acetylation of these compounds and then removal of the protecting groups and oxidative splitting off of the excess one-carbon unit gave the N,O-diacetyl reducing sugars. These were methyl-glycosylated to obtain methyl 3-acetamido-4-O-acetyl-2,3,6-trideoxy-α-L-*xylo*-hexo-pyranoside **(948)**, methyl N,O-diacetyl-α-L-acosaminide **(210)**, and methyl N,O-diacetyl-α-L-ristosaminide **(209)**, respectively.

The synthesis strategy elaborated by Hamada and collaborators for the production of derivatives of L-daunosamine **(12)** (555, 556), D-ristosamine **(20)** (557, 559), and L-vancosamine **(15)** (558), is based on the formation of an optically active 5-substituted-4-methoxycarbonyl-oxazole derivative by the direct C-acylation of methyl isocyanoacetate with optically active carboxylic acids aided by diphenyl phosphoroazidate (DPPA), to produce the C_3—C_4 bond of the target aminohexoses (Scheme 156). According to this methodology, O-methoxymethyl-L-lactic acid **(949)** was treated with methyl isocyanoacetate to obtain the oxazole **950** in 70% yield. Methanolysis of **950** quantitatively produced the unsaturated lactone **951**, which after N-protection was stereoselectively hydrogenated over rhodium-on-alumina catalyst to afford 2-*tert*-butoxycarbonylamino-2,5-dideoxy-L-*lyxono*-1,4-lactone **952**, the key-intermediate in the 3-amino-2,3,6-trideoxyhexose syntheses.

DIBAL reduction of **952** led to the lactol **953**, which was subjected to the introduction of the C-1 unit by a Wittig reaction to give the enolether **954**, from which L-daunosamine **(12)** was obtained (556) by acid hydrolysis in an overall yield of 24% — based on L-lactic acid.

To obtain the vancosamine derivatives (558) from **952**, the C-2 carbon was deprotonated with lithium diisopropylamide and then treated with methyl iodide to yield a 96:4 mixture of the 2-C-methyl-L-*lyxono*- **(955)** and *xylono*- **(956)** lactones. Lactone **955** was converted into peracetyl-L-vancosamine **(957)** by an analogous sequence of reactions as shown for **952**, (Scheme 156). The inversion of the configuration at C-5 of **952** was necessary for the preparation of D-ristosamine analogs (557, 559). Upon silylation **952** was converted into the 3-O-*tert*-butyldimethylsilyl ether **958** and then into the amino acid **959**. Mitsunobu reaction of **959** produced the D-*ribono*-lactone **960**, which can be converted into methyl N,O-diacetyl-α-D-ristosaminide **(961)** by means of a one-carbon chain building as shown above.

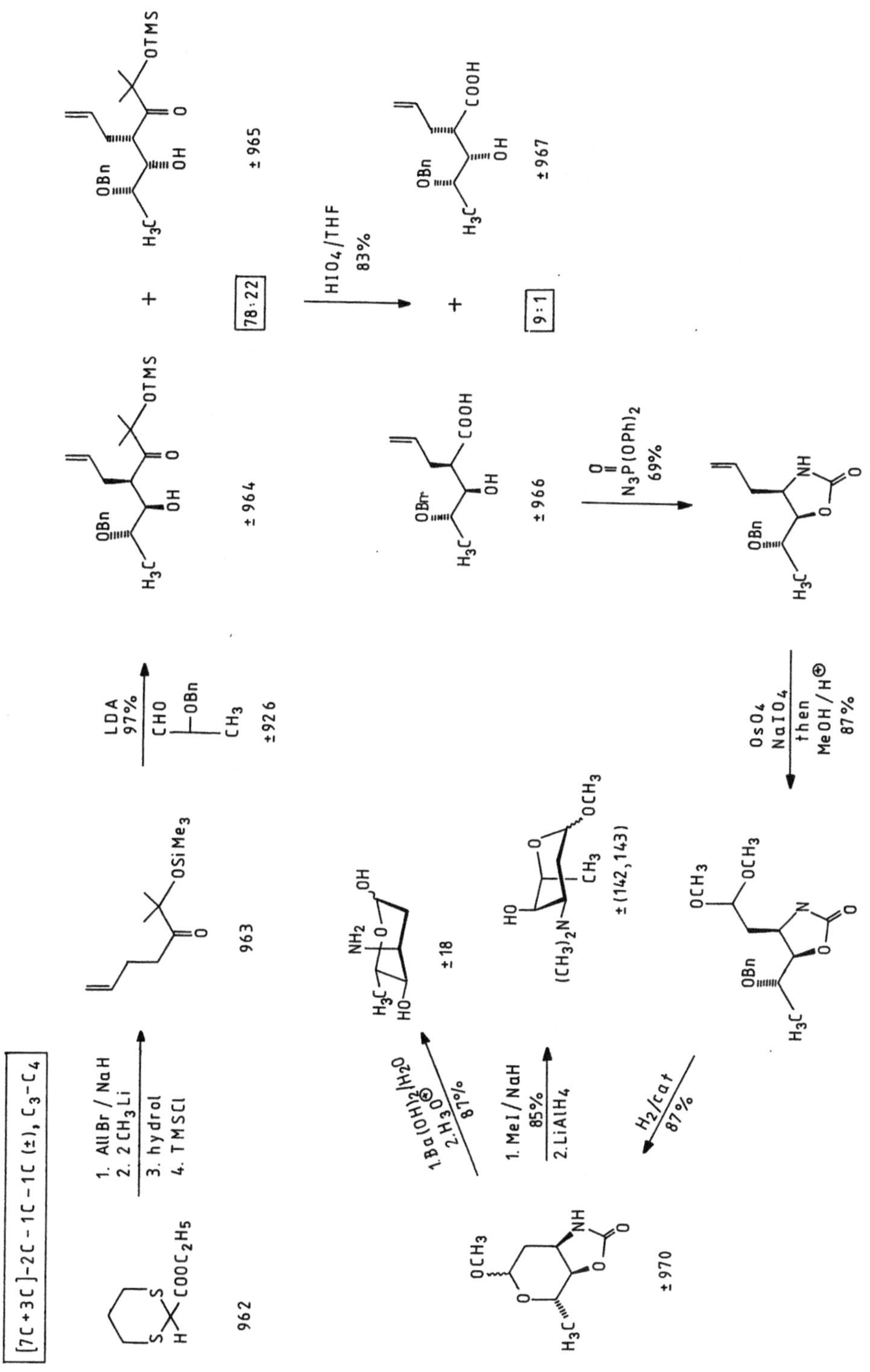
[7C + 3C]-2C - 1C - 1C (±), C3-C4
962
1. AllBr / NaH
2. 2 CH3Li
3. hydrol
4. TMSCl
963
LDA
97%
±926
±964
±965
78:22
HIO4/THF
83%
±966
±967
9:1
N3P(OPh)2
69%
±968
OsO4
NaIO4
then
MeOH/H⊕
87%
±969
H2/cat
87%
±970
1. Ba(OH)2/H2O
2. H3O⊕
87%
±18
1. MeI/NaH
85%
2. LiAlH4
±(142,143)

Scheme 157

3.2.5 Miscellaneous methods

The crucial step in the ingenious achiral ristosamine synthesis of Heathcock and Montgomery (560) is the diastereoselective aldol-addition of the ketone **963**, prepared from the glyoxylate **962** by allylation and addition of methyl anion, and O-benzyl-($\pm$)-lactaldehyde **($\pm$926)** to give a 78:22 mixture of the epimeric aldols $\pm$**964** and $\pm$**965** (Scheme 157).

Periodate oxidation of this mixture yielded the β-hydroxy acids $\pm$**966** and $\pm$**967** in a 9:1 ratio. The mixture could be readily separated by chromatography. Modified Curtius degradation of the major product $\pm$**966** gave the oxazolidinone $\pm$**968**, which was transformed into the acetal $\pm$**969** by Lemieux-Johnson oxidation and subsequent methanolysis. Hydrogenolytic removal of the benzyl group then provided the racemic α,β-methyl glycoside $\pm$**970**, which was deprotected to give D,L-ristosamine **($\pm$18)**. To synthesize the derivatives of the racemic megosamine (N,N-dimethyl-D,L-ristosamine), the oxazolidinone $\pm$**970** was methylated and then reduced with lithium aluminum hydride to yield a 7:3 mixture of racemic methyl α- **($\pm$142)** and β- **($\pm$143)** megosaminides. The trimethylsilyloxy-isopropyl group and the carbonyl function, present in the ketone **963**, might appear to be "unnecessary" for the construction of the carbon framework of the target aminodeoxy hexose because they are removed in the subsequent steps by means of oxidative destruction and a rearrangement reaction, respectively. But these two functions play decisive roles in the stereocontrol of the aldol addition reaction **963** → **964**.

4. Miscellaneous Functionalized Derivatives of 3-Amino-3-Deoxyhexoses of Antibiotics and Their Conversion into Other Organic Substances

The numerous synthetic routes worked out during the past twenty years for the production of 3-amino-2,3,6-trideoxyhexoses offered the challenge of further derivatization of these sugars and the intermediates in their syntheses. Such research was apparently stimulated and promoted by the goal of discovering semisynthetic anthracycline antibiotics that exhibit the same or higher antitumor activity along with diminished toxic side effects, thereby extending the limited therapeutic utility of the parent antibiotics. In addition, based on analogies taken from other fields of biologically active carbohydrate substances, structurally related derivatives containing a daunosamine-type aminodeoxy hexose moiety have also been prepared. The widening interest and research on the utilization of the derivatives of such aminosugars as "chiral templates" (561) in natural products syntheses apparently demonstrate the importance of this group of highly functionalized carbohydrates both chemically and biologically.

In the following sections a survey of the syntheses of diversely substituted and modified 3-aminodeoxy hexose derivatives is given[1] together with a discussion of their transformation reactions leading, in some cases, to important biologically active organic substances.

4.1 Thio-, nitrogen- and C-glycosides of 3-amino-2,3,6-trideoxyhexoses

4.1.1 1-Thio derivatives

The synthesis of the first thio-glycoside analog **(213)** of a 3-amino-2,3,6-trideoxyhexose was based (189–191) upon the stereoselective Michael addition of phenylmethanethiol across the double bond of 1,5-anhydro-2,6-dideoxy-L-*erythro*-hex-1-en-3-ulose **(204)** to result in benzyl 2,6-dideoxy-1-thio-α-L-*erythro*-hexopyranosid-3-ulose (Sect. 2.2.3, Scheme 38). Transformation of **205**, by the reduction of the oxime **212**, into benzyl 1-thio-2,3,6-trideoxy-3-trifluoroacetamido-α-L-*arabino*-hexopyranoside **(213)**, a protected 1-thio analog of L-acosamine **(8)**, is discussed in Sect. 2.2.3. (cf. Scheme 40).

[1] For the sake of methodological treatment of the synthetic strategies, the methods elaborated for the production of the 2-halogeno-derivatives of 3-amino-2,3,6-trideoxyhexoses are discussed separately in Sect. 2.4.

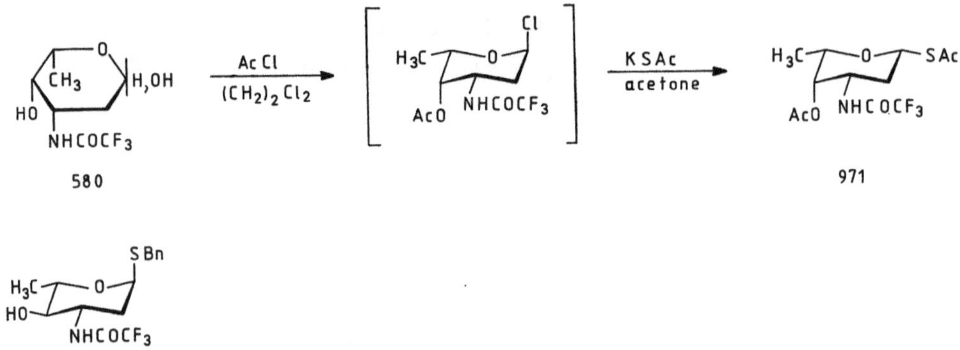

580 971 213

Scheme 158

4-O-Acetyl-1-S-acetyl-N-trifluoroacetyl-β-1-thio-L-daunosamine **(971)** was synthesized (190) directly from N-trifluoroacetyl-L-daunosamine **(580)** (Scheme 158). Treatment of this latter sugar with acetyl chloride in dichloroethane at room temperature gave the corresponding 4-O-acetyl-glycosyl chloride, which was treated (without isolation) with potassium thioacetate in acetone to produce the required 1-thio-daunosamine derivative **971** in 71% overall yield. The β-configuration of the anomeric 1-thioacetyl group in **971** was unequivocally shown by ^{1}H-NMR measurements. Attempts to remove the S-benzyl protecting group of **213** failed under a variety of conditions (562); only extensive decomposition of the molecule was observed. Thus, the desired utilization of **213** in glycosylations of more complex aglycones — including anthracyclinones — could not be realized. However, a recent synthesis of 7-thio-daunomycinone by Sallam et al. (563) might offer a successful route to the C-7 sulfur-bridged anthracyclinone glycosides.

In 1986 Hauser et al. (511) described an alternative method for the preparation of phenyl 1-thio-N-trichloroacetyl-α,β-D,L-daunosaminide **(±972)** from non-carbohydrate precursors (Scheme 159). Thus, (*E*)-1-(phenylthio)-1-acetoxy-3-(trichloroacetamido)-4-hexene **(757)**, prepared as shown in Sect. 3.1 from sorbic alcohol **(749)** via the sequence **749 → 750 → 751 → 756a, b → 757** (Scheme 127), was *cis*-hy-

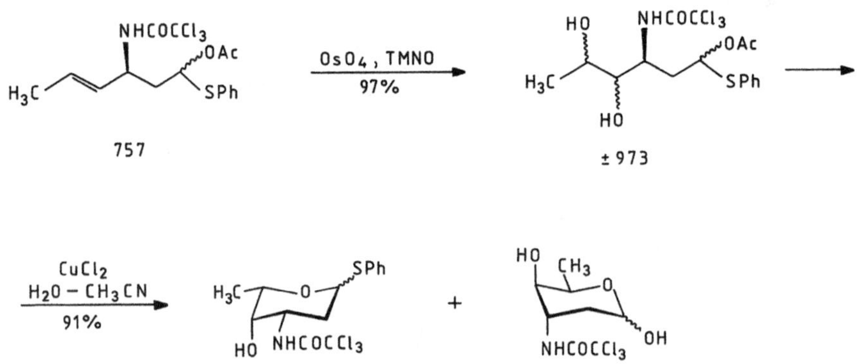

757 ±973

Scheme 159 ±972 6:4 ±760

droxylated with osmium tetroxide-trimethylamine oxide to give a 97% yield of the diastereoisomeric diols ±**973**. The hydrolysis of ±**973** with cupric chloride in acetone-water yielded a readily separable 6:4 mixture of the racemic phenyl thioglycoside ±**972** and the antipodes of the free *xylo* sugar ±**760**. This result indicated that the *lyxo* acetoxy sulfide diastereoisomers could adopt transition state geometries favorable to the selective displacement of the 1-acetoxy moiety, whereas those with the *xylo* stereochemistry could not.

4.1.2 Nitrogen-glycoside derivatives of daunosamine

Attempts to prepare (562) nitrogen glycosides of 3-amino-2,3,6-trideoxyhexoses by the Michael addition of secondary amines across the double bond of the enone **204**, followed by the introduction of the amino group at position C-3 of the sugar via the reduction of an oxime, as was done with the 1-O- and 1-S-hexopyranosid-3-uloses **109** and **205**, respectively (cf. Sect. 2.2.3, Schemes 38–40), were unsuccessful. The reaction of **204** with amines under a variety of conditions (562) resulted in open-chain enamino ketones, such as **974** in accordance with the results of Paulsen and Bünsch (294) (Scheme 160).

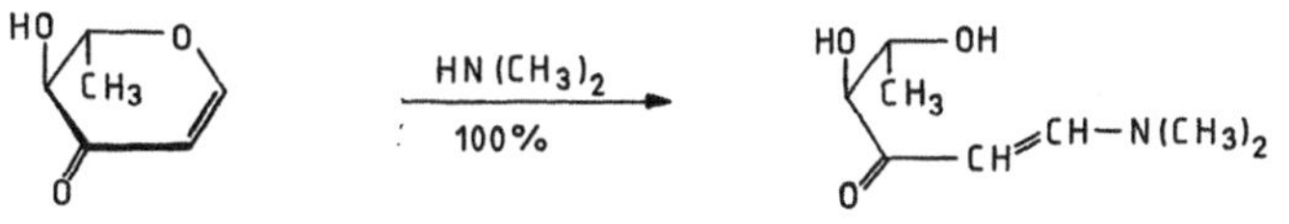

204 974

Scheme 160

The Arcamone group (564) employed 3-N-acetyl-1,4-di-O-acetyldaunosamine **(780)** and N,O-di-trifluoroacetyl-α-L-daunosaminyl chloride **(299)**, both readily available from L-daunosamine with simple chemical transformations, in pursuit of the synthesis of purine and pyrimidine nucleoside analogs of daunosamine (including the respective puromycin derivative) that represent complex N-glycosides of this sugar.

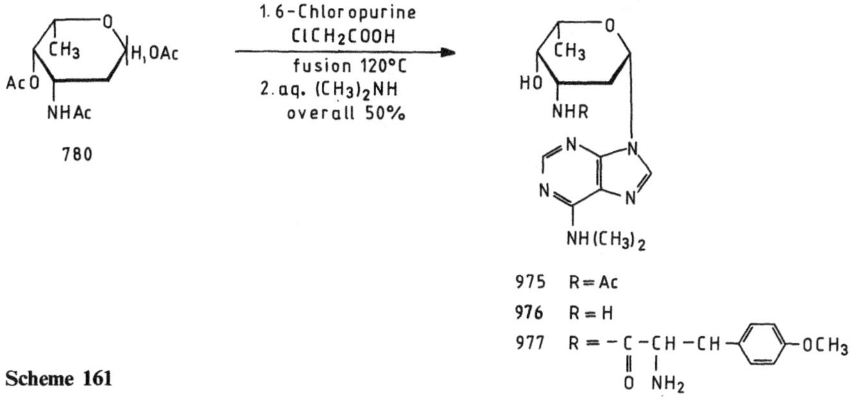

780

Scheme 161

The fusion of the peracetate **780** (α:β ratio = 3:7) with 6-chloropurine at 120 °C in the presence of a chloroacetic acid catalyst and treatment of the product with aqueous dimethylamine resulted in the β-nucleoside **975** (50%). N-Deacetylation with hot barium hydroxide yielded 9-β-daunosaminyl-6-dimethylaminopurine **(976)**, from which the puromycin derivative **977** was prepared by acylation with N-benzyloxy-carbonyl-O-methyltyrosine and subsequent deprotection (Scheme 161).

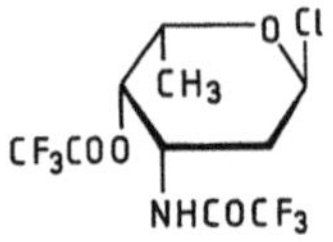

299

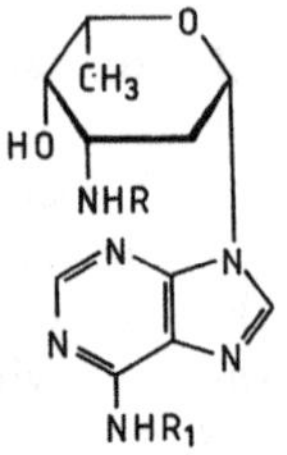

978 R = $COCF_3$
R_1 = Bz

979 R = R_1 = H

Scheme 162

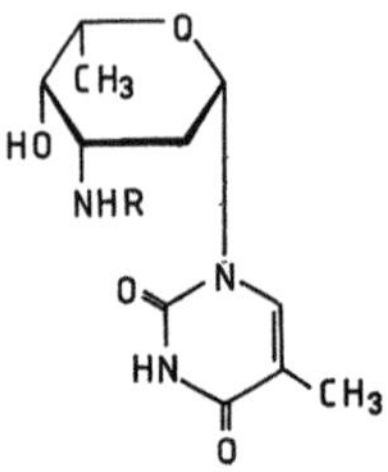

980 R = $COCF_3$
982 R = H

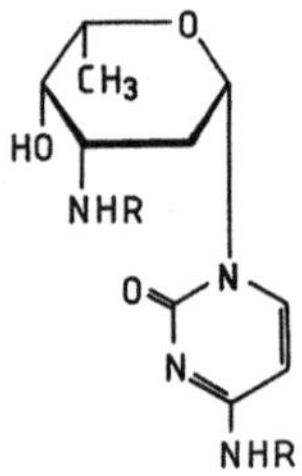

981 R = $COCF_3$; R_1 = Bz
983 R = R_1 = H

Scheme 163

The reaction of the 1-chloro compound **299** with N^6-benzoyl-9-chloromercuryadenine in boiling xylene gave only a 10% yield of the desired 6-benzoyl-9-(N-trifluoroacetyl-β-L-daunosaminyl)-adenine **978**. This low yield was attributed to the concurrent conversion of the halide **299** into N-trifluoroacetyl-L-daunosamine **(580)** and the respective glycal derivative **303** (cf. Sect. 2.2.6, Fig. 7). When N^6-benzoyladenine was used the yield of **978** was about 30%. The treatment of **978** with ammonia in methanol gave 9-β-daunosaminyladenine (**979**, Scheme 162).

The synthesis (564) of pyrimidine nucleoside derivatives of L-daunosamine was accomplished by the fusion method using the chloride **299**, bis(trimethylsilyl)thymine, and N^4-benzoyl-bis(trimethylsilyl)cytosine to obtain the protected β-nucleosides **980** and **981** (Scheme 163) in 55% and 21% yields, respectively. Removal of the protecting groups of **980** and **981** with ammonia in methanol yielded 1-β-daunosaminylthymine **(982)** and 1-β-daunosaminylcytosine **(983)**, respectively. The β-anomeric configuration of each of the prepared daunosaminyl nucleosides was shown by thorough ^{1}H-NMR studies (564).

4.1.3 C-Glycoside analogs of daunosamine

The Wittig reaction of 4-O-*p*-nitrobenzoyl-3-N-trifluoroacetyl-α-L-daunosamine **(984)** with ethoxycarbonylmethylenetriphenylphosphorane in refluxing DMF for two days (Scheme 164) gave (565) a 54% yield of ethyl 2-(4-O-*p*-nitrobenzoyl-2,3,6-trideoxy-3-trifluoroacetamido-α-L-*lyxo*-hexopyranosyl)-acetate **(986)** and a 15% yield of the corresponding β-anomer **(987)**; these were separable by chromatography. The uncyclized Wittig-product **985** could be also isolated and characterized at intermediate stages of the reaction. The use of DMF for the reaction was necessary to achieve ring closure; in refluxing THF the reaction did not proceed at all. Reduction of **986** with 3.5 equivalents of DIBAL at 0 °C produced the alcohol **988** and partially hydrolyzed the trifluoroacetyl group. The methyl ester derivative **989** of **986** was also prepared by transesterification.

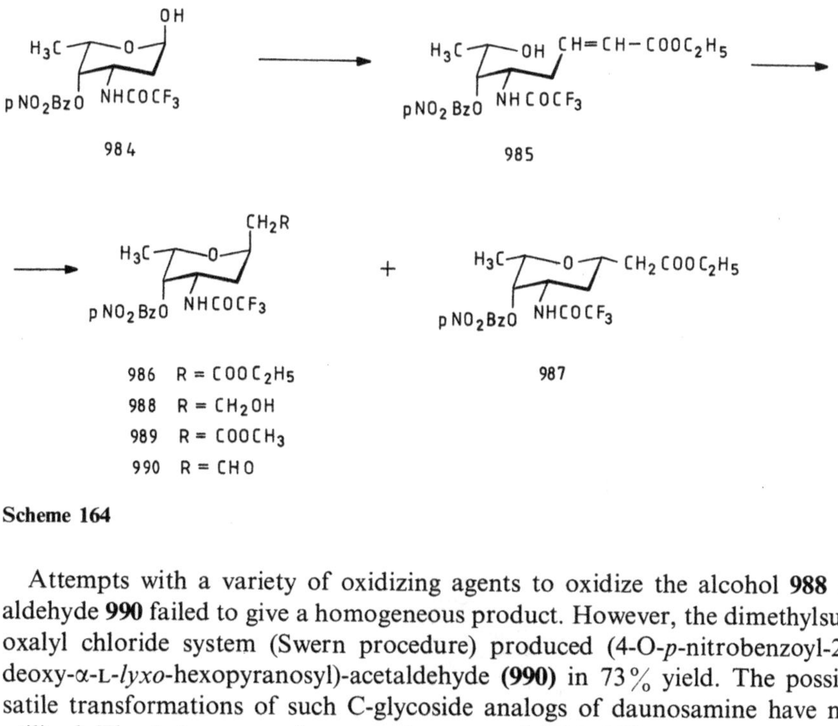

Scheme 164

Attempts with a variety of oxidizing agents to oxidize the alcohol **988** into the aldehyde **990** failed to give a homogeneous product. However, the dimethylsulfoxide-oxalyl chloride system (Swern procedure) produced (4-O-*p*-nitrobenzoyl-2,3,6-trideoxy-α-L-*lyxo*-hexopyranosyl)-acetaldehyde **(990)** in 73% yield. The possible versatile transformations of such C-glycoside analogs of daunosamine have not been utilized. The Acton group has recently prepated (566, 567) unsaturated C-glycosides of daunosamine and employed these in Diels-Alder reactions to obtain the methylene-bridged C-glycoside analogs of anthracycline antibiotics.

Thus, C-allylation of N-trifluoroacetyl-1,4-di-O-*p*-nitrobenzoyl-α-L-daunosamine **(991)** with allyltrimethylsilane in acetonitrile and in the presence of boron trifluoride etherate gave the C-1 allyl compound **992** as the exclusive product in 84% yield (Scheme 165). The analogous reaction of **991** with three equivalents of (*E*)-2,4-pentadienyltrimethylsilane and boron trifluoride etherate yielded 1-O-(4-O-*p*-nitrobenzoyl-N-trifluoroacetyl-α-L-daunosaminyl)-(*E*)-2,4-pentadiene **(993)**, as the sole product (yield >90%). Deacylation of **993** at O-4 with methanolic diisopropylamine

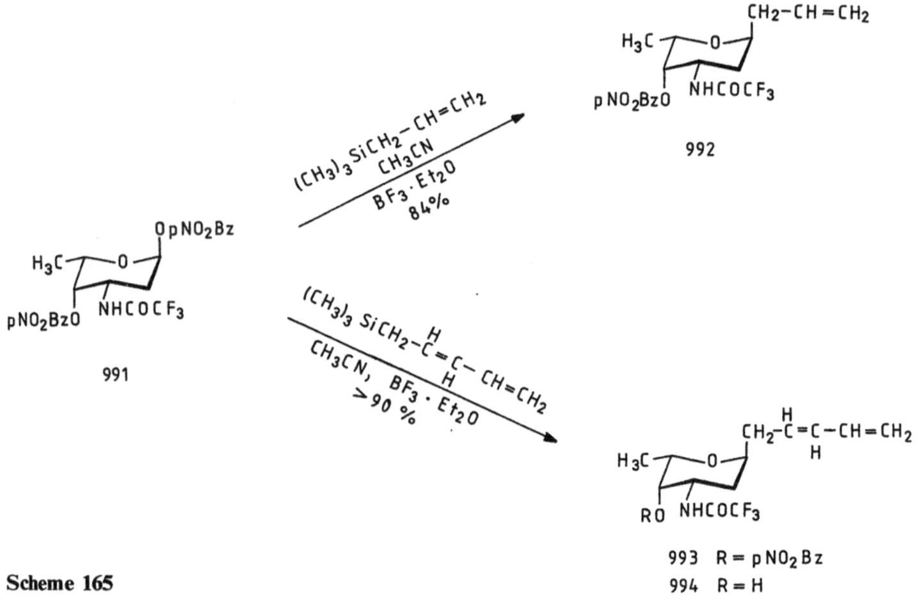

Scheme 165

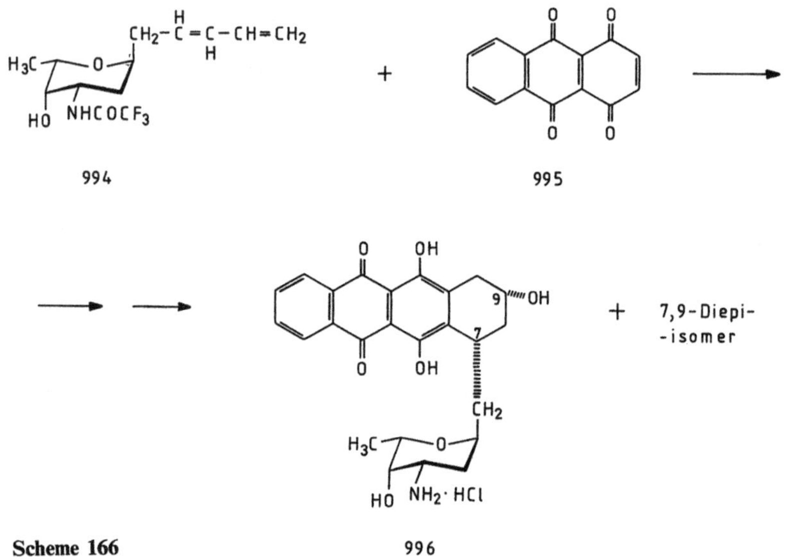

Scheme 166

produced a 77% yield of **994**. The Diels-Alder cycloaddition of **994** with quinizarin quinone **(995)**, followed by functionalization of ring A, provided (567) the C-glycosyl isostere of 4-demethoxy-9-deacetyldaunomycin **(996)** and its 7,9-diepi isomer (Scheme 166). Note that both glycosides exhibited cytotoxic effects in preliminary biological tests. The N-benzoyl- and N-trifluoroacetyl derivatives of the 1-C-methyl analog of L-daunosamine (**997** and **998**, Scheme 167) have been synthesized by Fuganti and his associates (547) according to a modified version of the method shown in Sect. 3.2.3 (Scheme 148).

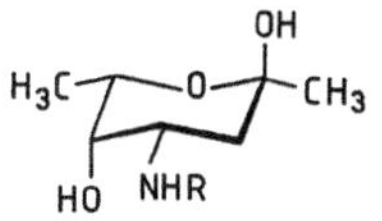

997 R = Bz
998 R = $COCF_3$

Scheme 167

4.2 N-Substituted derivatives of daunosamine-type aminodeoxy hexoses

Among the numerous N-substituted analogs of 3-amino-2,3,6-trideoxyhexoses, those prepared by the direct chemical transformation of the intact anthracycline glycoside molecule are the most important from both synthetic and biological points of view. The most successful approach for achieving such modifications involves the reductive alkylation of the amino group of the daunosamine moieties of daunomycin **(1)** (568) and adriamycin **(2)** (59) with 2,2′-oxybis-[acetaldehyde] **(999)** and sodium cyanoborohydride. These reactions led to the formation of the morpholino derivatives **1000** and **1001** of daunomycin and adriamycin, respectively. In these the nitrogen atom of the aminosugar portion is incorporated in the morpholino ring (Scheme 168). The respective diastereoisomeric 3′-deamino-(3-cyano-4-morpholinyl) analogs **1002** and **1003** were also isolated in moderate yields as by-products of the alkylation process. The product **1003**, obtained from adriamycin (**2**, doxorubicin), was found to be 100 to 1000 times as potent as the parent antibiotic against tumors in cell cultures and in mice. In view of this exceptional cytotoxic activity, anthracycline **1003** is one of the most potent antitumor agent ever investigated (59).

Other 3′-N-alkyl derivatives of the anthracycline antibiotics, synthesized by means of reductive alkylation (569) or through the hydrolysis (570) of the above cyano-

1 R = H
2 R = OH

1000 R = R_1 = H
1001 R = OH; R_1 = H
1002 R = H; R_1 = CN
1003 R = OH; R_1 = CN

Scheme 168

morpholinyl analogus, were either inactive or far less effective than compound **1003** and the unsubstituted parent antibiotics.

In search of other types of antitumor agents Fujiwara et al. (571) prepared numerous C-3-nitrosoureido-carbohydrate analogs of streptozotocin **(1004)**. For example, methyl α-L-daunosaminide **(260)** was first treated with methylisocyanate, and the resulting ureidoglycoside **1005** was nitrosated to obtain the nitrosourea **1006** (Scheme 169). As shown by biological tests, many of the prepared analogs, but the daunosaminide derivative **1006** exhibited similar or improved antileukemic activity, as compared to that of streptozotocin **(1004)**. In contrast the (nitrosoureido)-hexopyranoside **1007**, derived from methyl α-D-acosaminide **(234)** in an essentially similar way (572), was reported to posess antitumor properties.

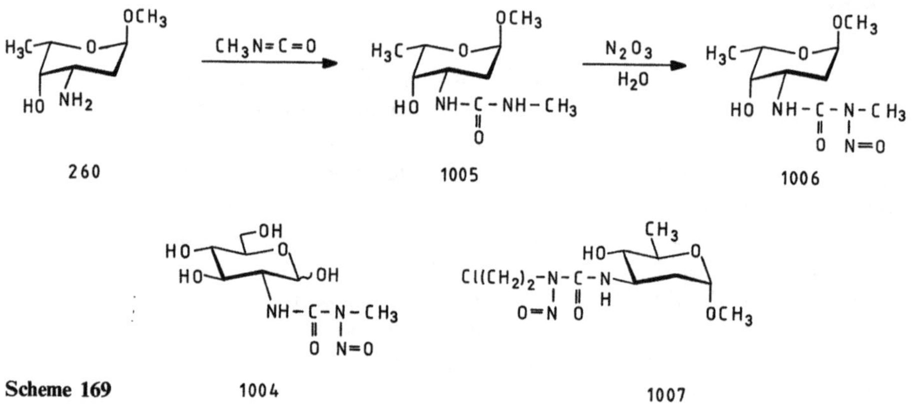

Scheme 169

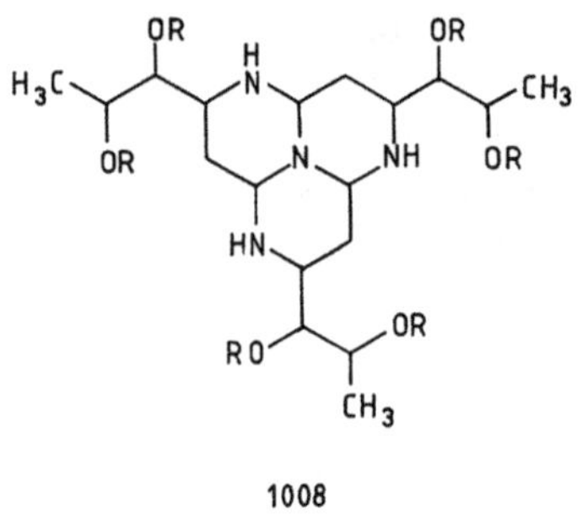

Scheme 170

The condensation product (**1008**, R=H) of three molecules of L-ristosamine **(18)** and one molecule of ammonia that was isolated (573) during the processing of the hydrolyzate of antibiotic avoparcin is depicted in Scheme 170 for the sake of curiosity. The structure of this tetrazaphenalene derivative was determined by X-ray crystallographic studies on the corresponding crystalline hexaacetate (**1008**, R=Ac). The formation of such a compound upon isolation of 3-amino-2,3,6-trideoxyhexoses from antibiotic substances, including the related work-up process (58) used for the separation of L-ristosamine from antibiotic ristomycin (ristocetin), had never before been observed.

4.3 4-Deoxy-, C-4 branched-chain and other C-4 substituted derivatives of 3-amino-2,3,6-trideoxyhexoses

4.3.1 Synthesis of 4-deoxy analogs

It has been already mentioned that 4'-deoxydaunomycin **(1009)** and 4'-deoxyadriamycin **(1010)** (Scheme 171), carrying 4-deoxy-L-daunosamine **(239)** as the aminodeoxy sugar moiety, are valuable antitumor agents (7, 8) and nearly free of cardiotoxic side-effects. Clinical studies have revealed that 4'-deoxyadriamycin exhibits (574) a higher activity against colon cancer than does the parent antibiotic.

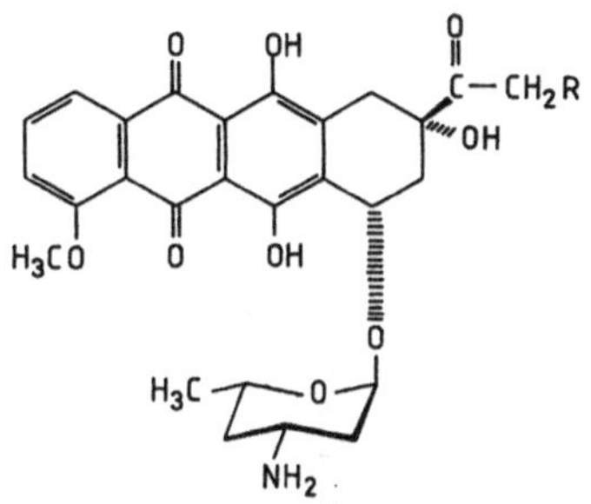

1009 R = H
1010 R = OH

Scheme 171

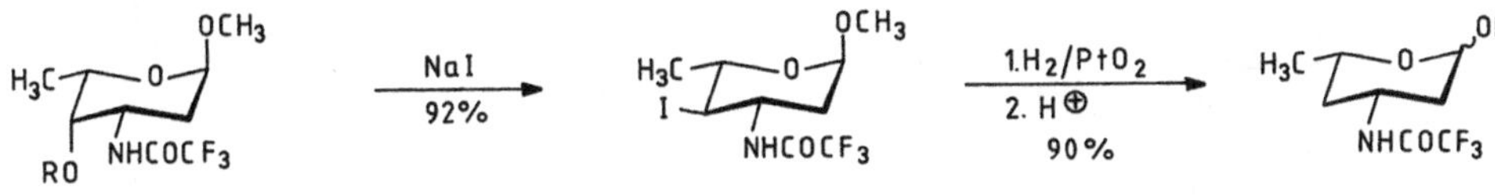

680 R = H
1011 R = p-Br-C_6H_4-SO_2

1012

1013

Scheme 172

The first synthesis of 4-deoxy-L-daunosamine **(239)** was performed by Arcamone et al. (575) (Scheme 172) by treating methyl N-trifluoroacetyl-α-L-daunosaminide **(680)** with 4-bromobenzenesulfonyl chloride in pyridine to obtain the 4-O-sulfonate **1011** (92%). Reaction of this latter substance with sodium iodide in refluxing methyl ethyl ketone gave an 84% yield of methyl 4-iodo-2,3,4,6-tetradeoxy-3-trifluoroacetamido-α-L-*arabino*-hexopyranoside **(1012)**. Hydrogenation of **1012** in the presence of Adams catalyst, followed by mild acid hydrolysis produced 2,3,4,6-tetradeoxy-3-trifluoroacetamido-L-*threo*-hexopyranose **(1013)** in 90% overall yield. This was then converted into the corresponding 1-chloro analog suitable for glycosylation of daunomycinone and adriamycinone to form the 4'-deoxy antibiotic analogs **1009** and **1010**, respectively.

Based on an ingenious thought of the Arcamone group (505), the intact glycosidic molecule of N-trifluoroacetyl-daunomycin **(704)** was subjected to a similar sequence of reactions (Scheme 173). However, when the 4-O-trifluoromethanesulfonyl ester **1014** of **704** was treated with tetrabutylammonium iodide the desired 4′-iodo-*arabino*-glycoside **(1015)** was not formed. Instead, a rearrangement involving the *trans*-elimination of the 4′-O-triflyl- and H-5′ substituents, and ring opening of the resulting enolic system occured, followed by the incorporation of the trifluoroacetamido moiety into a new six-membered ring-system **1016**.

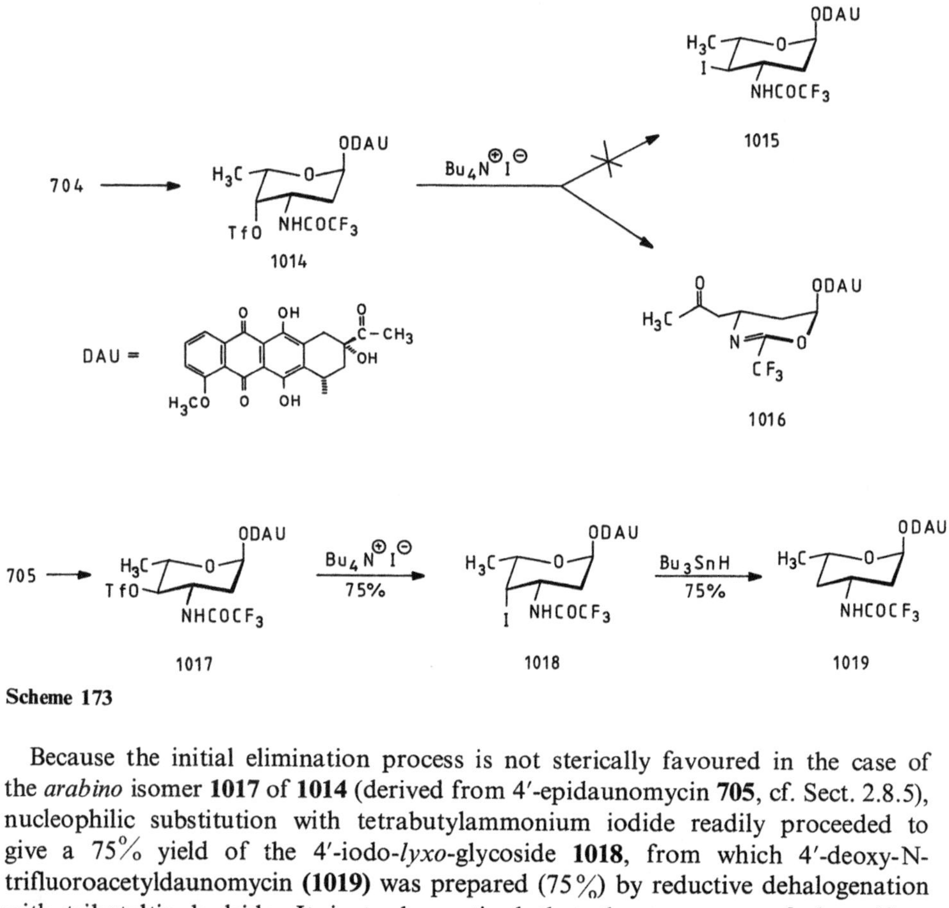

Scheme 173

Because the initial elimination process is not sterically favoured in the case of the *arabino* isomer **1017** of **1014** (derived from 4′-epidaunomycin **705**, cf. Sect. 2.8.5), nucleophilic substitution with tetrabutylammonium iodide readily proceeded to give a 75% yield of the 4′-iodo-*lyxo*-glycoside **1018**, from which 4′-deoxy-N-trifluoroacetyldaunomycin **(1019)** was prepared (75%) by reductive dehalogenation with tributyltin hydride. It is to be noticed that the treatment of the triflate ester **1017** with tetrabutylammonium bromide, followed by mild acid hydrolysis of the N-trifluoroacetyl group, resulted (576) in 4′-bromo-4′-deoxy-daunomycin, which is active at low concentrations against certain adriamycin-resistant tumor cells.

For the sake of methodological treatment of the construction of aminodeoxy hexose molecules, an alternative route, developed by Baer et al. (315, 316) and involving 3-nitrosugar intermediates, to 4-deoxy-L-daunosamine **(239)** and 4-deoxy-L-ristosamine **(238)** is discussed in Sect. 2.2.4 (see Scheme 46). Derivatives of the two important 3-amino-polydeoxyhexoses **239** and **238** were also found to be readily available through the C-4 deoxygenation of the 3-trifluoroacetamido glycals **1020** and **1021**, prepared by Florent et al. (577) according to the straightforward method outlined in Chapter 2.2.6. Applying the Barton procedure (206), **1020** and **1021** were first converted into the 4-O-thiocarbonates **1022** and **1023**. The subsequent reduction of these with tributyltin hydride produced the N-trifluoroacetyl-pentadeoxy-hex-1-enopyranoses **1024** and **1025**, corresponding to 4-deoxy-L-daunosamine **(239)** and 4-deoxy-L-ristosamine **(238)**, respectively, in near 70% overall yield (Scheme 174). An apparent advantage of this method is that the products **1024** and **1025** are suitable for the direct glycosylation of anthracyclinoses with high stereoselectivity (cf. Sect. 2.2.6) without any further modification.

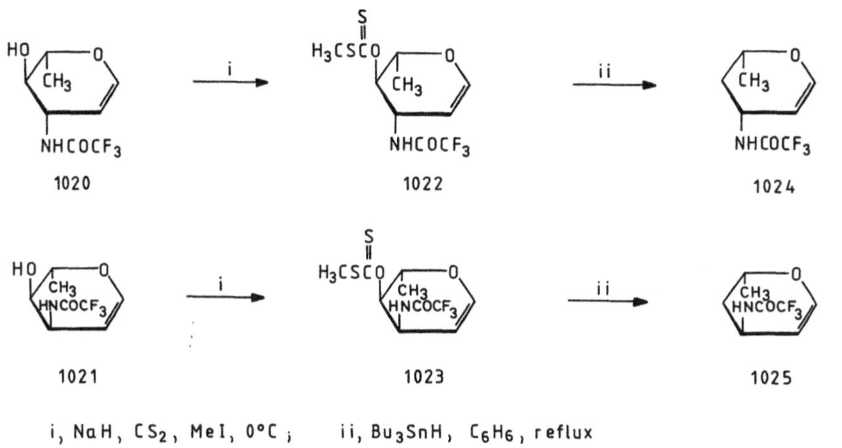

Scheme 174

In contrast to these successful Barton deoxygenations, Sztaricskai et al. (578) observed that the reduction of 3-azido-2,3,6-trideoxy-4-O-[(methylthio)-thiocarbonyl]-α-L-*ribo*-hexopyranoside **(1026)** with tributyltin hydride resulted in the oxazoline **1028**. This experimental finding can be explained by the initial formation of the 3-amino compound **1027**, the C-4 thiocarbonyl moiety of which is attacked by the C-3 amino group to form the oxazoline ring via the elimination of hydrogen sulfide (Scheme 175). Apart from these carbohydrate-based syntheses of 4-deoxy-L-daunosamine, three additional methods have also been reported for the preparation of the racemic form of this sugar from non-carbohydrate starting materials.

In the course of attempted saturation (515, 516) of the oxazolo-α-pyrone **777** (cf. Sect. 3.2.1, Scheme 130) by hydrogenation over Adams catalyst, a concomitant

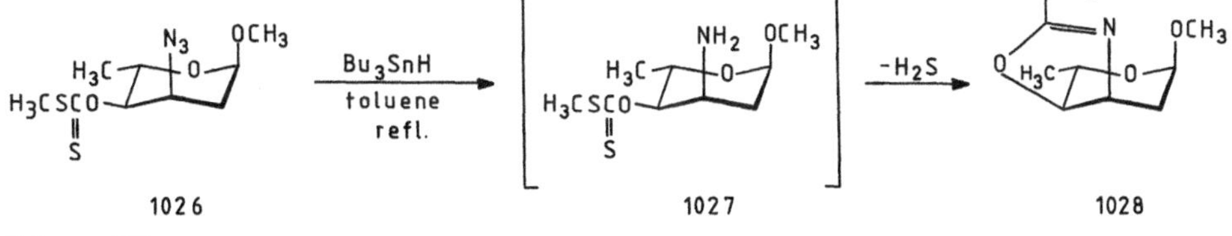

Scheme 175

hydrogenolytic cleavage of the oxazoline ring led to the formation of the N-benzoyl-tetradeoxy-D,L-*threo*-lactone **1029**, a precursor to 4-deoxy-D,L-daunosamine (Scheme 176).

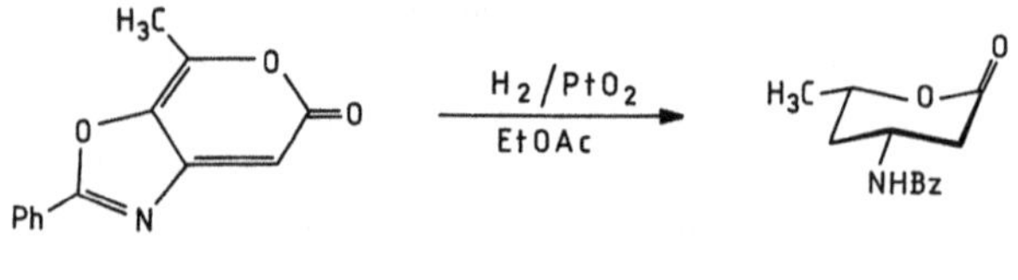

777 ±1029

Scheme 176

Another method (579) (Scheme 177) involves the mild acid hydrolysis of the unsaturated acetal **1030**, available with Diels-Alder cyclization (580) from 1-methoxy-1,3-butadiene and butyl glyoxylate, producing butyl (*E*)-2-hydroxy-6-oxo-hex-4-enoate **(1031)**. The addition of hydrazoic acid to **1031** resulted (579, 581) in the quantitative formation of the mixture of stereoisomeric butyl 3-azido-2,3,4-trideoxy-D,L-hexopyranuronates (±**1032**). Treatment of this mixture with methyl orthoformate and methanolic hydrogen chloride gave rise to all four stereoisomeric methyl (methyl 3-azido-2,3,4-trideoxy-D,L-hexopyranosid)-uronates (±**1033**). Equilibration

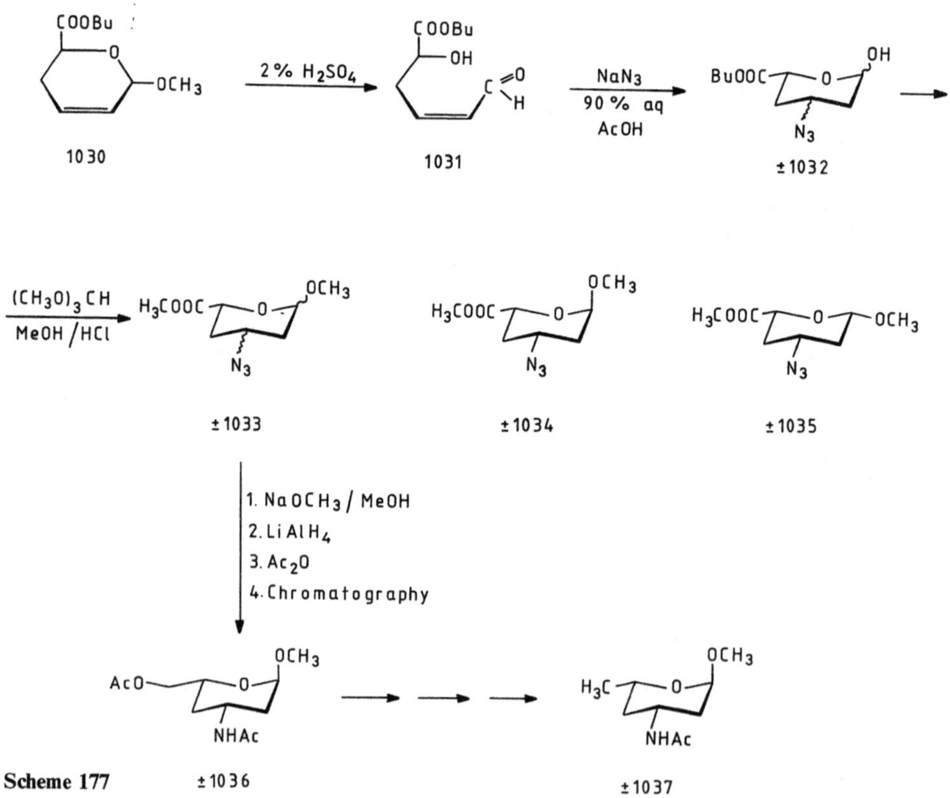

Scheme 177

with sodium methoxide in methanol led to the enrichment (up to 70%) of the thermodynamically more stable α- (±**1034**) and β- (±**1035**) -D,L-*threo* isomers via the epimerization of the corresponding *erythro* compounds at C-5. Reduction of the equilibrated mixture ±**1033** using lithium aluminum hydride, followed by acetylation and chromatographic separation, allowed the isolation of methyl 3-acetamido-6-O-acetyl-2,3,4-trideoxy-α-D,L-*threo*-hexopyranoside (±**1036**) in 45% yield. O-Deacylation of this compound and subsequent deoxygenation at the primary position, through the 6-O-tosylate and 6-iodo derivative according to the general strategy outlined in Sect. 2.7.1, yielded methyl N-acetyl-4-deoxy-α-D,L-daunosaminide (±**1037**). The mixture **1033** was also utilized in the preparation of racemic precursors for the synthesis of the antibiotic negamycin (see Sect. 4.5.2, Scheme 197). In recent work (542) aimed at using the *hetero*-Diels-Alder cyclization (Sect. 3.2.2, Scheme 144) to prepare daunosamine-type 3-aminodeoxy sugars, the D,L-hex-4-enopyranoside ±**868**, was made by this reaction and hydrogenated over Raney nickel in methanol to give ethyl N-benzoyl-4-deoxy-α-D,L-daunosaminide (±**1038**) in 65% yield (Scheme 178).

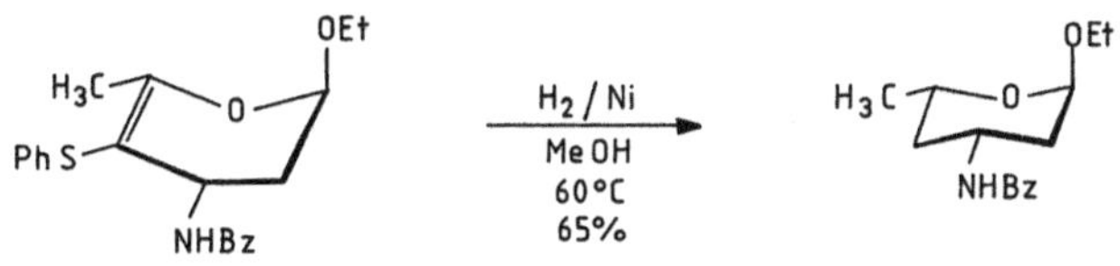

±868 ±1038

Scheme 178

4.3.2 C-4 Branched-chain derivatives

To synthesize (582–584) C-4 branched-chain 3-amino-2,3,6-trideoxy-L-hexoses, structurally related to L-daunosamine **(12)** and L-acosamine **(8)**, methyl 2,3,6-trideoxy-3-trifluoroacetamido-α-L-*threo*-hexopyranosid-4-ulose **(676)** — available by the oxidation of either the *arabino*-glycoside **211** (Sect. 2.8.1, Table 15 entry 3), or the *lyxo* compound **680** (Sect. 2.8.2, Scheme 117) — was treated with methyllithium and methylmagnesium iodide (Scheme 179). The reaction of **676** with ethereal methyllithium at —78 °C produced a 55% yield of methyl 4-C-methyl-2,3,6-trideoxy-3-trifluoroacetamido-α-L-*lyxo*-hexopyranoside **(1039)** as the exclusively isolable product. The steric course of the addition of methylmagnesium iodide onto the ketone **676** was found to be greatly dependent upon the reaction conditions (solvent and temperature); in ehtereal solution at —78 °C, a 95% yield of the *lyxo*-glycoside **1039** was produced. The best yield (75%) for the *arabino* 4-C-methyl compound **1040** was obtained when the reaction was conducted in THF at reflux temperature. The boron trifluoride-catalyzed methylation of **1039** and **1040** with diazomethane provided the 4-C,4-O-dimethyl derivatives **1041** and **1042** in yields of 65% and 75%, respectively.

Treatment of the C-4 ulose **676** with diazomethane in methanol gave a 92% yield of the spiro-epoxide **1045**, the configuration of the branch-point of which was

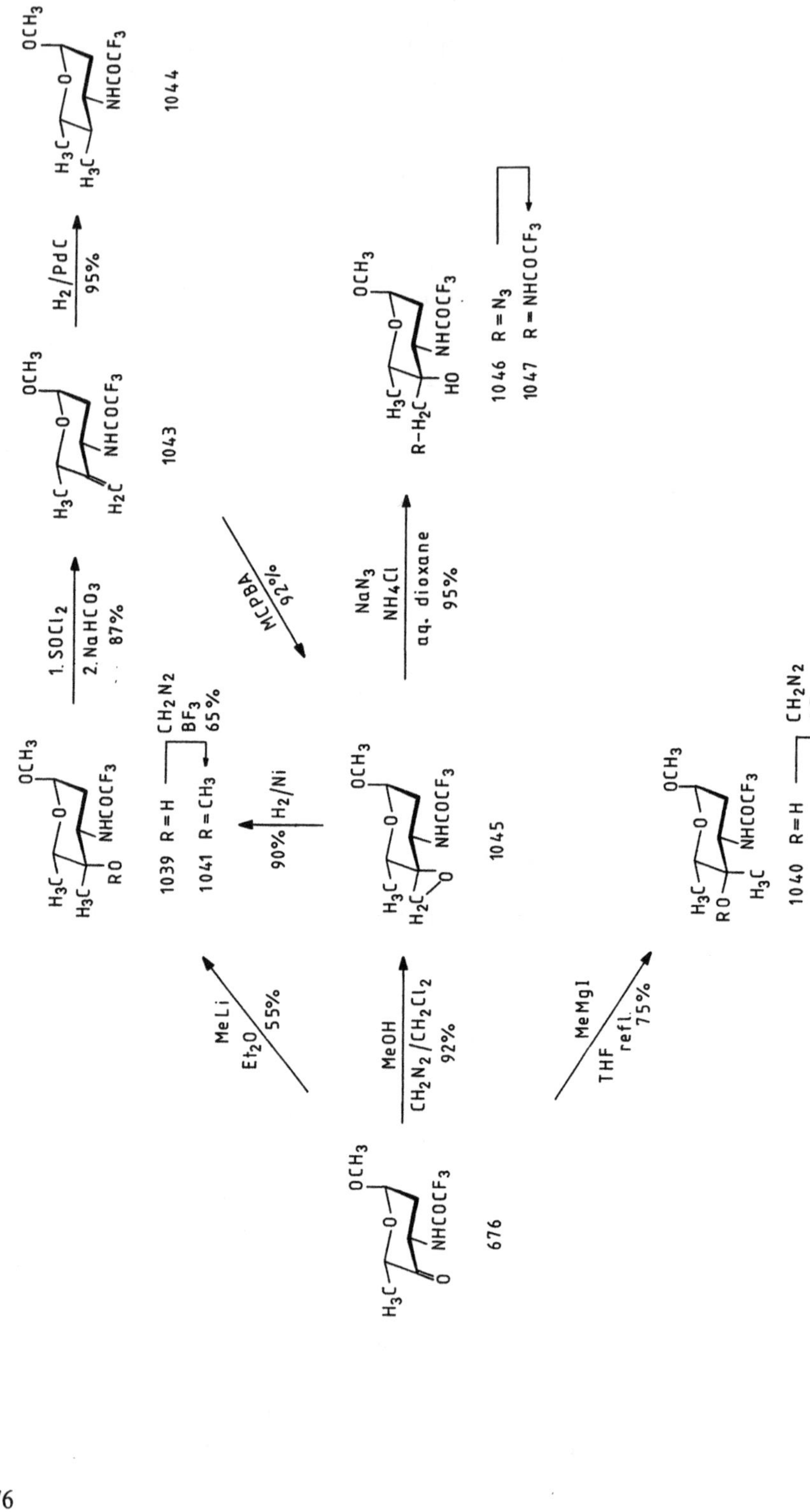

Scheme 179

shown by hydrogenolytic ring-opening into the *lyxo* 4-C-methyl glycoside **1039**. The spiro-oxirane **1045** could also be prepared in high yield by the epoxidation of the 4-exo-methylene compound **1043** (obtained from **1039** by sequential reactions with thionyl chloride and sodium hydrogen carbonate) with m-chloroperbenzoic acid. Catalytic addition of hydrogen the double bond of **1043** proceeded with high stereoselectivity to produce a 95% yield of methyl 4-deoxy-4-C-methyl-N-trifluoro-acetyl-α-L-daunosaminide **(1044)**.

Ring cleavage of the oxirane function of **1045** with sodium azide-ammonium chloride led to the α-L-*lyxo*-4-azidomethyl glycoside **1046**, from which methyl 2,3,6-trideoxy-3-trifluoroacetamido-4-C-(trifluoroacetamidomethyl)-α-L-*lyxo*-hexopyranoside **(1047)** was prepared by successive hydrogenation and N-trifluoroacetylation. The 4-C-methyl-branched aminosugars have been used in form of their respective 1-chloro derivatives (583, 584) to glycosylate anthracyclinones.

The conversion (233) of methyl 3,6-imino-N-*p*-toluenesulfonyl-2,3,6-trideoxy-α-D-*arabino*-hexopyranoside (**608**, cf. Sect. 2.7.4, Table 10) into the corresponding 4-C-nitromethyl derivative **579** involved the oxidation to form the C-4-ulose hydrate **1048**, which was then condensed with nitromethane in the presence of sodium hydride, as shown on Scheme 180.

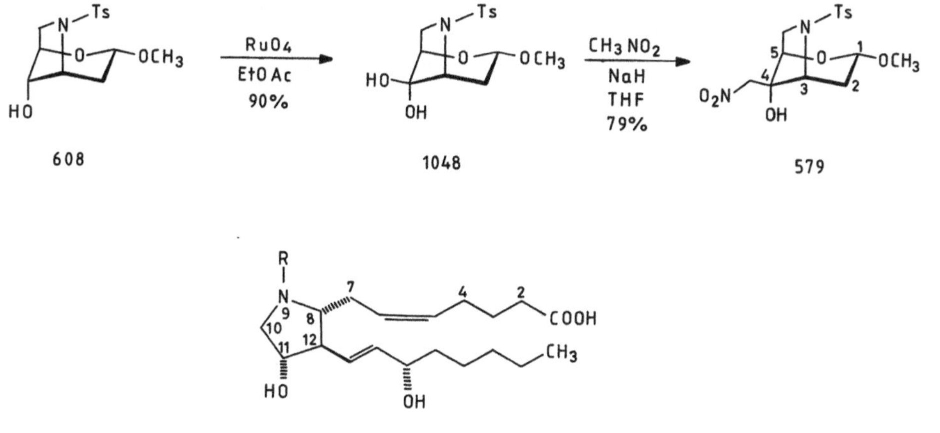

608

1048

579

1049

Scheme 180

Although deoxygenation of **579** at C-4 remained unsuccessful under a variety of conditions, this 3,6-iminohexopyranoside may serve as a possible precursor for 9-azaprostaglandin derivatives, such as **1049**, because the absolute stereochemistry of the chiral centers C-3, C-4, and C-5 in **579** corresponds to that of C-8, C-12, and C-11, respectively, of the prostaglandine **1049**.

4.3.3 Syntheses of C-4 substituted 3-amino-2,3,6-trideoxyhexoses susceptible to transformations into beta-lactam antibiotic (thienamycin) analogs

In the years 1982–1984 several approaches were described (224, 234, 269, 270, 585) for the chiral synthesis of the so-called Melillo lactone **(1051)** (586), a key intermediate in the total synthesis of (+)-thienamycin **(1050)** (Scheme 181) from

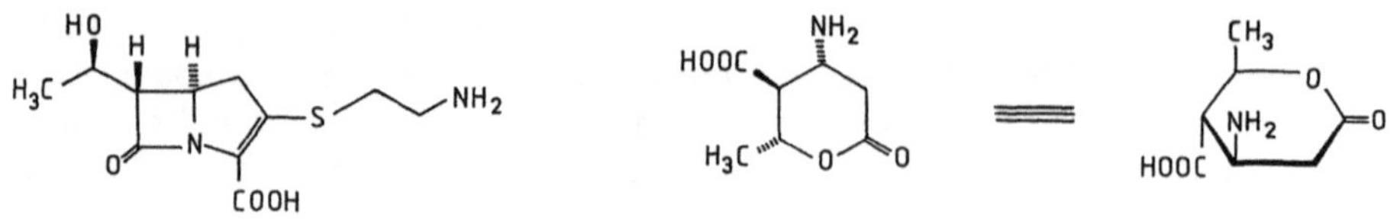

1050 1051

Scheme 181

3-azido-2,3,6-trideoxy-D-*arabino*-hexose precursors ("chiral templates") (561), readily available from D-glucose.

In the first method (Scheme 182) elaborated by Hanessian et al. (224), methyl 3-azido-2,3,6-trideoxy-α-D-*arabino*-hexopyranoside **(135)**, prepared from the α-D-*ribo* 3-O-mesylate **120** via **121** as shown in Sect. 2.2.1 (Scheme 35), was converted into the C-4 ulose **1052** by oxidation with pyridinium chlorochromate. The reaction of this azidoketone with the anion of O,O-dimethyl formylphosphonate-S,S-dimethyl dithioacetal (generated with butyllithium) gave the ketene dithioacetal derivative **1053** in 60% yield (not optimized). The reduction of **1053** with lithium aluminum hydride proceeded with high stereoselectivity via the attack of the hydride ion on the β-face that was presumably due to a prior coordination with the newly generated C-3 amino group. After trifluoroacetylation, the N-trifluoroacetyl dimethyl mercaptal derivative **1054** with the *arabino* configuration was obtained. Demercaptalation of **1054** with mercury oxide–mercury chloride in aqueous acetone and subsequent

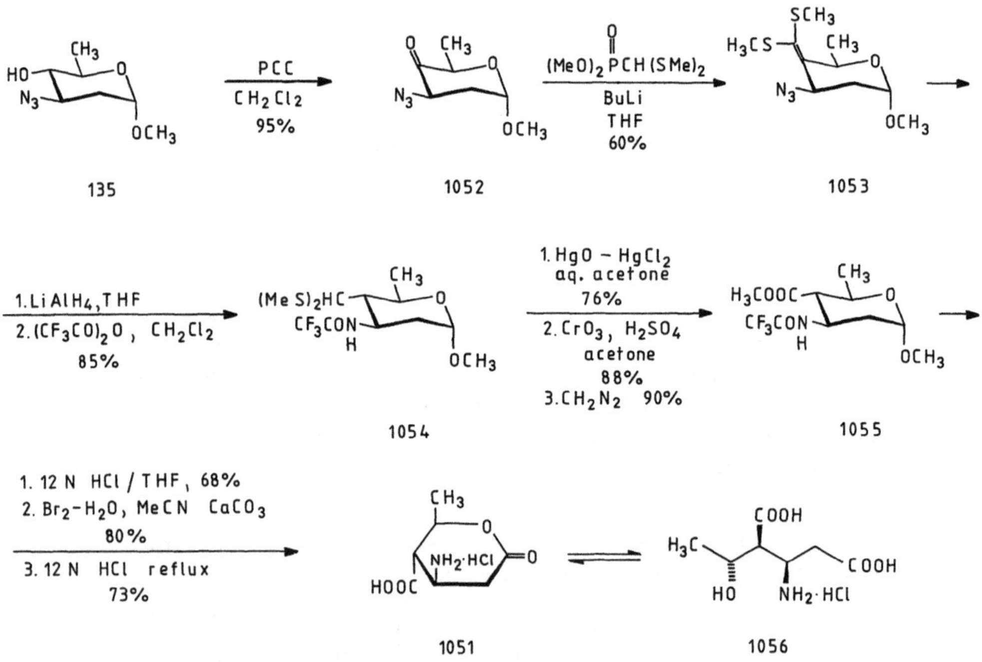

Scheme 182

oxidation of the liberated formyl group into carbonyl were followed by esterification with diazomethane to give the C-4 methyl carboxylate **1055**. Selective acid hydrolysis of the anomeric acetal and oxidation with bromine in water produced the "Melillo lactone" **1051** after vigorous acid hydrolysis to deprotect the carboxyl group. Based on NMR studies, **1051** is present in aqueous solution as a 1:6 equilibrium mixture with the amino acid **1056**, which had previously been used (in racemic form) for a total synthesis of (±)-thienamycin by Melillo and associates (586). In parallel work by Koga et al. (234, 585), the *threo*-hexopyranosid-4-ulose **1057** was prepared from the 3-aminoglycoside **572**. This latter compound was synthesized from the 4-O-mesylate **144** (Sect. 2.2.1, Scheme 27) via the azide **146** (cf. Sect. 2.7.2, Table 7 entry 1).

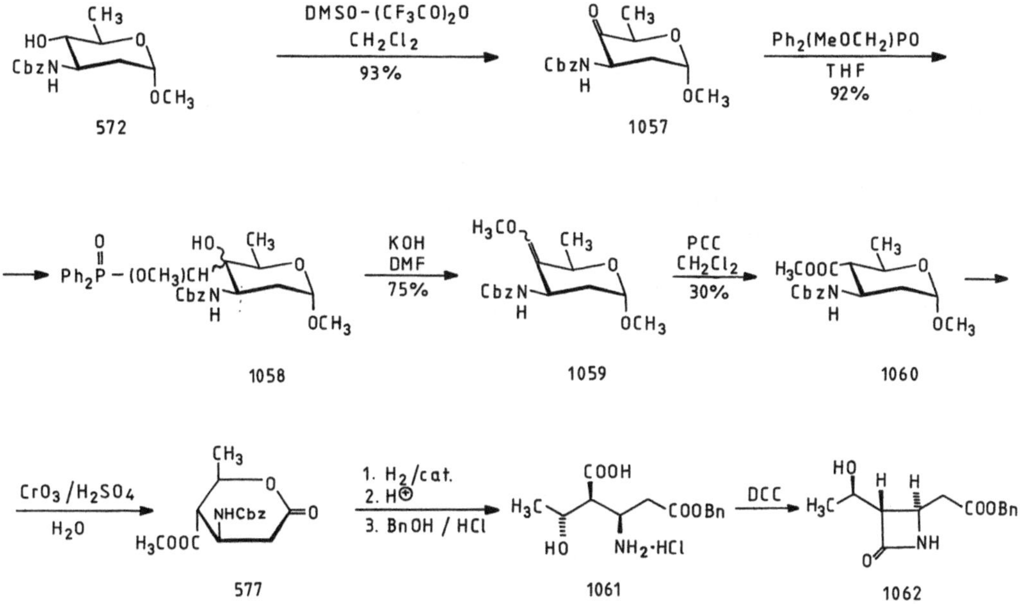

Scheme 183

The Wittig reaction of **1057** with the lithium derivative of methoxymethyldiphenylphosphine oxide gave the adduct **1058** (92%) consisting of two diastereoisomers in a ratio of ca. 27:10, separable by column chromatography (Scheme 183). Alkaline hydrolysis of the major product produced the 4-methoxymethylene derivative **1059**, which was oxidized with pyridinium chlorochromate in dichloromethane to obtain methyl (3-benzyloxycarbonyl)amino-4-C-(methoxycarbonyl)-2,3,4,6-tetradeoxy-α-D-*arabino*-hexopyranoside **1060** in 30% yield. Acid hydrolysis of **1060** and subsequent oxidation gave the lactone **577**, from which the benzyl ester **1061** of the amino acid **1056** was prepared. Cyclization of **1061** with dicyclohexylcarbodi-imide gave the beta-lactam **1062**, an established substrate (586) for conversion into (+)-thienamycin.

Although a closely related work published by Stewart and Williams (270) and starting with the 3-azidosugar **135** showed a considerable improvement of the yield from the oxidation step **1059** → **1060** (69%) when the Swern method was used even

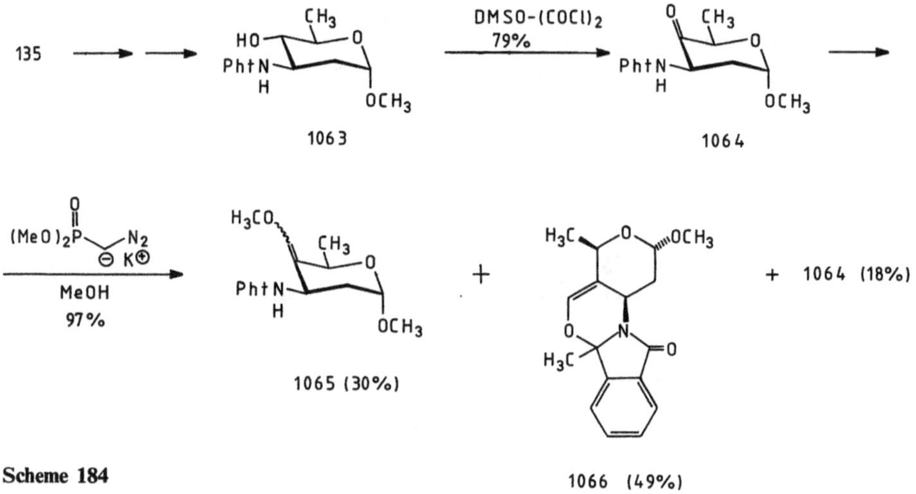

Scheme 184

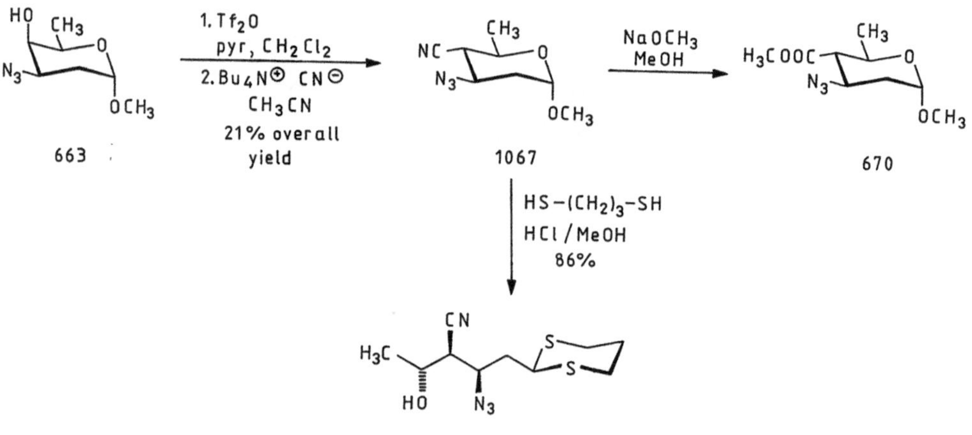

Scheme 185

with an N-phthalimido protecting group **(1063 → 1064)** (Scheme 184), the C-4 homologation of **1064** with potassium dimethyl-diazomethylphosphonate did not bring satisfactory results. Presumably due to an intramolecular trapping of the diazo intermediate, the major product of the reaction was a phthalimido enolether derivative **(1066)**, and the desired diastereoisomeric 4-methoxymethylene compound **1065** was prepared in only 30% yield.

According to a completely different strategy (Scheme 185) for the introduction of the carbonyl group to C-4 of a suitable carbohydrate precursor, Durette (269) employed 3-azido-2,3,6-trideoxy-α-D-*lyxo*-hexopyranoside **(663)**, prepared by means of the inversion of the configuration at C-4 of the *arabino* azide **135** as discussed in Sect. 2.8.1 (Table 14 entry 2). The azido alcohol **663** was converted into the 4-O-trifluoromethanesulfonate which was allowed to react, without isolation, with tetrabutylammonium cyanide, providing a cyanide ion with enhanced nucleophilicity, to obtain the *arabino*-nitrile **1067** still in moderate yield (21%). Treatment of **1067** with methanolic sodium methoxide yielded methyl 3-azido-4-C-(methoxycarbonyl)-2,3,6-trideoxy-α-D-*arabino*-hexopyranoside **(670)**.

The reaction of the azido-nitrile **1067** with 1,3-propanedithiol in the presence of acid gave an 86% yield of 3-azido-4-C-cyano-2,3,4,6-tetradeoxy-D-*arabino*-hexose trimethylene dithioacetal **(1068)**, an additional "chiral template" susceptible to conversion into (+)-thienamycin. Related synthetic strategies may also be effective for the preparation of other carbocyclic beta-lactam antibiotics.

4.4 Furanose-ring analogs of 3-amino-2,3,6-trideoxyhexoses

In order to extend the scale of potentially active semisynthetic antitumor anthracycline antibiotics, several research groups have studied the possibilities of preparing the furanose derivatives of the daunosamine-type aminodeoxy hexoses as novel candidates for the glycosylation of anthracyclinones. The syntheses allowing the direct obtainment of two furanoses, methyl 3-acetamido-5-O-benzyl-2,3,6-trideoxy-α-L-*lyxo*-hexofuranoside **(161)** (135) and methyl 3-amino-2,3,6-trideoxy-α,β-L-*lyxo*-hexofuranoside **(533)** (248), are discussed in Sect. 2.2.1 (Scheme 30) and Sect. 2.6.2 (Scheme 99), respectively.

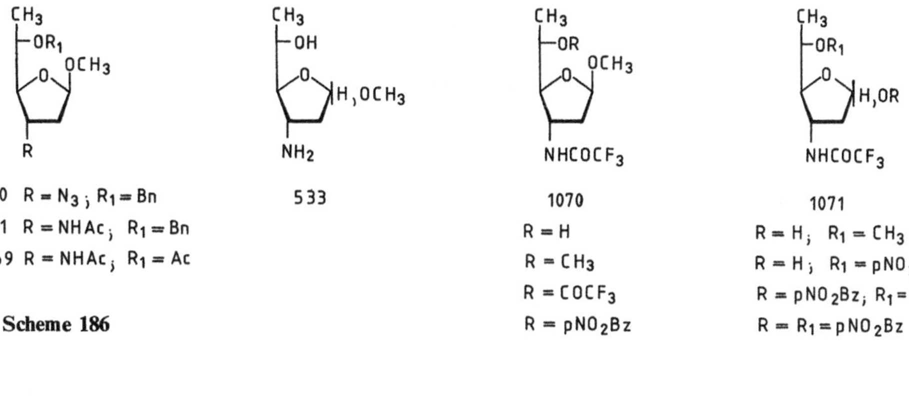

160 R = N_3; R_1 = Bn
161 R = NHAc; R_1 = Bn
1069 R = NHAc; R_1 = Ac

533

1070
R = H
R = CH_3
R = $COCF_3$
R = pNO_2Bz

1071
R = H; R_1 = CH_3
R = H; R_1 = pNO_2Bz
R = pNO_2Bz; R_1 = CH_3
R = R_1 = pNO_2Bz

Scheme 186

By using the 3-azido-α-L-*lyxo*-hexofuranoside **160** (cf. Scheme 30) as a key intermediate, the Kuszmann group has synthesized (135, 246) the 3,5-diacetyl compound **1069** and various N-trifluoroacetyl-L-daunosamino-furanose derivatives that are depicted by the general formulas **1070** and **1071** (Scheme 186).

The five-membered ring-isomers of the 3-amino-2,3,6-trideoxyhexoses often appear — mainly in the form of γ-lactones — in syntheses involving non-carbohydrate precursors as starting materials. The enantiomerically pure and racemic furanose derivatives of the foregoing type of aminodeoxy sugars, used as intermediates in related procedures, are summarized (cf. Chapt. 3) in Scheme 187.

Two alternative methods of transforming 3-amino-2,3,6-trideoxyhexopyranoses into the corresponding hexofuranose derivatives have been elaborated. In one of these procedures (Scheme 188), Arcamone et al. (584) treated N-trifluoroacetyl-L-

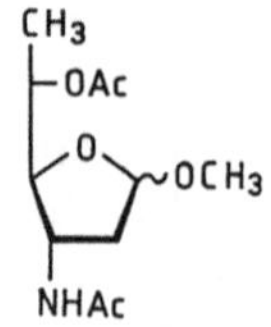

825
(Scheme 136)

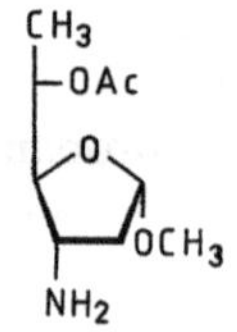

914
(Scheme 150)

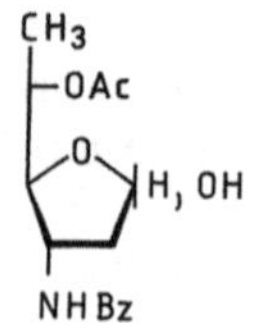

±784
(Scheme 131)

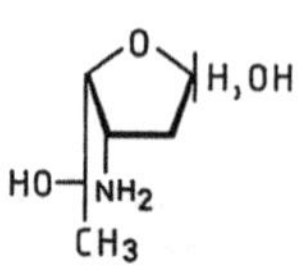

941
(Scheme 154)

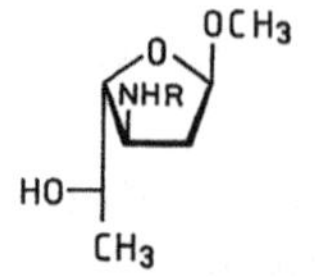

±762 R = COCCl$_3$
(Scheme 127)

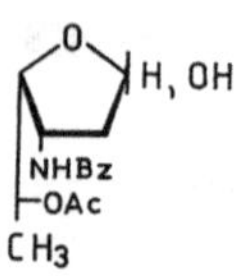

±785
(Scheme 131)

Scheme 187

ristosamine **(1072)** with methanol in the presence of Dowex 50 (H^+) resin, to obtain a separable mixture of the α- (**1073**, R=H) and β- (**1074**, R=H) -L-*ribo*-hexofuranosides. These were first converted into the 5-O-*p*-nitrobenzoates **1073** (R = *p*-NO_2Bz) and **1074** (R=*p*-NO_2Bz) and then into the 1-chloro compound **1075**, a suitable glycosylating agent for the preparation of the 3-amino-2,3,6-trideoxy-L-*ribo*-hexofuranosyl anthracyclinones. An oxidation-reduction sequence was used to convert the α-L-*ribo*-hexofuranoside **1073** (R=H) via the hexofuranosid-5-ulose **1076** into the C-5 epimeric β-D-*lyxo*-hexofuranoside **1077**, a five-membered ring derivative of D-daunosamine.

An alternative route that makes similar 3-aminohexofuranosides readily accessible from the corresponding pyranose derivatives has been reported by El Khadem and associates (247, 587). This approach (Scheme 189) involves, as the first step, the conversion of the N-protected aminodeoxy sugar, e.g. N-trifluoroacetyl-L-daunosamine **(580)**, into the diethyl dithioacetal **162** with ethanethiol in the presence of hydrochloric acid. Mercury-induced cyclization of **162** gave a separable mixture of the methyl α- **(1078)** (37%) and β- **(1079)** (29%) furanosides together with methyl N-trifluoroacetyl-α,β-L-daunosaminide **(1080)** (7%) and the dimethyl acetal **1081** (5%). Para-nitrobenzoylation of **1078** and **1079** yielded the 5-O-*p*-nitrobenzoates **1082** and **1083**, respectively. The acetolysis of these with acetic anhydride and acetic acid-sulfuric acid in ethyl acetate produced 1-O-acetyl-5-O-(*p*-nitrobenzoyl)-2,3,6-trideoxy-3-trifluoroacetamido-L-*lyxo*-hexofuranose **(1084)** in an overall yield of 42%, based on **580**. The 1-O-acetate **1084** was coupled with daunomycinone (587) using either stannic chloride or *p*-toluenesulfonic acid. Deblocking gave 3-amino-2,3,6-trideoxy-α-L-*lyxo*-hexofuranosyl daunomycinone **(1085)**. The syntheses of **1085**, the corresponding 5-O-methyl analog, and the structurally related 7-O-(3-azido-2,3,6-trideoxy-5-O-methyl-α-L-*lyxo*-hexofuranosyl)-carminomycinone were reported by Kuszmann and his associates (246).

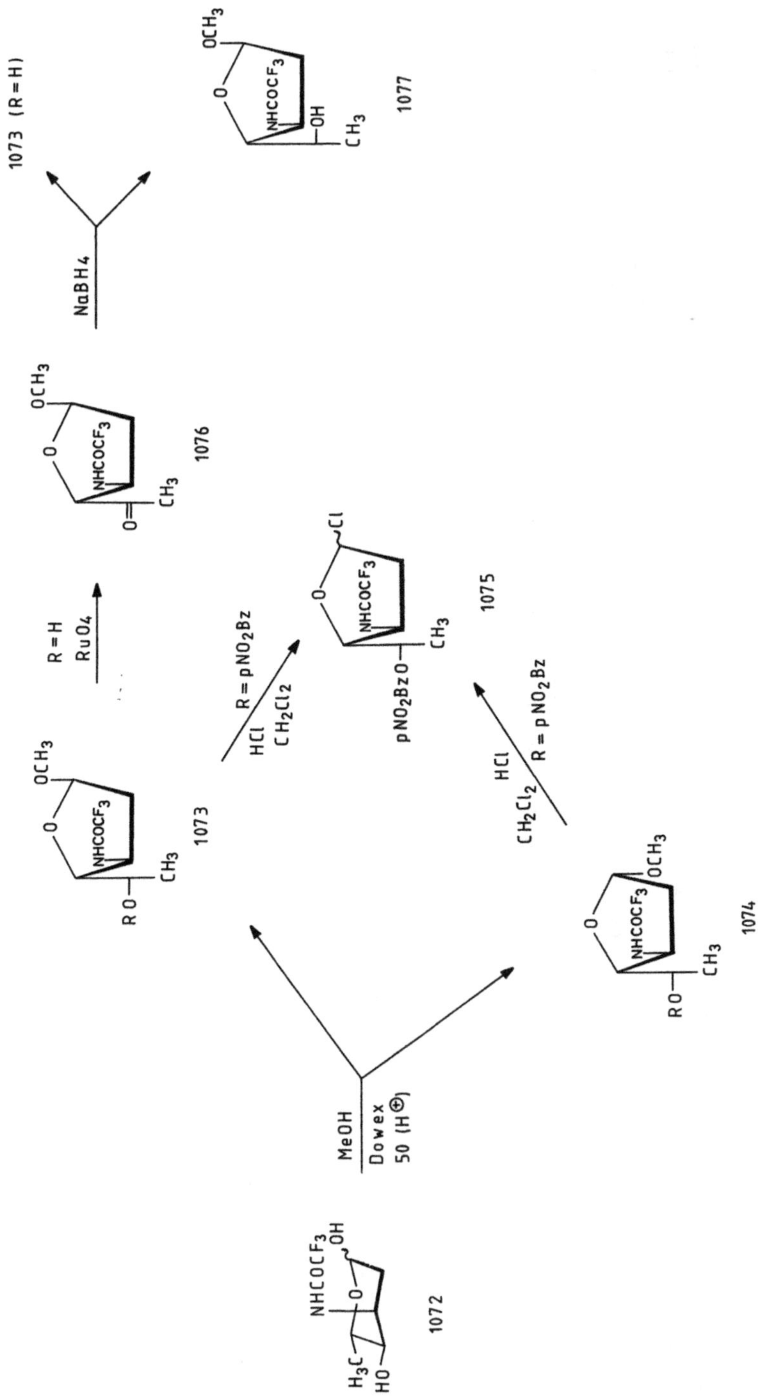

Scheme 188

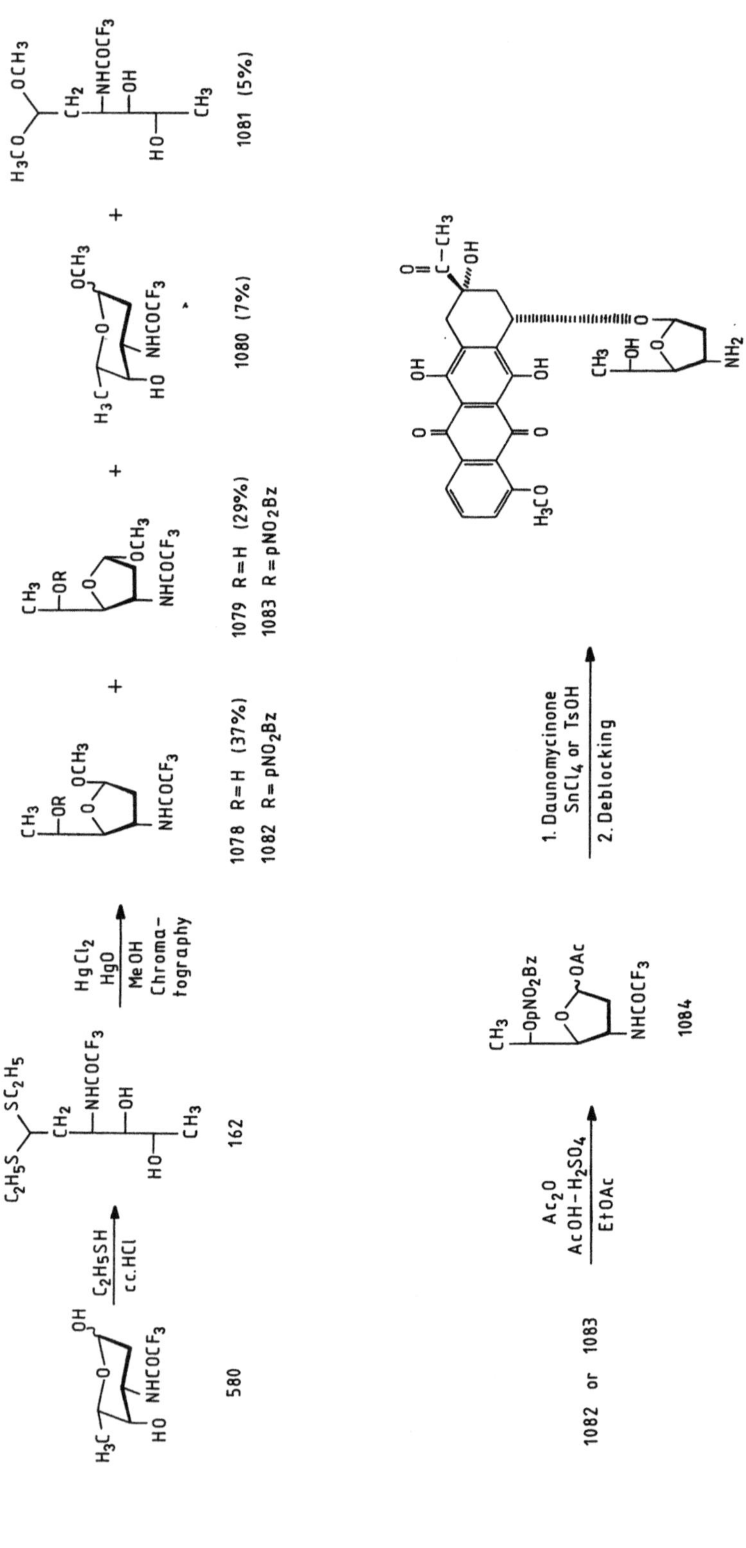

Scheme 189

The N-benzoyl analog of the diethyl dithioacetal **162** was prepared by Iida et al. (539) from the O-protected masked acetal **851** (Sect. 3.2.2, Scheme 142) by boron trifluoride-catalyzed mercaptalation. An essentially similar reaction sequence starting with N-acetyl-D-ristosamine **(1086)** was used (247, 587) to prepare 1-O-acetyl-5-O-(*p*-nitrobenzoyl)-2,3,6-trideoxy-3-trifluoroacetamido-D-*ribo*-hexofuranose **(1095)** and its N-acetyl derivative **1096** via the respective methyl α- **(1087–1090)** and β- **(1091–1094)** -D-*ribo*-hexofuranosides (Scheme 190). The acetolysis of the N-protected methyl α- and β-glycofuranosides into the 1-O-acetates was accompanied by extensive degradation when a 5-O-acetyl protecting group was used.

In order to avoid the formation of the undesired methyl pyranosides and dimethyl acetals (such as **1080** and **1081**, respectively, cf. Scheme 189) upon cyclization of the diethylmercaptals, an improved protecting group strategy and modified reaction conditions (Scheme 191) have been introduced (587).

Starting with a 3-N,4-O-diacyl-aminosugar, e.g., N-acetyl-O-benzoyl-D-ristosamine **1097**, the corresponding diethly dithioacetal **(1098)** was prepared as previously

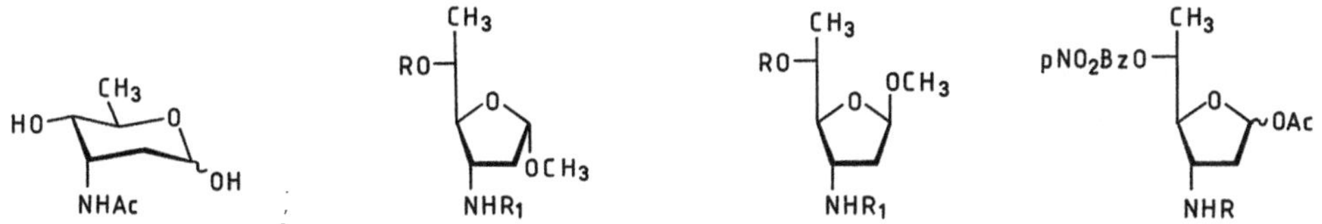

1086

1087 R = H; R_1 = Ac
1088 R = H; R_1 = $COCF_3$
1089 R = pNO_2Bz; R_1 = Ac
1090 R = pNO_2Bz; R_1 = $COCF_3$

1091 R = H; R_1 = Ac
1092 R = H; R_1 = $COCF_3$
1093 R = pNO_2Bz; R_1 = Ac
1094 R = pNO_2Bz; R_1 = $COCF_3$

1095 R = $COCF_3$
1096 R = Ac

Scheme 190

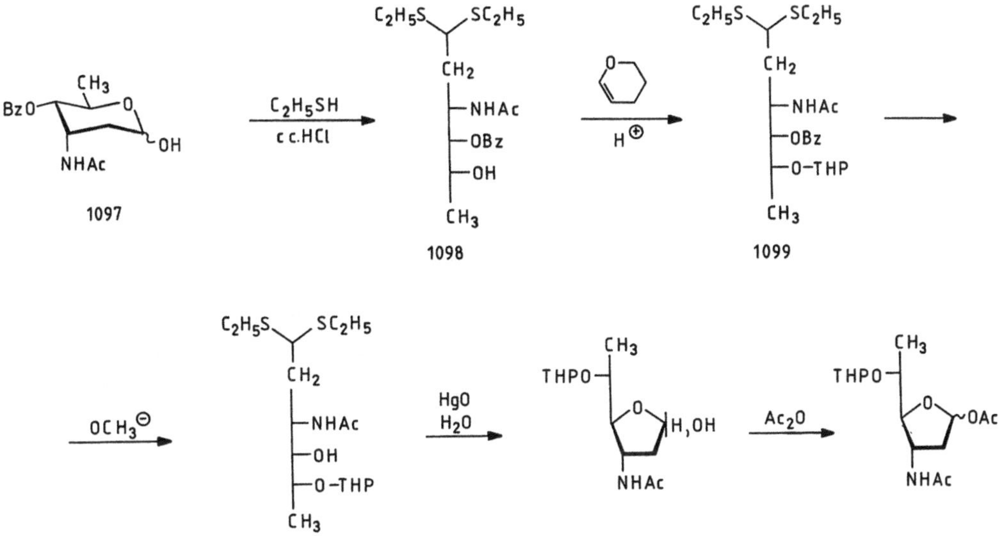

1097 1098 1099

1100 1101 1102

Scheme 191

shown. The mercaptal **1098** was transformed into the 5-O-tetrahydropyranyl ether **1099** and the C-4 hydroxyl group was liberated by means of Zemplén deacylation **(1100)** before cyclization into the furanose derivative. The reaction was conducted in water instead of methanol to avoid competitive production of the dimethyl acetal upon mercury-induced cyclization of **1100** into **1101**, and the furanose sugar **1101** was obtained practically without any by-product. Subsequent acetylation of **1101** then gave the 1-O-acetate **1102** as a mixture of the α- and β-anomers, which was used in boron trifluoride- or stannic chloride-catalyzed glycosylations of anthracyclinones. Apparently, this method can be uniformally employed to make the other stereoisomeric 3-aminodeoxy hexofuranoses.

Kuszmann et al. (246) used methyl 2,3:5,6-di-anhydro-β-D-mannofuranoside **(1103)** the synthesize a unique 5-amino-furanoside analog of daunosamine (Scheme 192). As described in Sect. 2.1.2 (cf. Scheme 13), the terminal oxirane function of **1103** was selectively cleaved by reduction with sodium borohydride to give an 83% yield of the 6-deoxy-β-D-mannofuranoside **1104**. The 2,3-epoxide function of the 5-O-tosyl derivative **(1105)** of **1104** was then split off by lithium aluminum hydride reduction, and the resulting crude, unstable 2-deoxyfuranoside **1106** was mesylated. The mixed 3,5-di-O-sulfonate **1107** obtained was also rather unstable, and thus it was immediately treated with sodium azide in DMF to produce methyl 3,5-diazido-2,3,5,6-tetradeoxy-α-L-*lyxo*-hexofuranoside **(1108)** in 62% overall yield, based on the mono-epoxide **1105**. Catalytic hydrogenation of **1108**, followed by trifluoroacetylation, gave the protected 5-aminofuranosyl derivative **1109** of L-daunosamine **(12)**, which was used in form of the respective 1-chloro derivative to glycosylate carminomycinone.

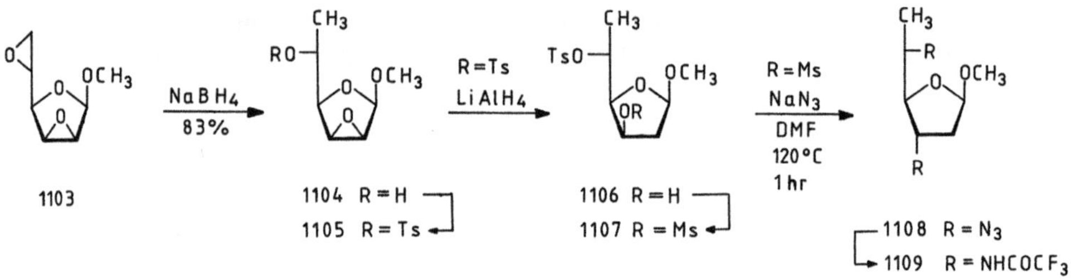

Scheme 192

In recent work by Welch et al. (588) methyl 3-acetamido-2,3,6-trideoxy-α,β-D,L-*arabino*-hexopyranoside, prepared according to the method of Dyong and and Bendlin (507) (cf. Sect. 3.1, Scheme 120), was treated with sulfur tetrafluoride in hydrogen fluoride. The resulting methyl 3-acetamido-5-fluoro-2,3,5,6-tetradeoxy-α,β-D,L-hexofuranoside was said to belong to the *ribo* series, but due to an apparent inconsistency between this configurational assignment and the chemical structure and perspective view of the molecule of the product, as depicted by X-ray crystallography, the steric outcome of this reaction is not clear to us.

4.5 Synthesis of the 6-azido- and 6-amino derivatives of 3-amino-di- and trideoxyhexoses

4.5.1 6-Azido and amino analogs of D-ristosamine

Research efforrs to prepare the 6-azido- and 6-amino derivatives of 3-amino-2,3-dideoxy-D-*ribo*-hexopyranose and to utilize such sugars that carry two nitrogen functions to glycosylate anthracyclinones were made by the groups of Horton (379, 589) and Bognár (590, 591). The starting material used by Horton and associates (379, 589) was methyl 3-acetamido-4-O-benzoyl-6-bromo-2,3,6-trideoxy-α-D-*ribo*-hexopyranoside **(565)**, prepared from methyl α-D-mannopyranoside by the reaction sequence **98** → **100** → **189** → **196** → **565** (cf. Scheme 19, Table 3 entry 2 and Table 7 entry 5 in Sects. 2.1.4, 2.2.3, and 2.7.2, respectively).

Treating **565** with sodium azide in DMF produced the 6-azide **1110** in 88% yield (589) (Scheme 193). This was deacylated **(1111)** and then reduced using lithium aluminum hydride in ether to obtain methyl 3,6-diamino-2,3,6-trideoxy-α-D-*ribo*-hexopyranoside **1113**. This latter compound is a regioisomer of methyl nebrosaminide (tobrosaminide), the α-methyl glycoside of an aminosugar component of the amino-cyclitol antibiotic tobramycin. Both the 4-O-acetyl- **(1114)** and 4-O-*p*-nitrobenzoyl- **(1115)** esters were prepared by the trifluoroacetylation of their amino functions. Acetolysis of **1114** and **1115** gave the α,β-1-O-acetates **1116** and **300**, respectively. The anomeric 1-O-acetyl group of **1116** could not be exchanged for chlorine; the addition of hydrochloric acid in dichloromethane gave a separable three-component mixture of the reducing sugar **1117** (57%), the 3,6-diaminoglycal **1118** (31%), and a bicyclic derivative **1119** [4-O-acetyl-1,6-anhydro-2,3,6-trideoxy-3,6-bis-(trifluoro-acetamido)-β-D-*ribo*-hexopyranose] (10%).

In order to enhance the yield of the diaminoglycal — susceptible to glycosylation of the anthracyclinones — the 1-O-acetyl-4-O-*p*-nitrobenzoyl compound **300** was heated in toluene with silica gel (cf. Sect. 2.2.6, Fig. 7) to produce 1,5-anhydro-2,3,6-trideoxy-3,6-bis-(trifluoroacetamido)-4-O-*p*-nitrobenzoyl-D-*ribo*-hex-1-enitol **304** in 78% yield. In agreement with the experimental findings discussed in Sect. 2.2.6, the glycal **304** was coupled with daunomycinone to form preponderantly an α-linked glycoside in near 80% yield.

In parallel work (590) carried out by the Bognár group, the β-methyl glycoside **1124** of **1113** was synthesized (Scheme 194) from methyl 3-azido-4-O-benzoyl-6-bromo-2,3,6-trideoxy-β-D-*ribo*-hexopyranoside **568** (cf. Sect. 2.7.2, Table 7 entry 8). O-Debenzoylation of **568** according to Zemplén's method gave **1120**, which was treated with sodium azide to obtain the 3,6-diazido-glycoside **1122** in 92% yield. Under the same conditions, the displacement of the bromine atom in the 4-O-benzoate ester **568** by azide gave only a moderate yield (<50%). Reduction of **1122** with lithium aluminum hydride gave methyl 3,6-diamino-2,3,6-trideoxy-β-D-*ribo*-hexopyranoside **(1124)**, a novel structural isomer of methyl nebrosaminide. The 4-O-*p*-nitrobenzoyl derivatives **1121** and **1123** of the 3-azido-6-bromo- **(1120)** and 3,6-diazido- **(1122)** sugars, as well as that of methyl 3-azido-2,3,6-trideoxy-α-L-*ribo*-hexopyranoside (**127**, Sect. 2.2.1, Scheme 26) have been used by Sztaricskai et al. (591, 592) to prepare semisynthetic azido-hexopyranosyl analogs of anthracycline antibiotics.

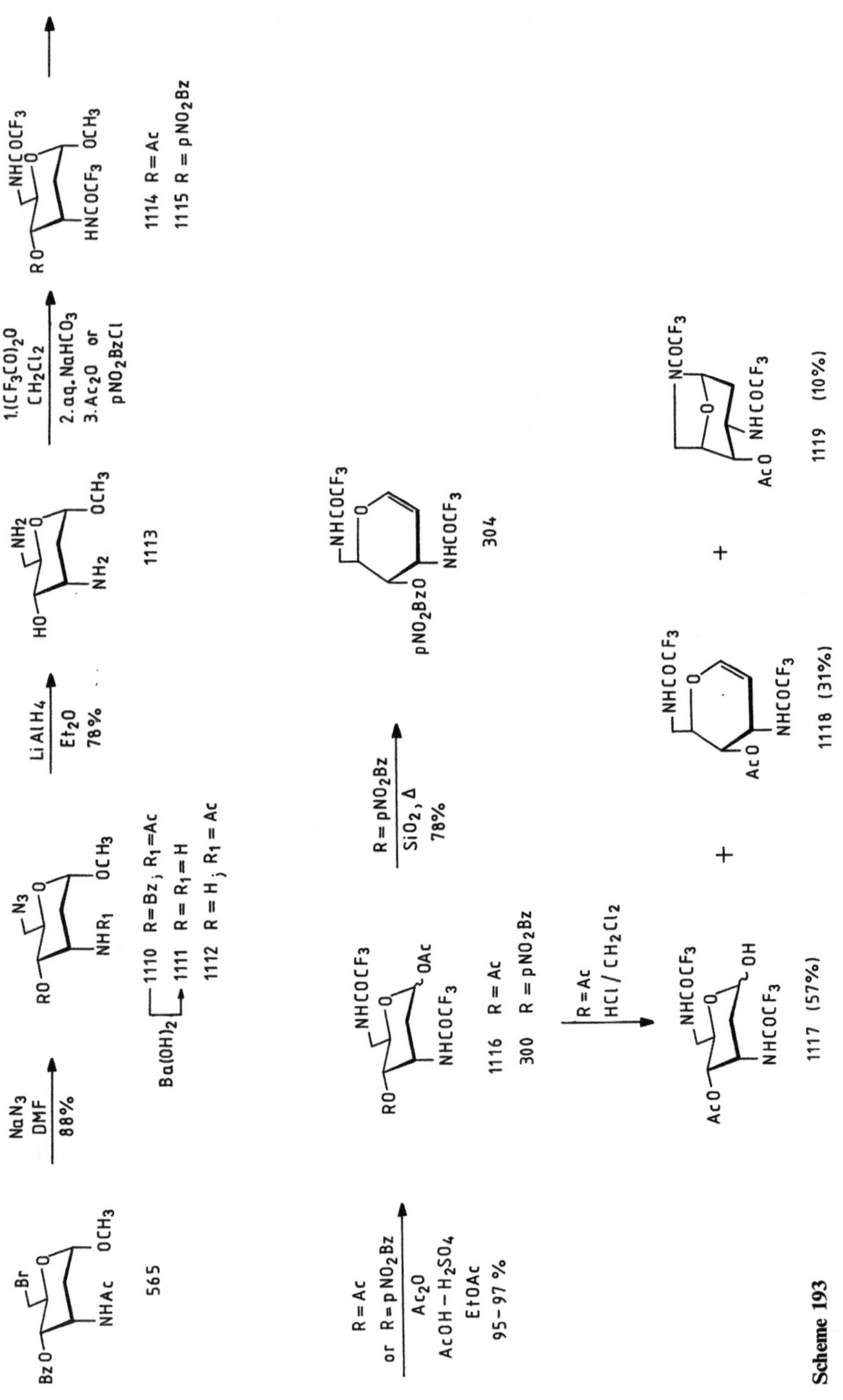

Scheme 193

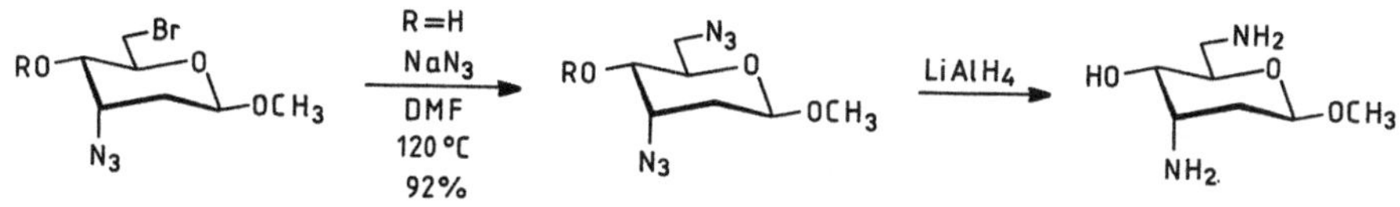

568 R = Bz
1120 R = H
1121 R = pNO$_2$Bz

1122 R = H
1123 R = pNO$_2$Bz

1124

Scheme 194

4.5.2 Approaches to 3,6-diaminohexose precursors for the synthesis of antibiotic negamycin

Synthetic routes that allow the preparation of 3(R),6-diacetamido-5(R)-hydroxyhexonolactone (**1133**, Scheme 195), which corresponds to the δ-hydroxy-β-lysine building unit of the antibiotic negamycin **(1125)** (593), and of the corresponding (3s,5s) antipode were elaborated by Umezawa and his associates (372) in 1972. In order to obtain the lactone **1133** the glycal **1126** — derived from D-galacturonic acid —

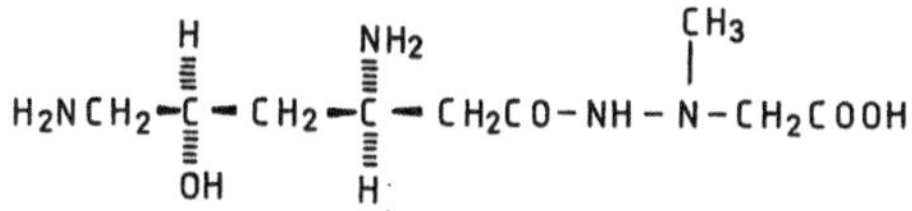

1125

was iodomethoxylated (cf. Sect. 2.1.1) and then dehalogenated by hydrogenolysis and methyl (methyl 3,4-di-O-acetyl-2-deoxy-α-D-*lyxo*-hexopyranosid) -uronate **(1127)** was isolated (Scheme 195).

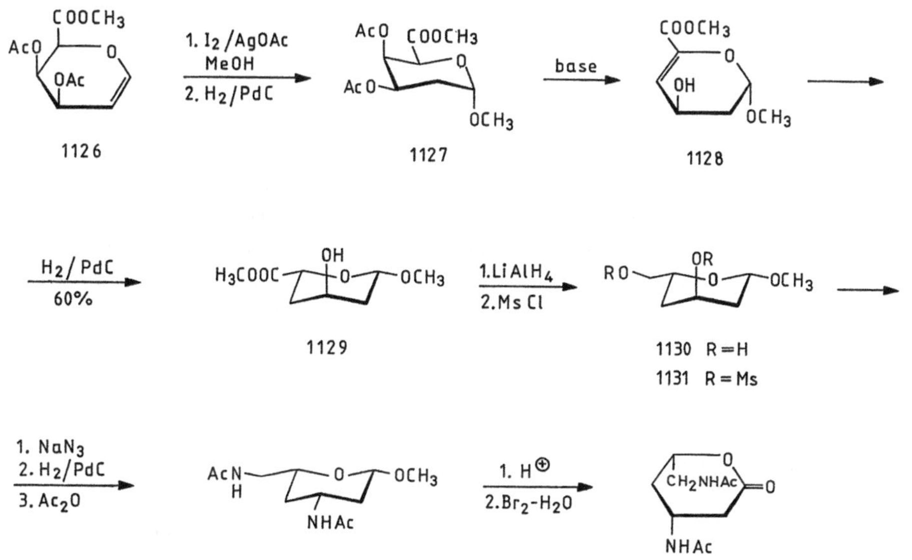

1126
1127
1128
1129
1130 R = H
1131 R = Ms
1132
1133

Scheme 195

Base-catalyzed β-elimination off the C-4 acetoxy substituent of **1127** gave the (hex-4-enopyranosid)-uronate **1128**, the double bond of which was saturated upon catalytic hydrogenation. As the major product, methyl (methyl 2,4-dideoxy-β-L-*erythro*-hexopyranosid)-uronate **(1129)** was obtained in 60% yield after chromatographic separation from the C-5 diastereoisomer, also produced in smaller quantities. Reduction of the carbonyl group of **1129** with lithium aluminum hydride yielded the diol **1130**, which was mesylated. The resulting 3,6-di-O-mesylate **1131** was treated with sodium azide. Catalytic hydrogenation over palladium-on-carbon catalyst and acetylation then gave methyl 3,6-diacetamido-2,3,4,6-tetradeoxy-β-L-*threo*-hexopyranoside **(1132)**. The free sugar, prepared by mild acid hydrolysis, was oxidized into 3(R),6-diacetamido-5(R)-hydroxyhexonolactone **(1133)** using aqueous bromine.

To synthesize (372) the enantiomer **(1137)** of the lactone **1133**, methyl 3-acetamido-4,6-di-O-acetyl-2,3-dideoxy-α-D-*arabino*-hexopyranoside **(518)** — prepared from the aminoglycal **296** as shown in Sect. 2.6.1 (Scheme 93) — was used. Zemplén O-deacylation followed by selective tosylation of the primary hydroxyl group gave the 6-O-tosylate **542** (Scheme 196). Displacement with sodium azide in DMF yielded the

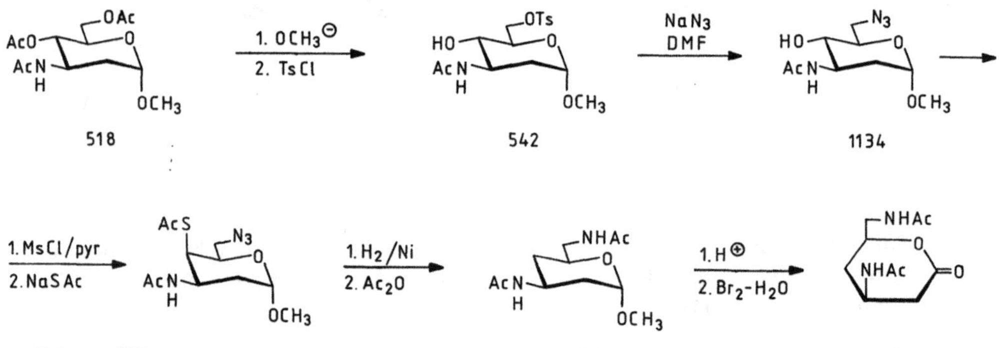

Scheme 196

C-6 azide **1134**, which was mesylated and then treated with sodium thioacetate to give the 4-S-acetyl-6-azidoglycoside **1135**. Simultaneous reduction of the two reducible functions of this latter over Raney nickel followed by acetylation of the crude product provided methyl 3,6-diacetamido-2,3,4,6-tetradeoxy-α-D-*threo*-hexopyranoside **1136**. Acid hydrolysis and oxidation with aqueous bromine then yielded the (3S,5S)-lactone **1137**, the optical antipode of the natural compound **1133**. The conversion of the lactones **1133** and **1137** into the N-hydroxysuccinimide ester of the respective δ-hydroxy-β-lysines, and the acetylation of N^1-methyl-hydrazino acetic acid with these "active" esters led to the preparation of negamycin **(1125)** and its enantiomer, respectively, the latter being ca. 30 times less active than the natural antibiotic.

Chmielewski et al. (581) elaborated three routes to the derivatives of the racemic 3,6-diacetamido-5-hydroxy-*threo*-hexonolactone **±1133**. In the most effective of these procedures (Scheme 197), the mixture of the 3-azido-uronates **±1033**, prepared as shown in Sect. 4.3.1 (Scheme 177) was treated (after equilibration with sodium methoxide in methanol) with ammonium hydroxide to obtain the uronic amide **±1138**. After reduction of this latter with lithium aluminum hydride and acetylation

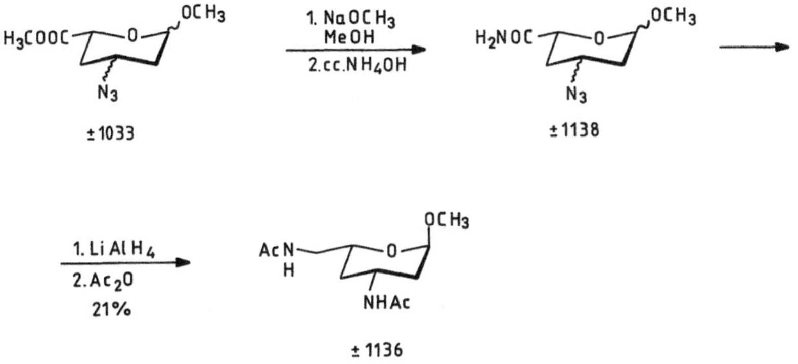

Scheme 197

of the diamino compounds produced, pure methyl 3,6-diacetamido-2,3,4,6-tetradeoxy-α-D,L-*threo*-hexopyranoside (±**1136**) was crystallized from the reaction mixture in 21% yield.

4.6 Synthesis of the uronic acid derivatives of 3-amino-di- and trideoxyhexoses

The applications of methyl 3-acetamido-6-azido-2,3,6-trideoxy-α-D-*ribo*-hexopyranoside (**1112**) and its 4-O-acetyl derivative **1139** in the synthesis of uronic acid derivatives of methyl α-D-ristosaminide was reported by Horton et al. (589) in 1979 (Scheme 198). Photolysis of the azide **1112** (involving the formation and rearrangement of an intermediate nitrene) gave the oligomeric form of the imine **1140**. Unambigous evidence for the structure of **1140** was obtained by converting it with 1,2-bis-(anilino)-ethane into the C-5 diphenylimidazolidine derivative **1141**. Mild acid hydrolysis of **1140** gave rise to methyl 3-acetamido-2,3-dideoxy-6-*aldehydo*-α-D-*ribo*-hexodialdo-1,5-pyranoside (**1142**) present in an intermolecularly associated form. The presence of the aldehyde function in **1142** was shown by converting it into the crystalline imidazolidine **1141**. Further chemical evidence for the hexodialdopyranoside structure of **1142** was given by oxidation with bromine-water and subsequent esterification with methyl iodide to produce methyl (methyl 3-acetamido-2,3-dideoxy-α-D-*ribo*-hexopyranosid)-uronate **1143**, a uronic acid derivative of D-ristosamine (**20**).

Irradiation of a benzene solution of the 4-O-acetyl-6-azido sugar **1139** gave the 4-O-acetyl derivative (**1144**) of **1140**, as demonstrated by its subsequent conversion into the C-5 imidazolidine **1141**. When an aqueous solution of **1144** was boiled under reflux, the α,β-unsaturated aldehyde was isolated as the sole product (in 57% yield, based on **1139**) as a result of β-elimination. The reaction of **1145** with 1,2- bis-(anilino)-ethane gave the dehydrated analog **1146** of the 1,3-diphenylimidazolidine **1141**. The oxime derivative **1147** of the aldehyde **1145**, a potential precursor of a regioisomer of the unsaturated diaminosugar (2,6-diamino-2,3,4,6-tetradeoxy-D-*glycero*-hex-4-enopyranose) present in the aminocyclitol antibiotic sisomycin (594),

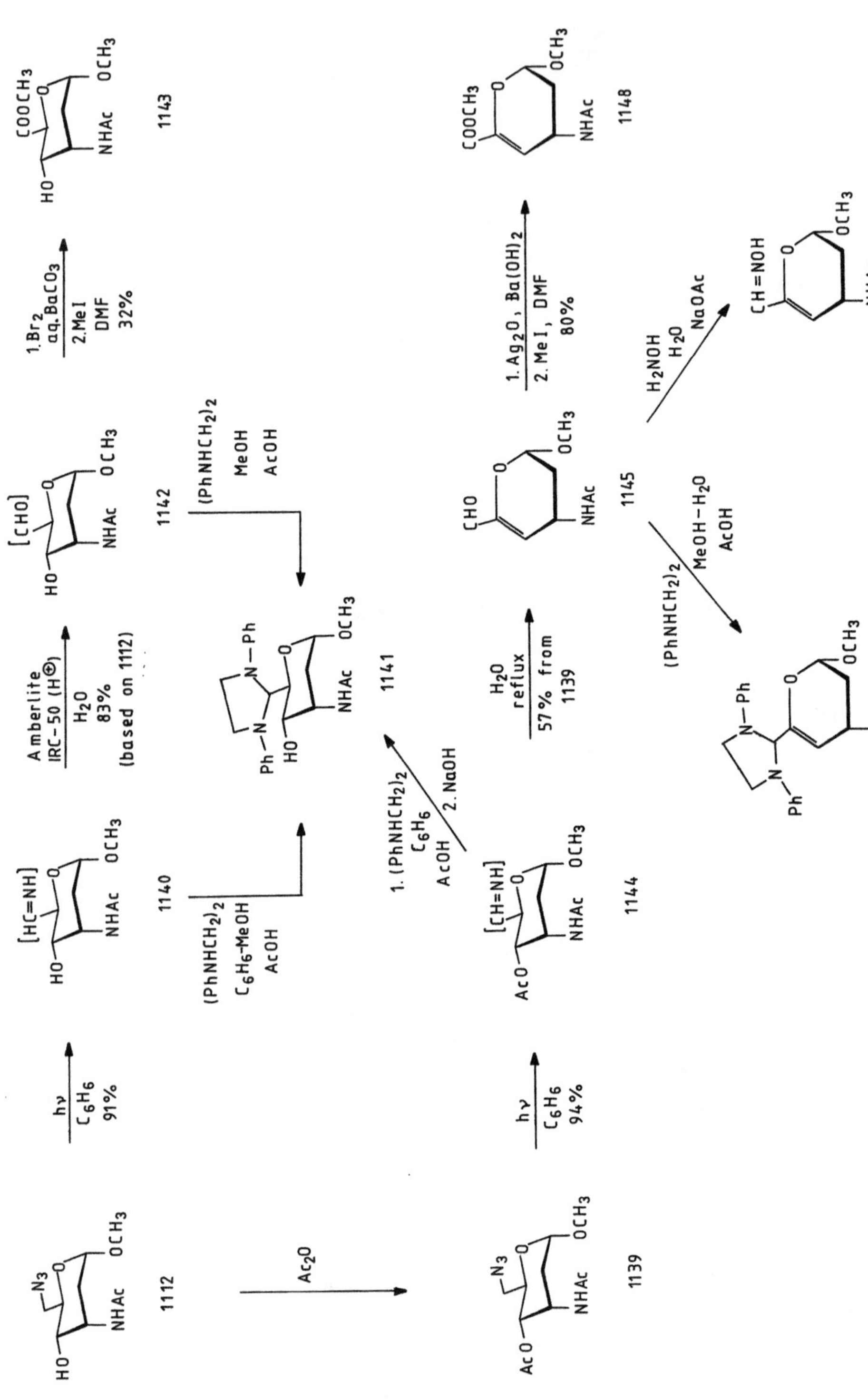
1112
Ac2O
1139
hν
C6H6
91%
hν
C6H6
94%
1140
1144
(PhNHCH2)2
C6H6-MeOH
AcOH
1. (PhNHCH2)2
C6H6
AcOH
2. NaOH
1141
Amberlite
IRC-50 (H⊕)
H2O
83%
(based on 1112)
H2O
reflux
57% from
1139
1142
(PhNHCH2)2
MeOH
AcOH
1145
(PhNHCH2)2
MeOH-H2O
AcOH
1146
H2NOH
H2O
NaOAc
1147
1.Br2
aq.BaCO3
2.MeI
DMF
32%
1143
1. Ag2O, Ba(OH)2
2. MeI, DMF
80%
1148

Scheme 198

was prepared by treatment with an aqueous solution of hydroxylamine. Oxidation of the hex-4-enodialdopyranoside **1145** with silver oxide, followed by esterification with methyl iodide, produced methyl (methyl 3-acetamido-2,3,4-trideoxy-α-D-*glycero*-hex-4-enopyranosid)-uronate **(1148)** in 80% yield.

4.7 Conversion of 3-amino- and 3-azido-2,3,6-trideoxyhexose derivatives into carbocyclic compounds

The application of Ferrier's mercury salt-catalyzed ring-transformation reaction (595) for 3-acylamino- and 3-azidodeoxyhex-5-enopyranosides (447, 448, 450) led to deoxyinosose derivatives carrying a nitrogen substituent at carbon C-3. The role of mercury in the reaction mechanism (596, 597) and in the stereochemical outcome (448) of the *trans* relationship between the C-3 substituent and the newly generated hydroxyl group at C-5 of the resulting cyclohexanone molecule, has been discussed in detail. Suitable chemical transformations of the carbonyl function of the highly functionalized cyclohexanones available by the Ferrier reaction provide access to 1,3-diaminocyclitol-type aglycones of the aminocyclitol antibiotics and to the carbocyclic analogs of 3-amino-2,3,6-trideoxyhexoses.

The reaction of methyl 3-azido-4-O-benzoyl-2,3,6-trideoxy-β-D-*erythro*-hex-5-enopyranoside (**618**, Sect. 2.7.4, Table 11 entry 7) with mercury chloride in hot aqueous acetone gave (447, 448) the cyclohexenone **582** in 78% yield by the elimination of the C-3 azide and the C-4 hydrogen functions of **618**, being *trans* to each other, presumably in form of hydrazoic acid. In contrast, in the case of the C-3 epimeric azidodeoxy α-D-*treo*-exo-methylene compound **610** (Sect. 2.7.4, Table 11 entry 1) no such elimination occured under equal conditions and an 82% yield of (2S,3R,5R)-2-benzoyloxy-3-azido-5-hydroxycylohexanone **(578)** was obtained (Scheme 199).

The Ferrier-reaction of the 3-acylamino-hex-5-enopyranosides **619** and **620** of the β-D-*erythro* series (Sect. 2.7.4, Table 11 entries 8 and 9) also proceeded (448)

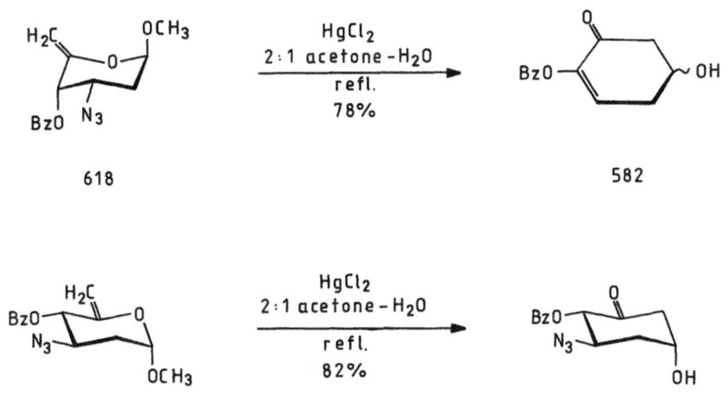

618 582 610 578

Scheme 199

normally and without an elimination reaction. This allowed the isolation of (2s,3s,5s)-2-benzoyloxy-3-benzoylamino-5-hydroxycyclohexanone **(583)** and the corresponding 3-trifluoroacetamido compound **584** in 81% and 86% yields, respectively (Scheme 200). As shown by high-resolution ^{1}H-NMR studies, both **583** and **584** exist in the form of a ca. 1:1 equilibrium mixture of the alternative $^2C_5$ and $^5C_2$ conformers in solutions for various solvents.

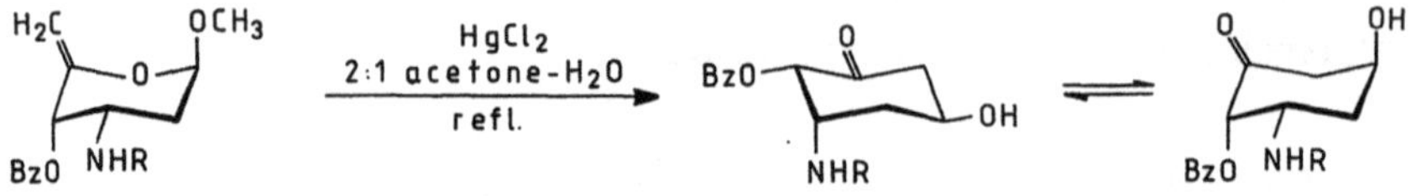

619 R = Bz
620 R = COCF3

583 R = Bz (81%)
584 R = COCF3 (86%)

Scheme 200

The first practically useful application of the Ferrier ring-transformation reaction came from the closely related research of the Lukacs group (450) involving the conversion of aminocyclohexanone **1149** into a carbocyclic L-daunosamine analog **581** (Scheme 201). The transformation of methyl 3-benzamido-4-O-benzyl-2,3,6-trideoxy-α-D-*erythro*-hex-5-enopyranoside (**616**, Sect. 2.7.4, Table 11 entry 5) into **1149** was carried out in acetone solution in the presence of dilute sulfuric acid and mercury(II)sulfate (a modification of the original reaction conditions) (595). The

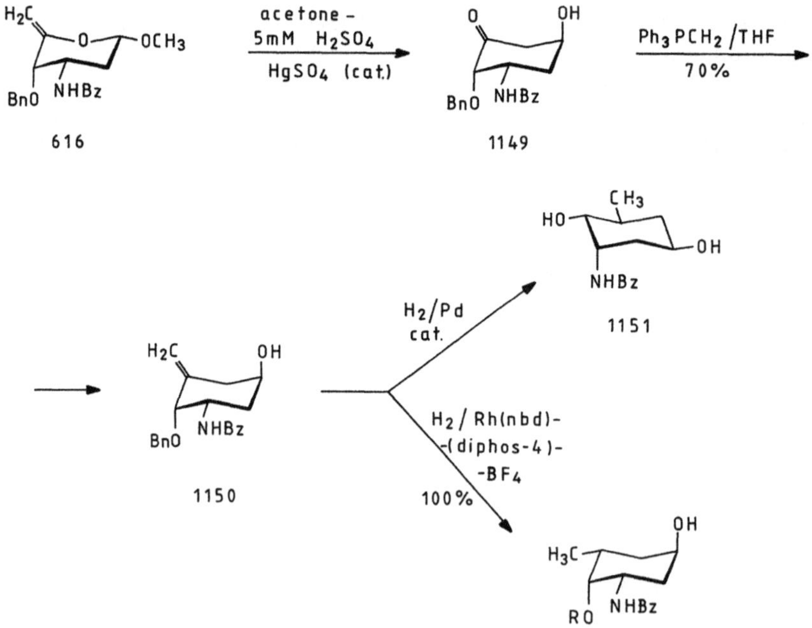

616 1149 1150 1151

581 R = H
1152 R = Bn

Scheme 201

ketone **1149** was treated with methylenetriphenylphosphorane to furnish the methylene compound **1150** in 70% yield.

The hydrogenation of **1150** under heterogeneous catalysis (Pd/C, Pd/$BaSO_4$, $PdCl_2$/C) and using a variety of conditions led to the *pseudo*-D-ristosamine derivative **1151**; the desired L-stereoisomer **581** was produced only in small quantities. Contrarily, hydroxyl-group-directed homogeneous hydrogenation with transition metal (rhodium and iridium) complex catalysts readily produced N-benzoyl-O-benzyl-*pseudo*-L-daunosamine **(1152)**. A complete stereoselectivity could be achieved when the rhodium complex [Rh(nbd) (diphos-4)]BF_4 was employed. Removal of the O-benzyl group of **1152** by catalytic hydrogenation then gave the target carbocyclic compound N-benzoyl-*pseudo*-α-L-daunosamine **(581)**.

The carbocyclic analogs of D-glucosamine and L-idosamine have been synthesized most recently by the Barton group (598) by applying the Ferrier reaction to D-glucosamine substrates.

5. Concluding Remarks

At the beginning of the 1980s we had been engaged in aminodeoxy sugar chemistry for almost ten years. We suspected that no more significant contributions would be made to the synthesis of daunosamine-type aminodeoxy hexoses in the ensuing years. By 1980 efficient procedures had been elaborated that used mainly carbohydrate starting materials to produce each of the stereoisomeric hexoses belonging to this series. Naturally, the derivatives of L-daunosamine were the most frequent targets of these syntheses along with L-acosamine, which is a component of the less cardiotoxic semisynthetic 4′-epi-anthracycline antibiotics. However, rapid development of numerous additional ingenious short and effective methods starting with both carbohydrates and non-sugar precursors, together with the expanding area of utilization of 3-aminodeoxyhexoses as "chiral templates" in the syntheses of natural products, disproved this assumption. Moreover, intense research on entirely new subjects, e.g., 2-halogeno derivatives of daunosamine, utilization of the allylic-azide rearrangement reaction, simple chiral syntheses involving carbon-carbon bond formation, and asymmetric introduction of the required functionalities, etc, has added new methods to the chemistry of aminodeoxy sugars. Also, previously less widely distributed procedures, such as C-5 configurational interchange via hex-5-enopyranosides, stereospecific glycosylations with glycals, introduction of a $CH_3-\overset{|}{\underset{|}{C}}-NH_2$ branching into carbohydrates, etc, have emerged as routine operations because of the research on daunosamine-type aminodeoxy sugars.

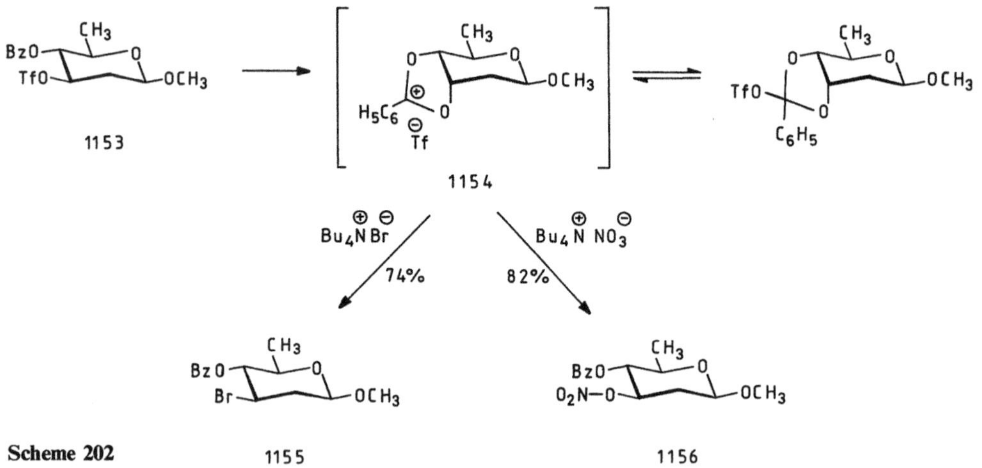

Scheme 202

We believe that there are still unexplored possibilities in the field of 3-amino-polydeoxy hexoses, as shown, for example, by the recent work of Binkley et al. (599, 600) (Scheme 202). These authors have recognized that the benzoxonium triflate **1154**, a rearrangement product of the 3-O-triflate **1153**, represents a cation similar to the intermediate *C* of the Hanessian-Hullar reaction (cf. Sect. 2.7.2, Scheme 111) but devoid of an effective nucleophile such as the bromide anion in *C*. Thus, as shown by the reported examples of the transformation of **1154** into the C-3 bromo- **(1155)** and O-nitro- **(1156)** sugars, a wide variety of displacement reactions can be executed, involving conversion into 3-amino derivatives, with added nucleophiles. Further research on this subject is expected to achieve such transformations.

The growing interest in elaborating chiral total synthesis routes from non-sugar precursors to daunosamine-type aminodeoxy hexoses is clearly demonstrated by fourteen related papers that appeared between 1985 and the beginning of 1987 (ref. Chap. 3). In this field highly diastereoselective intramolecular *hetero*-Diels-Alder cycloaddition reactions might also be considered as promising new approaches to such sugars.

The treasury of fantasy in the imaginations of synthetic organic chemists is almost limitless. Consequently, the biological importance and wide practical utility of 3-aminodeoxy hexoses as "chiral templates" in natural products syntheses will most likely encourage scientists to continue related research in the years to come. It is hoped that this volume will contribute to the forthcoming challenges.

Appendix

In this section the hitherto prepared functionalized derivatives of 3-amino-2,3-dideoxyhexopyranoses (Table 19), 3-amino-2,3,6-trideoxyhexopyranoses (Table 20) and the respective 1,5-anhydro-3-amino-hex-1-enitol (glycal) analogs (Table 21) are collected. Compounds without the satisfactory physical characterization that is essential for their identification are not included. It was intended to provide readily available references to primary journals, but patent literature was also involved whenever necessary.

Because we lack a computerized data-base system, we ask for your indulgence and tolerance in the event of any omission. At the same time we hope that despite some possible shortcomings concerning the derivatives included the collection of more than 400 substituted analogs of the foregoing sugars will be a useful source of information and will save the reader a laborious literature search.

Table 19. Derivatives of 3-amino-2,3-dideoxyhexopyranoses

L-Arabino:

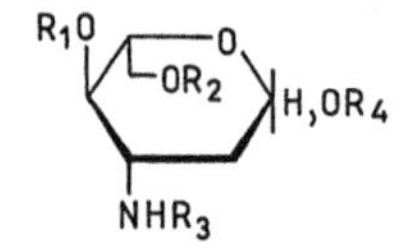

R_1	R_2	R_3	R_4	Refs.
H	H	H	H	235, 236
H	H	TFAc	H	235, 236
H	H	H	Me(α)	235, 236
H	H	TFAc	Me(α)	236, 601
H	Tr	TFAc	Me(α)	236, 601
*p*NBz	*p*NBz	TFAc	Me(α)	236
benzylidene		H	Me(α)	236
benzylidene		TFAc	Me(α)	236

D-Arabino

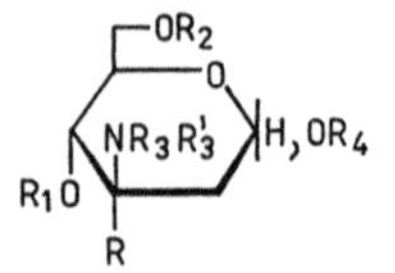

R	R_1	R_2	R_3	R'_3	R_4	Refs.
H	H	H	H	H	H	310, 311
H	H	H	Ac	H	H	383

(Continued)

Table 19 (continued)

R	R_1	R_2	R_3	R_3'	R_4	Refs.
H	H	H	H	H	Me(α)	312
H	H	H	H	H	Me(β)	310, 311
H	H	H	H	H	Me(α, β)	426
H	H	H	Ac	H	Me(α)	99, 229, 312, 372, 383, 426
H	H	H	Ac	H	Me(β)	310
H	H	Ts	Ac	H	Me(α)	229, 372
H	H	Ts	Ac	H	Me(β)	318
H	Ac	Ac	Ac	H	Me(α)	99–101
H	Ac	Ac	Ac	H	Me(β)	310
H	Ac	Ac	Ac	H	Et(α)	99–101
H	Ac	Ac	Ac	H	Ip(α)	99–101
H	Ac	Ac	Ac	H	c-Hex(α)	99–101, 360
H	Ac	Ac	Ac	H	t-Bu(α)	99–101
H	Ac	Ts	Ac	H	Me(β)	318
H	Ms	Ms	Ac	H	Me(β)	310
H	benzylidene		H	H	Me(α)	178, 228, 237
H	benzylidene		H	H	Me(β)	178
H	benzylidene		CHO	H	Me(α)	228
H	benzylidene		Ac	H	Me(α)	172, 227–229, 276, 292, 383, 602, 603
H	benzylidene		Cbz	H	Me(α)	383
H	benzylidene		Cbz	Me	Me(α)	383
H	benzylidene		Ts	H	Me(α)	233
Me	benzylidene		H	H	Me(α)	177, 180
Me	benzylidene		Ac	H	Me(α)	177, 180, 181, 418
Me	benzylidene		Ts	H	Me(α)	354
CHO	benzylidene		Ts	H	Me(α)	354
CHO	benzylidene		TCAc	H	Me(α)	354
CH_2OH	benzylidene		Ts	H	Me(α)	354
CH_2OTs	benzylidene		Ts	H	Me(α)	354
CH=N—NHTs	benzylidene		Ts	H	Me(α)	354
$CH{=}CH_2$	benzylidene		H	H	Me(α)	354
$CH{=}CH_2$	benzylidene		TCAc	H	Me(α)	354
$CH{=}CH_2$	benzylidene		Ts	H	Me(α)	354

L-Lyxo: 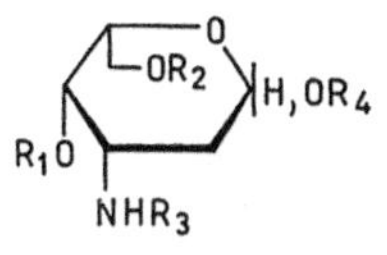

R_1	R_2	R_3	R_4	Refs.
H	H	H	H	601
H	H	TFAc	H	601
H	Bn	Cbz	H	548
H	H	TFAc	Me(α)	236
H	Tr	TFAc	Me(α)	236, 601
*p*NBz	*p*NBz	TFAc	*p*NBz(α, β)	601

(Continued)

Table 19 (continued)

D-Lyxo:

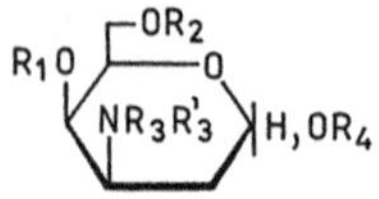

R_1	R_2	R_3	R'_3	R_4	Refs.
H	H	H	H	H	317
H	H	H	H	Me(β)	317
H	H	CHO	H	Me(α)	228
H	H	Ac	H	Me(β)	317
H	H	Bz	H	Me(β)	317
Ac	Ac	Ac	H	Me(α)	312
Ac	Ac	Ac	H	Me(β)	317
Ms	Ms	Ac	H	Me(β)	317
Ac	Ts	Ac	H	Me(β)	318
benzylidene		CHO	H	Me(α)	228
benzylidene		Me	Me	Me(α)	228

L-Ribo:

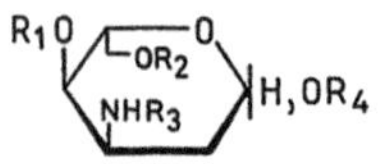

R_1	R_2	R_3	R_4	Refs.
*p*NBz	*p*NBz	TFAc	H	604
H	H	H	Me(α)	236
H	H	TFAc	Me(α)	184, 236, 604
*p*NBz	*p*NBz	TFAc	Me(α)	236, 604
Ms	Ms	TFAc	Me(α)	236, 605
*p*NBz	*p*NBz	TFAc	*p*NBz(α, β)	604
benzylidene		H	Me(α)	236

D-Ribo:

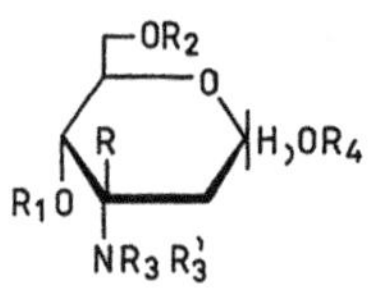

R	R_1	R_2	R_3	R'_3	R_4	Refs.
H	H	H	H	H	Me(α)	228, 388
H	H	H	Ac	H	Me(α)	276, 388
H	Ac	Ac	Ac	H	c-Hex(α)	360
H	Ts	Ts	Me	Me	Me(α)	229
H	Ac	TBDMS	Ac	H	Bn(α)	351
H	benzylidene		H	H	Me(α)	178, 228, 237, 388
H	benzylidene		H	H	Me(β)	178
H	benzylidene		CHO	H	Me(α)	228
H	benzylidene		Ac	H	Me(α)	172, 276, 424, 602, 603
H	benzylidene		TFAc	H	Me(α)	179
H	benzylidene		Bz	H	Me(β)	448

(Continued)

Table 19 (continued)

R	R_1	R_2	R_3	R_3'	R_4	Refs.
H	benzylidene		TFAc	H	Me(β)	448
H	benzylidene		Me	H	Me(α)	228
H	benzylidene		salicylidene		Me(α)	228
H	cyclohexylidene		Ts	H	Me(α)	277
Me	benzylidene		H	H	Me(α)	177
Me	benzylidene		Ac	H	Me(α)	177, 182, 414
Me	benzylidene		TFAc	H	Me(α)	182, 380, 414
$CH_2{=}CH$	benzylidene		TCAc	H	Me(α)	354

L-Xylo: 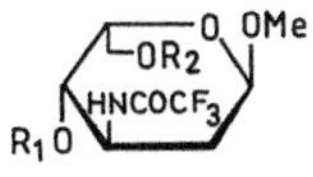

R_1	R_2	Refs.
H	H	236, 605
Bz	Bz	236, 605
*p*NBz	*p*NBz	236, 605

D-Xylo: 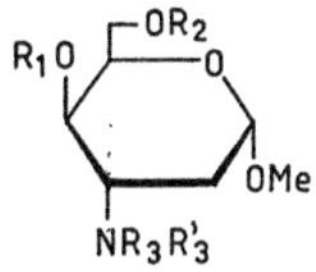

R_1	R_2	R_3	R_3'	Refs.
H	H	CHO	H	228
H	Ts	CHO	H	228
benzylidene		CHO	H	228
benzylidene		Me	Me	228

Table 20. Derivatives of 3-amino-2,3,6-trideoxyhexopyranoses

L-Arabino:

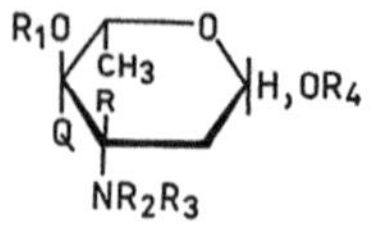

R	Q	R_1	R_2	R_3	R_4	Refs.
H	H	H	H	H	H	265
H	H	H	Ac	H	H	187, 507; DL: 507, 510, 588, 606
H	H	H	TFAc	H	H	189, 190, 498, 502, 607
H	H	H	Bz	H	H	514, 544, 545
H	H	Ac	Ac	H	H	554
H	H	Me	TFAc	H	H	608
H	H	H	H	H	Me(α)	155, 264–266, 496, 497
H	H	H	H	H	Me(β)	264
H	H	H	H	H	Me(α, β)	496
H	H	H	Ac	H	Me(α)	187, 265, 266; DL: 588
H	H	H	TFAc	H	Me(α)	189–191, 496, 497, 502
H	H	H	Bz	H	Me(α)	266
H	H	H	MOC	H	Me(α)	496, 497
H	H	H	Me	H	Me(α)	264
H	H	H	Me	H	Me(β)	264
H	H	H	Me	Me	Me(α)	264
H	H	H	Me	Me	Me(β)	264
H	H	Me	Ac	H	Me(α)	265
H	H	MOM	Ac	H	Me(α)	187
H	H	Me	TFAc	H	Me(α)	608
H	H	Ac	Ac	H	Me(α)	187–190, 265, 359, 554
H	H	Ac	Ac	H	c-Hex(α)	360
H	H	*p*NBz	TFAc	H	Me(α)	362
H	H	*p*NBz	TFAc	H	Me(β)	362
H	H	H	MOC	$CH(CH_3)Ph$	Me(α)	496, 497
H	H	Ms	MOC	H	Me(α)	496, 497
H	H	Ms	Ac	H	Me(α)	123, 187
H	H	*p*NBz	TFAc	H	c-Hex(α)	362
H	H	*p*NBz	TFAc	H	c-Hex(β)	362
H	H	TFAc	TFAc	H	TFAc(α, β)	607
H	H	Me	TFAc	H	*p*NBz(α, β)	608, 609
Me	H	H	Bz	H	H	499, 500
Me	H	H	H	H	Me(β)	326
Me	H	H	Ac	H	Me(α)	181, 405
Me	H	H	Ac	H	Me(β)	181, 405
Me	H	Ac	Ac	H	Me(α)	181, 183, 405
Me	H	Ac	Ac	H	Me(β)	181, 183, 405
Me	H	Ac	Ac	H	Me(α, β)	374
Me	H	Me	Ac	H	Me(α)	180, 181, 186, 405, 610
Me	H	MOM	Ac	H	Me(α)	183
Me	H	Me	Ac	H	Me(β)	378, 405, 610
Me	H	Me	Ac	H	Me(α, β)	611
Me	H	Me	Ac	Me	Me(β)	405
Me	H	Ms	Bz	H	Me(α)	499, 500
H	Me	Me	TFAc	H	H	583

(Continued)

Table 20 (continued)

R	Q	R_1	R_2	R_3	R_4	Refs.
H	Me	H	TFAc	H	Me(α)	582, 583
H	Me	Me	TFAc	H	Me(α)	582, 583
H	Me	Me	TFAc	H	*p*NBz(α, β)	583

D-Arabino:

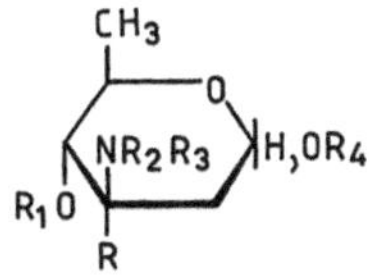

R	R_1	R_2	R_3	R_4	Refs.
H	H	H	H	H	174, 314, 318
H	H	Ac	H	H	174, 318
H	H	TFAc	H	H	174, 379
H	H	Bz	H	H	174
H	H	Me	Me	H	314
H	*p*NBz	TFAc	H	H	363
H	H	H	H	Me(α)	270, 314, 363
H	H	Ac	H	Me(α)	12, 174, 229, 231
H	H	Ac	H	Me(β)	318
H	H	TFAc	H	Me(α)	174, 363, 379
H	H	Bz	H	Me(α)	174
H	H	Cbz	H	Me(α)	234, 270
H	H	Me	Me	Me(α)	55
H	H	Me	Me	Me(β)	55
H	H	phthalimido		Me(α)	270
H	Ac	Ac	H	Me(α)	12, 174, 231, 232
H	*p*NBz	TFAc	H	Me(α)	363
H	Ac	Ac	H	Me(β)	12
H	Ac	Ac	H	Me(α, β)	318
H	Bz	Ac	H	Me(α)	174, 232
H	TFAc	TFAc	H	Me(α)	379
H	Ms	Ac	H	Me(β)	318
H	*p*NBz	TFAc	H	Ac(α)	363
H	*p*NBz	TFAc	H	Ac(β)	363
H	Ac	Me	Me	Ac(α)	14
H	Ac	Me	Me	Ac(α, β)	14
H	*p*NBz	TFAc	H	*p*NBz(α)	379
H	TFAc	TFAc	H	TFAc(α)	379
Me	Me	H	H	Me(α)	180
Me	H	Ac	H	Me(α)	180, 181, 418
Me	Me	Ac	H	Me(α)	180, 181, 186, 418, 610
Me	Me	Ac	H	Me(β)	180, 186, 610
Me	Bz	Ac	H	Me(α)	180

(Continued)

Table 20 (continued)

L-Lyxo Daunosamine derivatives

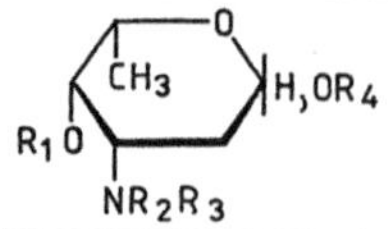

R_1	R_2	R_3	R_4	Refs.
H	H	H	H	135, 155, 172, 210, 248, 373, 550 555, 556, 616; DL: 347, 517, 549
H	Ac	H	H	21, 172, 538
H	TFAc	H	H	178, 179, 190, 191, 498, 538, 564, 609, 617–619
H	TCAc	H	H	DL: 511
H	Bz	H	H	19, 21, 172, 518, 536, 539, 540 544, 545, 551; DL: 509, 517, 518
H	Ts	H	H	527
*p*NBz	TFAc	H	H	565, 620
Me	TFAc	H	H	608
H	H	H	Me(α)	19, 21, 123, 155, 348, 496, 497; DL: 537
H	Ac	H	Me(α)	21, 123, 187, 346, 349
H	Ac	H	Me(β)	172, 373
H	TFAc	H	Me(α)	190, 191, 238, 502
H	TFAc	H	Me(β)	179, 615, 616
H	TFAc	H	Me(α, β)	619
H	TCAc	H	Me(α, β)	DL: 511
H	Bz	H	Me(α)	DL: 347
H	Bz	H	Me(β)	172
H	Bz	H	Me(α, β)	619
H	MOC	H	Me(α)	496, 497
H	Ts	H	Me(α)	527, 528
H	Ac	H	Bn(α)	351
H	TFAc	H	Bn(α)	613
H	TFAc	H	Bn(α, β)	620
H	DCAc	H	Bn(α)	351
H	Me	Me	Me(β)	238
H	Me	Me	Ac(α)	33
H	Me	Me	Ac(β)	33
H	CON(Me)NO	H	Me(α)	571
Ac	Ac	H	Me(α)	19, 21, 123, 346, 348, 349, 528, 529, 535; DL: 534
Ac	Ac	H	Me(β)	172, 543; DL: 543
Ac	Ac	H	Bn(β)	351
Ac	TFAc	H	Me(β)	238
TCAc	TCAc	H	Me(α)	511; DL: 511
Bz	Bz	H	Me(α)	DL: 347
Bz	Bz	H	Me(β)	395
Ms	TFAc	H	Me(α)	503, 621
Me	TFAc	H	Me(α)	608
*p*NBz	TFAc	H	Me(α)	565
$pBrC_6H_4SO_2$	TFAc	H	Me(α)	575
*p*NBz	TFAc	H	Et(α)	622
*p*NBz	TFAc	H	Et(α, β)	620
Ac	Ac	H	Ip(α, β)	DL: 526

(Continued)

Table 20 (continued)

R_1	R_2	R_3	R_4	Refs.
*p*NBz	TFAc	H	Bn(α)	622
*p*NBz	TFAc	H	Bn(α, β)	620
Ac	Ac	H	Ac(β)	21
Ac	Ac	H	Ac(α, β)	19, 21; DL: 515
*p*NBz	TFAc	H	Ac(α, β)	620
TFAc	TFAc	H	TFAc(α, β)	607, 612, 614
*p*NBz	TFAc	H	*p*NBz(α)	617
*p*NBz	TFAc	H	*p*NBz(α, β)	618
Me	TFAc	H	*p*NBz(α, β)	608, 609

L-Lyxo — Branched-chain derivatives 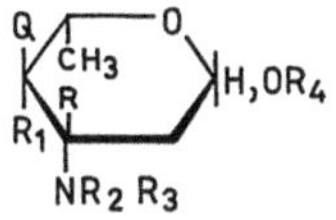

R	Q	R_1	R_2	R_3	R_4	Refs.
Me	H	OH	Bz	H	H	499, 500
Me	H	OH	Ts	H	H	DL: 533
Me	H	OH	Me	Me	H	15–17
Me	H	OBz	TFAc	H	H	380
Me	H	OH	H	H	Me(α)	47, 406, 623
Me	H	OH	Ac	H	Me(α)	406
Me	H	OH	Ac	H	Me(β)	177
Me	H	OH	TCAc	H	Me(α)	DL: 511
Me	H	OH	Bz	H	Et(α)	47, 623
Me	H	OH	Ts	H	$(CH_2)OH$(α, β)	DL: 533
Me	H	OAc	Ac	H	Me(α)	406
Me	H	OAc	Ac	H	Me(α, β)	558
Me	H	OMe	Ac	H	Me(α)	194
Me	H	OBz	Ac	H	Me(β)	177
Me	H	OBz	TFAc	H	Me(β)	380
Me	H	OBz	Bz	H	Me(α)	47, 406, 623
Me	H	OBz	Bz	H	Me(β)	47, 623
Me	H	OBz	Bz	H	Et(α)	47, 623
Me	H	OBz	Bz	H	Et(β)	47, 623
Me	H	OAc	Ts	H	Me(α)	623; DL: 533
Me	H	OAc	Ts	H	Me(β)	DL: 533
Me	H	OTs	Ts	H	Me(α)	623
Me	H	OTs	Ts	H	Me(β)	623
Me	H	OAc	Ac	H	Ac(β)	DL: 532
Me	H	OAc	Ts	H	Ac(β)	DL: 533
Me	H	OAc	Ac	H	Ac(α, β)	47
H	Me	OMe	TFAc	H	H	583
H	Me	H	TFAc	H	Me(α)	582, 583
H	Me	OH	TFAc	H	Me(α)	582, 583
H	CH_2N_3	OH	TFAc	H	Me(α)	582
H	$CH_2NHTFAc$	OH	TFAc	H	Me(α)	582
H	Me	OMe	TFAc	H	Me(α)	582, 583
H	Me	OMe	TFAc	H	*p*NBz(α, β)	583

(Continued)

Table 20 (continued)

D-Lyxo:

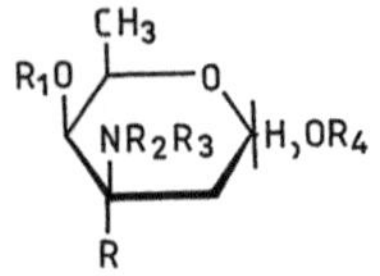

R	R_1	R_2	R_3	R_4	Refs.
H	H	Ac	H	H	318
H	H	Bz	H	H	229, 230
H	*p*NBz	TFAc	H	H	363
H	H	H	H	Me(α)	363
H	H	Ac	H	Me(α)	229–231
H	H	Ac	H	Me(β)	318
H	H	TFAc	H	Me(α)	363
H	H	Bz	H	Me(α)	229, 230
H	H	Me	Me	Me(α)	55
H	H	Me	Me	Me(β)	55
H	Ac	Ac	H	Me(α)	229–231, 318
H	*p*NBz	TFAc	H	Me(α)	363
H	*p*NBz	TFAc	H	Ac(α)	363
H	*p*NBz	TFAc	H	Ac(β)	363
Me	H	Ac	H	Me(α)	183

L-Ribo:

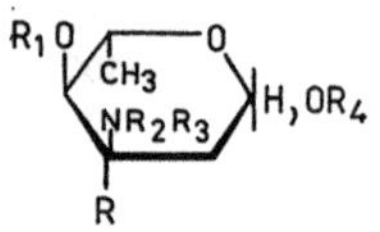

R	R_1	R_2	R_3	R_4	Refs.
H	H	H	H	H	122, 226, 560, 624
H	H	Ac	H	H	170, 171, 554
H	H	Bz	H	H	48, 122, 225, 231, 504, 545; DL: 510
H	*p*NBz	TFAc	H	H	624
H	H	H	H	Me(α)	48, 50, 122, 226, 345, 347
H	H	H	H	Me(α, β)	225
H	H	Ac	H	Me(α)	344, 349
H	H	Ac	H	Me(β)	231
H	H	TFAc	H	Me(α)	122, 184, 503, 624
H	H	TCAc	H	Me(α)	344, 349; DL: 511
H	H	Bz	H	Me(α)	48, 122
H	H	Bz	H	Me(β)	231
H	H	Bz	H	Me(α, β)	225
H	H	Cbz	H	Me(α)	578
H	H	Me	Me	Me(α)	55, 625
H	H	Me	Me	Me(β)	55, 625
H	H	Me	Me	Me(α, β)	560
H	H	Me	Me	Ac(α)	DL: 626
H	Ac	Ac	H	Me(α)	48, 50, 188, 226, 344, 345, 349, 359, 554
H	Ac	Ac	H	Me(β)	231
H	Ac	Ac	H	Me(α, β)	560

(Continued)

Table 20 (continued)

R	R_1	R_2	R_3	R_4	Refs.
H	Ac	Ac	H	Bn(α)	351, 365
H	Ac	Ac	H	c-Hex (α)	360
H	MOM	Ac	H	Me(α)	171
H	MEM	Ac	H	Me(α)	170, 171
H	TFAc	TFAc	H	Me(α)	122
H	Bz	Bz	H	Me(α)	122
H	Ms	H	H	Me(α)	578
H	Ms	TFAc	H	Me(α)	503
H	*p*NBz	TFAc	H	Me(α)	184, 362, 624
H	*p*NBz	TFAc	H	Me(β)	362
H	*p*NBz	TFAc	H	c-Hex(α)	362
H	*p*NBz	TFAc	H	*p*NBz(α)	624
Me	H	Bz	H	H	500
Me	H	H	H	Me(α)	419
Me	H	H	H	Me(β)	326
Me	H	Ac	H	Me(α)	185
Me	H	Bz	H	Me(α)	500
Me	Ac	Ac	H	Me(α)	185
Me	Me	H	H	Me(β)	186, 610
Me	Me	Ac	H	Me(α)	186, 610
Me	Me	Ac	H	Me(β)	186, 610
Me	MOM	Ac	H	Me(α)	185

D-Ribo:

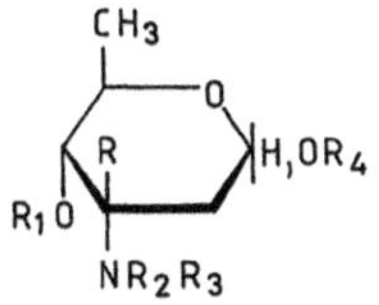

R	R_1	R_2	R_3	R_4	Refs.
H	H	H	H	H	114, 175, 240
H	H	Ac	H	H	175
H	H	Bz	H	H	114, 175, 240
H	H	H	H	Me(α)	176
H	H	Ac	H	Me(α)	175
H	H	TFAc	H	Me(α)	176
H	H	Bz	H	Me(α)	175
H	H	Bz	H	Me(α)	114, 239, 240, 485
H	H	Me	Me	Me(α)	55
H	H	Me	Me	Me(α, β)	55
H	Ac	Ac	H	Me(α)	175, 557, 559
H	Bz	Ac	H	Me(α)	175, 239, 485
H	Bz	Bz	H	Me(β)	114
H	*p*NBz	TFAc	H	Me(α)	176
H	*p*NBz	TFAc	H	Ac(α, β)	176
Me	H	H	H	Me(α)	352
Me	H	Ac	H	Me(α)	177, 182, 352, 414
Me	H	TFAc	H	Me(α)	182, 414
Me	Bz	Ac	H	Me(α)	177, 182
Me	Bz	TFAc	H	Me(α)	182, 380, 414

(Continued)

Table 20 (continued)

L-Xylo: 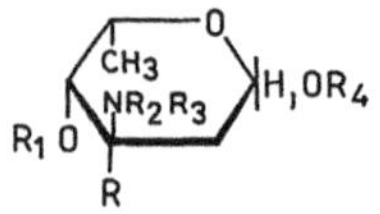

R	R_1	R_2	R_3	R_4	Refs.
H	H	Ac	H	H	223, 554
H	H	Bz	H	H	523, 544, 545; DL: 509, 517, 518
H	H	TFAc	H	H	223
H	*p*NBz	TFAc	H	H	363, 628
H	H	H	H	Me(α)	264
H	H	H	H	Me(β)	223, 231, 264, 627
H	H	Ac	H	Me(β)	223, 231, 232
H	H	TFAc	H	Me(α)	122, 503
H	H	TFAc	H	Me(β)	223, 363
H	H	Me	Me	Me(α)	264
H	H	Me	Me	Me(β)	264
H	Ac	Ac	H	Me(α)	554
H	Ac	Ac	H	Me(β)	223, 231, 232; DL: 543
H	Ac	Ac	H	Bn(α)	351, 365
H	Prop	Ac	H	Bn(α)	365
H	Bz	Ac	H	Me(β)	223
H	*p*NBz	TFAc	H	Me(α)	628
H	*p*NBz	TFAc	H	Me(β)	363
H	*p*NBz	TFAc	H	c-Hex(α)	628
H	*p*NBz	TFAc	H	Ac(α)	363
Me	H	Bz	H	H	499,500
Me	H	Ts	H	H	DL: 533
Me	Bz	Bz	H	H	56
Me	H	H	H	Me(α)	56
Me	H	Ac	H	Me(α)	185
Me	H	Bz	H	Me(α)	56,500
Me	H	Ts	H	$(CH_2)_2OH$(α)	DL: 533
Me	Ac	Ts	H	Me(α)	DL: 533
Me	Ac	Ts	H	Me(β)	DL: 533
Me	Bz	Bz	H	Me(α)	56, 192, 499, 500
Me	Bz	Bz	H	Me(β)	56
Me	Bn	Bz	H	Me(β)	192
Me	MOM	Bz	H	Me(β)	192
Me	Me	H	H	Me(α)	194
Me	Me	Ac	H	Me(α)	194
Me	Ac	Ts	H	Ac(α)	DL: 533
Me	Ac	Ts	H	Ac(β)	DL: 533

(Continued)

Table 20 (continued)

D-Xylo 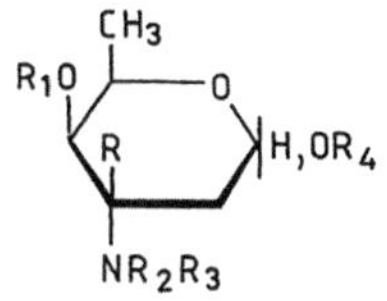

R	R_1	R_2	R_3	R_4	Refs.
H	H	TCAc	H	H	DL: 511
H	H	Bz	H	H	518,523
H	H	Ts	H	H	527
H	H	Ac	H	Me(α)	527, 529
H	H	TCAc	H	Me(α)	DL: 511
H	H	Ts	H	Me(α)	529
H	H	Ac	Ac	Me(α)	529
H	H	Me	H	Me(α)	228
H	H	Me	Me	Me(α)	55
Me	H	Ac	H	Me(α)	182, 414
Me	H	TFAc	H	Me(α)	182, 414, 416
Me	Me	Ac	H	Me(α)	182, 414
Me	Ms	TFAc	H	Me(α)	182, 416

Table 21. 3-Amino-2,3-dideoxy- and -2,3,6-trideoxy-hex-1-enopyranose derivatives

L-Arabino: 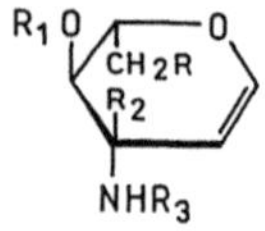

R	R_1	R_2	R_3	Refs.
H	H	H	H	362
H	H	H	TFAc	362, 577
H	*p*NBz	H	TFAc	362
H	CSSMe	H	TFAc	577
H	Ac	Me	Ac	374, 405
O*p*NBz	*p*NBz	H	TFAc	235

D-Arabino: 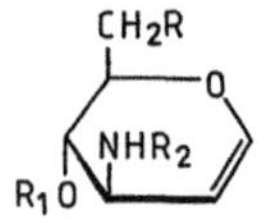

R	R_1	R_2	Refs.
H	H	H	363
H	*p*NBz	TFAc	363
OH	H	H	79
OAc	Ac	Ac	79, 99, 370–372
OAc	Ac	TFAc	377
OAc	Ac	EOC	370, 371, 629
benzylidene		Ac	80

L-Lyxo 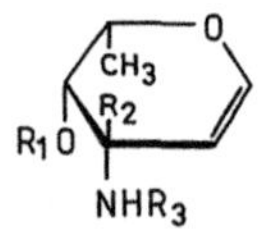

R_1	R_2	R_3	Refs.
H	H	TFAc	614
Ac	H	Ac	373
Ac	H	TFAc	616
TFAc	H	TFAc	378, 564
Bz	Me	TFAc	380

D-Lyxo: 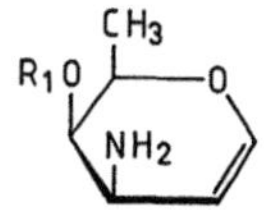 No derivatives have been prepared

(Continued)

Table 21 (continued)

L-Ribo: 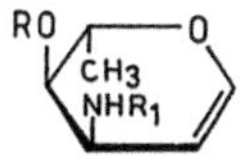

R	R_1	Refs.
H	H	362
H	TFAc	362
*p*NBz	TFAc	362

D-Ribo: 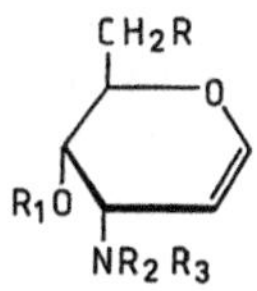

R	R_1	R_2	R_3	Refs.
H	H	H	H	363
H	*p*NBz	TFAc	H	176, 363
OH	H	H	H	79
OH	H	$=CMe_2$		79
OAc	Ac	Ac	H	79
OAc	Ac	$=CMe_2$		79
OAc	Ac	EOC	H	370, 371
NHTFAc	Ac	TFAc	H	379
NHTFAc	*p*NBz	TFAc	H	379
benzylidene		Ac	H	80

L-Xylo: 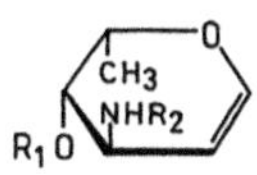

R_1	R_2	Refs.
H	TFAc	627
*p*NBz	TFAc	627

D-Xylo: 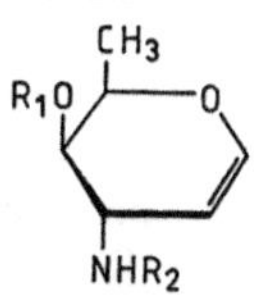

R_1	R_2	Refs.
H	H	363
H	TFAc	363
*p*NBz	TFAc	363

Note Added in Proof

Since the manuscript of this volume was closed to further additions, a number of papers dealing with the synthesis and chemical transformations of 3-amino- and nitro-polydeoxyhexoses have appeared. These results are briefly summarized in the present section (which covers the literature through December 1987), and organized according to the respective previous chapter/section numbers.

Chapter 1 (Table 1)

The novel glycopeptide-type antibiotics actaplanin (A-4696) (630, 631) and chlorosporins A, B and C (632) have been shown to contain an L-ristosamine **(18)** aminosugar building unit.

Section 2.1.1 (Addition of halogens)

In contrast with results obtained upon chlorination of 3,4,6-tri-O-acetyl- and 3,4,6-tri-O-benzyl-D-glucal (see pages 10–11) Sakakibara et al. (633) have reported that the chlorination of 3-O-acetyl- and 3-deoxy-4,6-O-benzylidene-D-glucal in carbon tetrachloride predominantly results in the β-D-*manno*- and β-D-*arabino*-1,2-dichloro adducts with 52% and 89% yield, respectively.

Section 2.2.2

Treatment of 2,3-anhydrohexopyranosides with N,N-diethyltrimethylsilylamine or N,N-dimethyltrimethylsilylamine in the presence of aluminum chloride catalyst has been shown (634) susceptible of the direct introduction of the C-2 or C-3 amino functions into carbohydrate molecules under mild conditions. Although moderate yields (41–48%) are reported, easy recovery (and recycling) of the unreacted starting epoxides makes this procedure efficiently applicable for the preparation of methyl 3-dimethylamino- **(164)** (or diethylamino)-3-deoxy-α-D-glucopyranoside, and of the corresponding 6-deoxy analogue (methyl micaminoside, **167**; R_1=H, Q=CH_3; Scheme 31).

Section 2.2.5

It has been shown (635) that conjugate addition of hydrazoic acid to hex-2-enopyranoses (pseudoglycals) leads to a predominant *equatorial* incorporation of the azido group into C-3 of the sugar molecule. Thus, reaction of 4-O-acetyl-6-deoxy-α,β-L-*erythro*-hex-2-enopyranose with sodium azide in aqueous acetic acid gave, after acetylation, a mixture of the 3-azido-1,4-di-O-acetyl-α- and β-L-*arabino*- and -*ribo*-

2,3,6-trideoxyhexoses. Glycosylation of this mixture with methanol or benzyl alcohol in the presence of K-10 montmorillonite allowed the convenient chromatographic isolation of methyl 4-O-acetyl-3-azido-2,3,6-trideoxy-α-L-*arabino*-hexopyranoside (48 %) or the corresponding benzyl glycoside (44 %), along with unsignificant amounts (1–10 %) of the respective β-L-*arabino*- and α- and β-L-*ribo*-glycosides. O-Deacetylation and catalytic hydrogenation of the major products then afforded (635) methyl α-L-acosaminide **(181)** or benzyl α-L-acosaminide in an overall yield of 35 % — based on 3,4-di-O-acetyl-L-rhamnal **(34)**.

Section 2.4

The Lukacs group (636) synthesized derivatives of 3-amino-2,2-difluoro-2,3,6-trideoxy-L-*lyxo*-hexopyranose (2,2-difluorodaunosamine) by means of the addition of trifluorofluoroxymethane to the double bond of 4-O-benzoyl-3,6-dideoxy-2-fluoro-3-trifluoroacetamido-L-galactal.

Section 2.5

Methyl 3-C-methyl-3-nitro-2,3,6-trideoxy-α-D-*ribo*-hexopyranoside (**491**, Scheme 86) has been directly applied (637) for the synthesis of methyl α-D-kijanoside (**437**, see Table 6, entry 10) via the 4-O-triflate ester, which was subjected to azide displacement, followed by chemoselective reduction of the azide function ($NaBH_4$—$NiCl_2$) and methoxycarbonylation of the resulting methyl 4-amino-3-C-methyl-3-nitro-2,3,4,6-tetradeoxy-α-D-*xylo*-hexopyranoside.

By means of the stereoselective α-C-methylation of N-protected methyl 3-amino-2,3,6-trideoxy-α-hexopyranosid-4-uloses (derived from readily available L- and D-ristosamine analogs), followed by reduction of the C-4 keto function with L-Selectride or sodium borohydride (to generate *axial*- or *equatorial* C-4 hydroxyl groups, respectively; cf. Sect. 2.8) a number of methyl 3-amino-3-C-methyl-2,3,6-trideoxyhexopyranosides with L-*lyxo*-, L-*ribo*-, L-*xylo* and D-*xylo* configurations have been synthesized by Klemer and Wilbers (638). The series of the prepared stereoisomeric 3-C-methyl-branched aminohexoses (see Table 23) thus involves several new

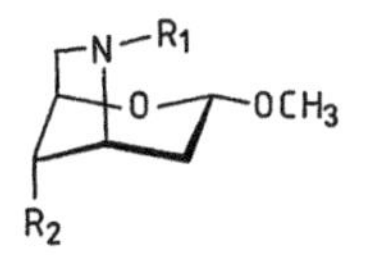

	R_1	R_2
1157	$COCF_3$	OBz
1158	$COCF_3$	OH
1159	H	OBz
1160	$COCF_3$	OMs
1161	Bn	OBz
1162	Bn	OH
1163	Bn	OTs
1164	Bn	CN
1165	Bn	CHO

derivatives of L-vancosamine **(15)** and those of the aminosugar **(21)** from antibiotic A 35512 B, as well as precursors to D-rubranitrose **(241)**, D-kijanose **(242)** and L-decilonitrose **(243)**. The nitrosugar D-rubranitrose was also synthesized.

Section 2.7.4 (Table 10)

The reaction of the hitherto unknown methyl 4-O-benzoyl-6-bromo-2,3,6-trideoxy-3-trifluoroacetamido-α-D-*arabino*-hexopyranoside with sodium hydride in DMF readily afforded (639) 86% of the substituted 3,6-iminopyranoside **1157** (cf. Table 10). By means of further chemical transformations a number of related bicyclic amides **(1158–1165)**, as potential chiral synthons for the preparation of 9-aza-9-deoxy-prostaglandine derivatives have been synthesized (cf. Sect. 4.3.2, Scheme 180).

Section 3.2.1

The "azetidinone method" (Scheme 131) established by Hauser and coworkers (517, 518) has been extended (640) for the preparation of 3-benzamido-2,3-dideoxy- and 3-benzamido-2,3-dideoxy-3-C-methyl-pentofuranoses with *erythro* and *threo* configurations, thus demonstrating the usefulness of the procedure for the general synthesis of the 3-amino-2,3-dideoxysugar framework.

Recent studies of Roush et al. (641) have shown that epoxidations of the 3-benzamido-hexenoate **738** (Scheme 131) or its N-trichloroacetyl analog (both derived from the azetidin-2-one **782**) with either peracids or the $Mo(CO)_6$-*tert*-butylhydroperoxide system were essentially non-stereoselective. In contrast, epoxidation of the 4-aminoheptadiene **1166** with MCPBA gave 42% of N-(*xylo*-5,6-epoxy-hept-1-en-4-yl)-trichloroacetamide ±**1167**, along with 12% of the corresponding *lyxo* isomer and 10% of a mixture of diepoxides (Scheme 203). α-Opening of the anhydro ring of ±**1167** afforded the oxazoline ±**1168**, which upon sequential hydrolysis, N-benzoylation and one-carbon ozonolytic descent (cf. Scheme 147) of the resulting ±**883** furnished N-benzoyl-D,L-ristosamine (±**700**).

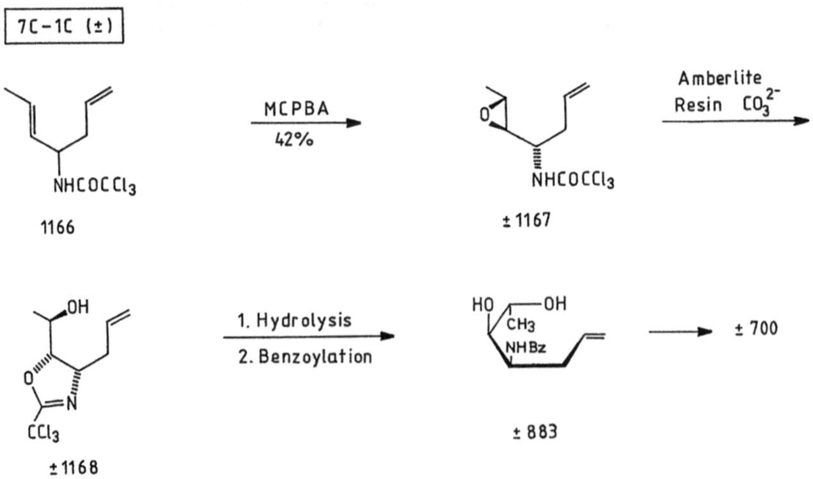

Scheme 203

Section 3.2.2

A full paper (642) describing the experimental details of the preparation of N-benzoyl-L-daunosamine (**833**, Schemes 142 and 143) (540) and N-benzoyl-L-acosamine (**863**, Scheme 143) (541) has been published by the Hiyama group.

Section 3.2.4

Further development of the previously reported routes to the derivatives of D,L- and L-daunosamine (509, 551) (Schemes 123, 151), D,L-acosamine (510) (Scheme 125), D,L-ristosamine (510) (Scheme 126) and to the corresponding D,L-*xylo* stereoisomer (509) (Scheme 124) has been described by the Hirama group. Thus, N-acetyl-L-acos-

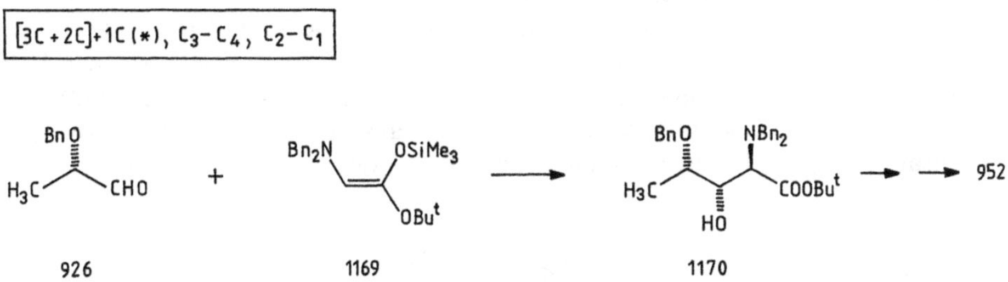

Scheme 204

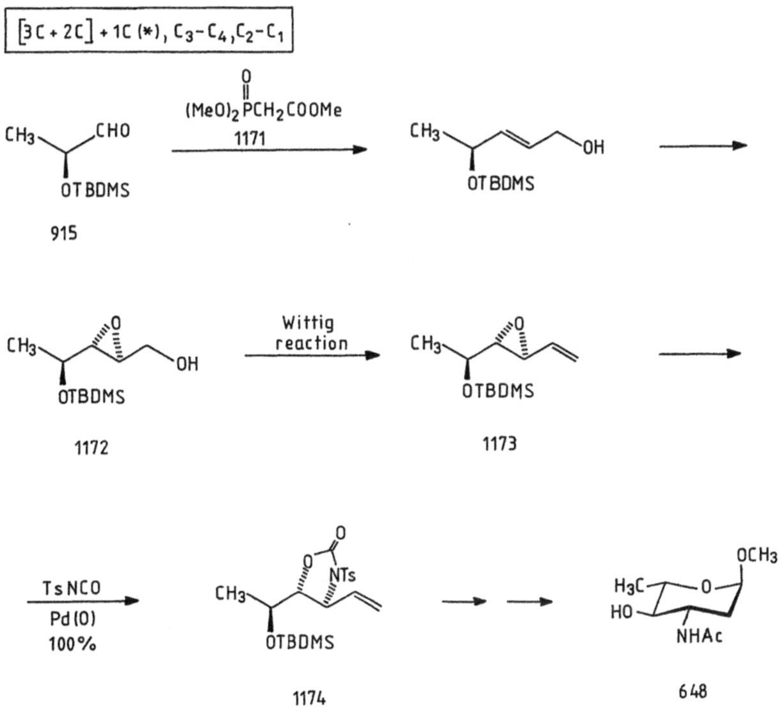

Scheme 205

amine **(138)** and N-benzoyl-L-ristosamine **(700)** were synthesized (643) by means of the stereoselective intramolecular conjugate addition of the appropriate carbamates of methyl (*Z*)-4,5-dihydroxy-hex-2-enoates, derived from L-lactaldehyde (cf. Scheme 151).

A new procedure for the preparation of the N-carbobenzyloxy analog of the 2-amino-2,5-dideoxy-L-*lyxono*-1,4-lactone (**952**, Scheme 156), used as the key intermediate in syntheses of L-daunosamine (555, 556), D-ristosamine (557, 559) and L-vancosamine (558) has been reported by Banfi et al. (644). The Lewis acid-catalyzed aldol-type condensation of O-benzyl-L-lactaldehyde **(926)** and the ketene silyl acetal **1169** (obtained from the ester enolate of *tert*-butyl N,N-dibenzylamino acetate) readily afforded the *tert*-butyl 2-amino-2,5-dideoxy-L-*lyxo*-pentanoate derivative **1170** (Scheme 204). By the application of standard procedures this latter compound was then transformed into the desired derivative of the 1,4-lactone **952.**

Palladium-mediated vicinal hydroxyamination (Scheme 205) of the vinyl epoxide **1173** (derived from the silylated L-lactaldehyde **915** and the phosphonoacetate **1171**, followed by a Wittig one-carbon ascent of the epoxy-pentanediol **1172**) gave exclusively the oxazolidine **1174.** This was then transformed into methyl N-acetyl-α-L-acosaminide **(648)** by means of a hydroboration-oxidation sequence (645).

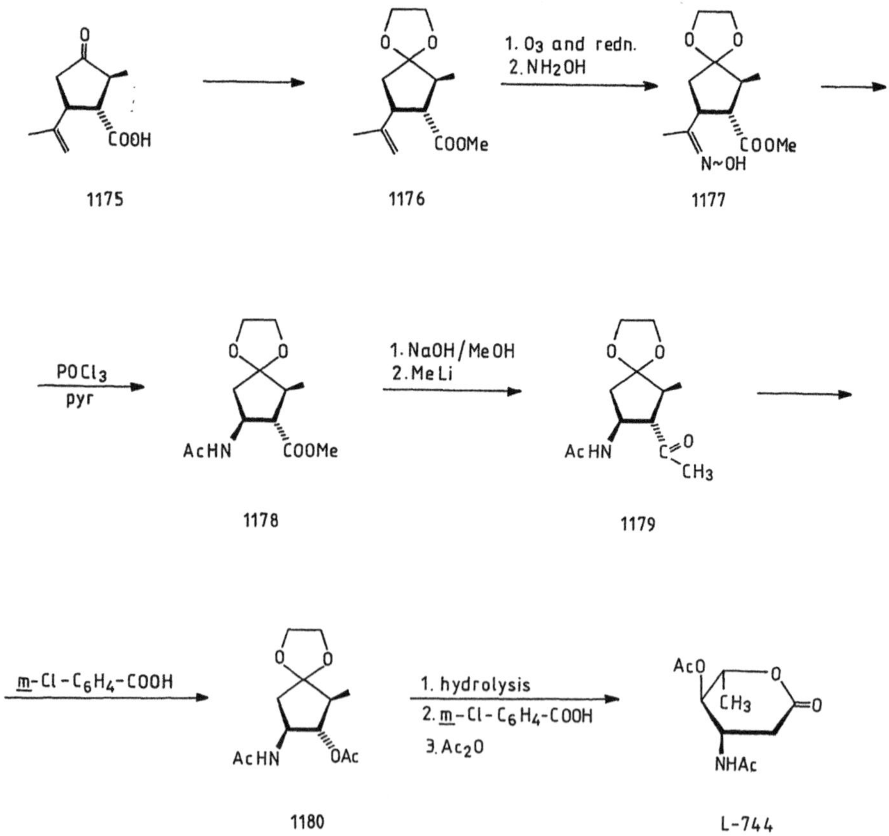

Scheme 206

Section 3.2.5

An ingenious chiral formal synthesis of N-acetyl-L-acosamine **(138)** was developed by Kametani et al. (646) by means of assembling the L-*arabino*-hexono-δ-lactone L-**744** (cf. Scheme 125) from the carbon skeletone of *S*-(+)-carvone. Thus, the latter natural compound was transformed into the cyclopentanone carboxylic acid **1175** (Scheme 206), and then into the acetalated keto-carboxylic ester **1176**. Beckmann rearrangement of the oxime **1177**, derived from **1176** by means of successive ozonolysis and oximation, afforded the amide **1178**, which was converted into the methylketone **1179**. Baeyer-Villiger oxidation of this latter gave the acetoxy derivative **1180**. An additional hydrolysis-oxidation procedure then transformed **1180** into the desired δ-lactone L-**744**, the racemic form of which has already been applied (507, 510) for the preparation of the target aminodeoxysugar ±**138** (cf. Schemes 120 and 125).

Multistep syntheses of methyl N-benzoyl-α-L-acosaminide and methyl N-(benzyloxyoxalyl)-α-L-daunosaminide have been accomplished (647) from a chiral 7-aza-3-oxabicyclo-[4.2.0]octan-8-one derivative, thus outlining a novel strategy of constructing 3-amino-2,3,6-trideoxyhexoses from beta-lactams (cf. Sect. 3.2.1).

Section 4.2

Syntheses of new N-alkyl-daunomycin derivatives via the Michael addition to substituted maleimides (648), as well as the preparation of novel morpholino-anthracyclines (cf. Scheme 168) derived from 13-deoxo-daunomycin and 13-deoxo-carminomycin have been published (649).

Condensation of daunomycin or adriamycin with "sugar-dialdehydes" in the presence of sodium cyanoborohydride has led to novel morpholino-anthracycline analogs carrying chiral substituents (OCH_3, CH_2OH) in the heterocyclic ring (650). Of the prepared derivatives some representatives were found to be more active than daunomycin in increasing the survival time of mice transplanted with leukemia ascites cells.

Section 4.4 (Scheme 187)

The furanose form of N-benzoyl-L-ristosamine has been described and characterized by the Hirama group (643).

Chapter 5

As we have predicted in the "Concluding Remarks" further achievement on the "triflate-rearrangement" reaction (cf. Scheme 202) has been reported by Binkley et al. (651). The presented novel results allow the convenient preparation of partially protected 2,6-dideoxyhexose derivatives in each of the D-*arabino*-, *lyxo*-, *ribo* and *xylo* series. As outlined throughout the previous sections, such compounds are widely used starting materials for conversions into 3-amino-2,3,6-trideoxy-D-hexose derivatives, or into the corresponding 4-amino-4-deoxy analogs.

Addendum to the Appendix

Up to the end of 1987 the following 3-amino-2,3-dideoxy- and 3-amino-2,3,6-trideoxyhexose derivatives have been synthesized (Tables 22 and 23).

Table 22. 3-Amino-2,3-dideoxyhexopyranose derivatives

D-Arabino

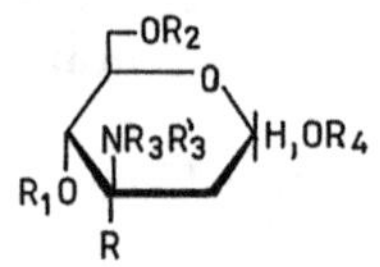

R	R_1	R_2	R_3	R_3'	R_4	Refs.
H	benzylidene		H	H	Me(α)	639
H	benzylidene		TFAc	H	Me(α)	639

D-Ribo

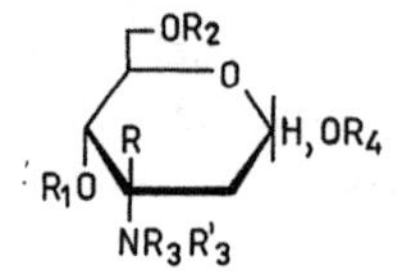

R	R_1	R_2	R_3	R_3'	R_4	Ref.
H	benzylidene		BOC	H	Me(α)	638

Table 23. Derivatives of 3-amino-2,3,6-trideoxyhexopyranoses

L-Arabino

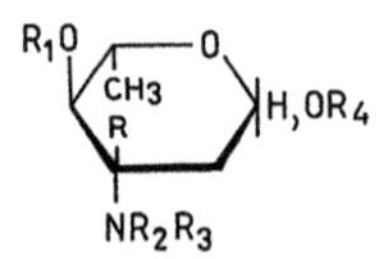

R	R_1	R_2	R_3	R_4	Refs.
H	H	Ac	H	H	643
H	H	Bz	H	H	642
H	H	H	H	Me(α)	635
H	H	H	H	Bn(α)	635
H	H	Ac	H	Me(α)	645
H	H	Bz	H	Me(α)	647

(Continued)

Table 23 (continued)

L-Lyxo

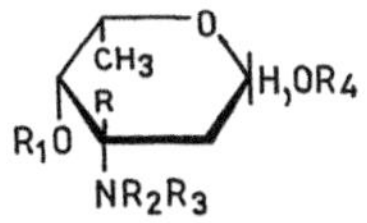

R	R_1	R_2	R_3	R_4	Refs.
H	H	Bz	H	H	642
H	H	COCOOBn	H	Me(α)	647
Me	H	BOC	H	Me(α)	638
Me	Bz	Bz	H	Me(α)	638

L-Ribo

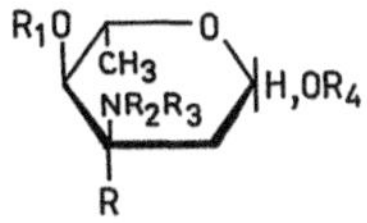

R	R_1	R_2	R_3	R_4	Refs.
H	H	Bz	H	H	D,L: 641
H	H	BOC	H	Me(α)	638
H	H	Cbz	H	Me(α)	638
Me	H	H	H	Me(α)	638
Me	H	Cbz	H	Me(α)	638

D-Ribo

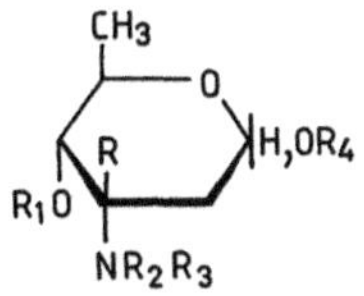

R	R_1	R_2	R_3	R_4	Ref.
H	H	BOC	H	Me(α)	638

L-Xylo

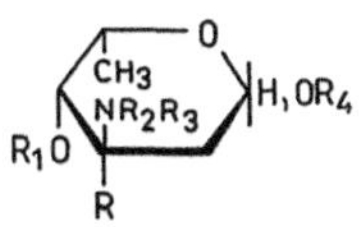

R	R_1	R_2	R_3	R_4	Refs.
Me	H	H	H	Me(α)	638
Me	H	BOC	H	Me(α)	638
Me	H	Cbz	H	Me(α)	638
Me	Bz	Bz	H	Me(α)	638
Me	H	BOC	Me	Me(α)	638

Table 23 (continued)

D-Xylo

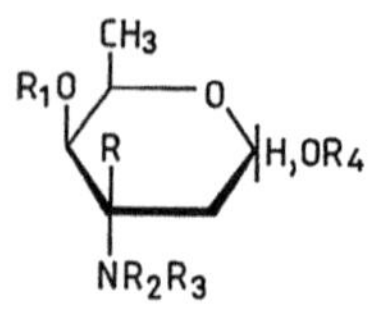

R	R_1	R_2	R_3	R_4	Refs.
Me	H	H	H	Me(α)	638
Me	H	BOC	H	Me(α)	638
Me	Me	BOC	H	Me(α)	638

References

1 Umezawa S (1974) Advan. Carbohydr. Chem. Biochem. 30: 111
2 Rinehart KL, Suami T (eds) (1980) Aminocyclitol antibiotics. Washington D.C. (ACS Symposium Series 125)
3 Arcamone F (1977) Lloydia 40: 45
4 Arcamone F (1978) Daunomycin and related antibiotics. In: Sammes P (ed) Topics in antibiotic chemistry, vol 2. Ellis Horwood, Chichester
5 Remers WA (1979) The chemistry of antitumor antibiotics, vol 1. John Wiley, New York
6 Crooke ST, Reich SD (eds) (1980) Anthracyclines-Current status and new development. Academic Press, New York
7 Arcamone F (1981) Doxorubicin. Anticancer antibiotics. Academic Press, New York (Medicinal chemistry, a series of monographs)
8 El Khadem HS (ed) (1982) Anthracycline antibiotics. Academic Press, New York
9 Lomakina NN, Spiridonova IA, Sheinker YuN, Vlasova TF (1973) Khim. Prir. Soedin. 9: 101, Chem. Abstr. (1973) 78: 148170
10 Spiridonova IA, Yurina MS, Lomakina NN, Sztaricskai F, Bognár R (1976) Antibiotiki 21: 304, Chem. Abstr. (1976) 85: 108925
11 Sztaricskai, F, Harris CM, Harris TM (1979) Tetrahedron Lett. 2861
Batta Gy, Sztaricskai F, Csanádi J, Komáromi I, Bognár R (1986) J. Antibiotics 39: 910
12 Harada KI, Ito S, Suzuki M (1979) Carbohydr. Res. 75: C-17
13 Kinumaki A, Suzuki M (1972) J. Antibiotics 25: 480
14 Brufani M, Keller-Schierlein W (1966) Helv. Chim. Acta 49: 1962
15 Furukawa M, Hayakawa I, Otha G, Litaka Y (1975) Tetrahedron 31: 2989
16 Sequin U, Furukawa M (1978) Tetrahedron 34: 3623
Zehnder M, Sequin U, Nadig H (1979) Helv. Chim. Acta 62: 2525
17 Kondo S, Miyamoto M, Naganawa H, Takeuchi T, Umezawa H (1977) J. Antibiotics 30: 1143
18 Okabe T, Nomoto K, Funabashi H, Okuda S, Suzuki H, Tanaka N (1985) J. Antibiotics 38: 1333
18a Rasmussen RR, Nuss ME, Scherr MH, Mueller SL, McAlpine JB, Mitscher LA (1986) J. Antibiotics 39: 1515
19 Arcamone F, Cassinelli G, Orezzi P, Franceschi G, Mondelli R (1964) J. Amer. Chem. Soc. 86: 5335
20 Arcamone F, Cassinelli G, Franceschi G, Orezzi P, Mondelli R (1968) Tetrahedron Lett. 3353
21 Arcamone F, Cassinelli G, Franceschi G, Mondelli R, Orezzi P, Penco S (1970) Gazz. Chim. Ital. 100: 949
22 Arcamone F, Franceschi G, Penco S, Selva A (1969) Tetrahedron Lett. 1007
23 Brazhnikova MG, Zbarsky VB, Ponomarenko VI, Potapova NP (1974) J. Antibiotics 27: 254
24 Arcamone F, Cassinelli G, Penco S, Tognoli L (1970) Ger. Offen. 1,923,885. Chem. Abstr. (1970) 72: 131086
25 Lunel J, Prendhomme J (1970) Ger. Offen. 1911240. Chem. Abstr. (1970) 72: 20603
26 Cassinelli G, Grein A, Masi P, Suarto A, Bernardi L, Arcamone F, DiMarco A, Casazza A, Pratesi G, Soranzo C (1978) J. Antibiotics 31: 178
27 Arcamone F, Cassinelli G, Matteo F, Forenza S, Ripamonti MC, Rivola G, Vigevani A, Clardy J, McCabe T (1980) J. Amer. Chem. Soc. 102: 1462
28 Takahashi Y, Nagasawa H, Takeuchi T, Umezawa H, Komiyama T, Oki T, Inuri T (1977) J. Antibiotics 30: 622

29 Fedorova GB (1970) Antibiotiki 15: 403
30 Ogawa Y, Sugi H, Fujikawa H, Mori M (1981) J. Antibiotics 34: 938
31 Oki T, Matsuzawa Y, Kiyoshima A, Yoshimoto H, Naganawa H, Takeuchi T, Umezawa H (1981) J. Antibiotics 34: 783
32 Oki T, Shibamoto N, Matsuzawa Y, Ogasawara T, Yoshimoto A, Kitamura I, Inui T, Naganawa H, Takeuchi T, Umezawa H (1977) J. Antibiotics 30: 683
33 Brockmann H, Spohler E, Waehneldt T (1968) Chem. Ber. 96: 2925
34 Brockmann H, Waehneldt T (1961) Naturwissenschaften 48: 717
35 Brockmann H, Waehneldt T, Niemeyer J (1969) Tetrahedron Lett. 415
36 Oki T, Matsuzawa Y, Kiyoshima A, Yoshimoto A, Numata K, Kitamura I, Hori S, Takamatsu A, Umezawa H, Ishizuka M, Naganawa H, Suda H, Hamada M, Takeuchi T (1975) J. Antibiotics 28: 830
37 Kitamura I, Oki T, Inui T (1978) J. Antibiotics 31: 919
38 Brockmann H, Niemeyer J (1968) Chem. Ber. 101: 1341
39 Brockmann H, Lenk W (1959) Chem. Ber. 92: 1904
40 Kumar V, Remers WA, Grulich RE (1977) J. Antibiotics 30: 881
41 Keller-Schierlein W, Richle W (1970) Antimicrob. Agents Chemother. 68
42 Richle W, Winkler EK, Hawley DM, Dobler M, Keller-Schierlein W (1972) Helv. Chim. Acta 55: 467
43 Mitschler LA, McCare W, Anders WW, Lowery JA, Bohonos N (1964) J. Pharm. Sci. 53: 1140
44 Nettleton Jr. DE, Bardner W, Bush JA, Coon AB, Mosely JE, Myllymaki RW, O'Herron FA, Schreiber RH, Vulcano AL (1977) J. Antibiotics 30: 525
45 Oki T, Yoshimoto A, Matsuzawa Y, Takeuchi T, Umezawa H (1981) J. Antibiotics 34: 916
46 Fujiwara A, Tazoe M, Hoshino T, Sekine Y, Masuda S, Nomura S (1981) J. Antibiotics 34: 912
47 Johnson AW, Smith RM, Guthrie RD (1972) J. Chem. Soc. Perkin Trans. I 2153
Weringa WD, Williams DH, Feeney J, Brown JP, King RW (1972) J. Chem. Soc. Perkin Trans. I 443
48 Bognár R, Sztaricskai F, Munk ME, Tamás J (1974) J. Org. Chem. 39: 2971
49 Sztaricskai F, Harris CM, Neszmélyi A, Harris TM (1980) J. Amer. Chem. Soc. 102: 7093
50 Spiridonova IA, Lomakina NN, Sztaricskai F, Bognár R (1974) Antibiotiki 19: 400
51 Williams DH, Rajananda V, Kalman JR (1979) J. Chem. Soc. Perkin Trans. I 787
52 McGahren WJ, Martin JH, Morton GO, Hargreaves RT, Leese RA, Lovell FM, Ellestad GA (1979) J. Amer. Chem. Soc. 101: 2237
53 McGahren WJ, Martin JH, Morton GO, Hargreaves RT, Leese RA, Lovell FM, Ellestad GA, O'Brien E, Holker JSE (1980) J. Amer. Chem. Soc. 102: 1671
54 Thang TT, Lukacs G, Omura S, Bartner P, Boxler DL, Brambilla R, Mallams AK, Morton JB, Reichert P, Sancilio FD, Supernant H, Tomalesky G (1978) J. Amer. Chem. Soc. 100: 663
55 Bartner P, Boxler DL, Brambilla R, Mallams AK, Morton JB, Reichert P, Sancillio FD, Supernant H, Tomalesky G, Lukacs G, Olesker A, Thang TT, Valente L, Omura S (1979) J. Chem. Soc. Perkin Trans. I 1600
56 Debono M, Molloy M (1980) J. Org. Chem. 45: 4685
57 Mallams AK (1969) J. Amer. Chem. Soc. 91: 7505
58 Sztaricskai F, Bognár R (1984) The chemistry of the vancomycin group of antibiotics. Publishing House of the Hungarian Academy of Sciences, Budapest (Recent Developments in the Chemistry of Natural Carbon Compounds, vol 10)
59 Acton EM, Tong GL, Mosher RL, Wolgemuth RL (1984) J. Med. Chem. 27: 638
60 Yoshimura J (1984) Advan. Carbohydr. Chem. Biochem. 42: 69
61 McGarvey GJ, Kimura M, Oh T, Williams JM (1984) J. Carbohydr. Chem. 3: 125
62 Hauser FM, Ellenberger S (1986) Chem. Rev. 86: 35
63 Ferrier RJ (1965) Advan. Carbohydr. Chem. 20: 67
64 Ferrier RJ (1969) Advan. Carbohydr. Chem. 24: 199
65 Sharma M, Brown, RK (1966) Can. J. Chem. 44: 2825
66 Ireland RE, Wilcox CS, Thaisrivongs S (1978) J. Org. Chem. 43: 786
67 Guthrie RD, Irwine RW (1979) Carbohydr. Res. 72: 285
68 Tsuda N, Yokota S, Kudo T, Mitsunobu O (1983) Chem. Lett. 289

69 Bovin NV, Zurabyan SE, Khorlin AYa (1983) J. Carbohydr. Chem. 2: 249
Csuk R, Fürstner A, Glänzer BI, Weidmann H (1986) J. Chem. Soc. Chem. Commun. 1149
70 Overend WG, Stacey M (1953) Advan. Carbohydr. Chem. 8: 45
71 Matthews AS, Overend WG, Safizadeh F, Stacey M (1955) J. Chem. Soc. 2511
72 Gonzalez F, Lesage S, Perlin AS (1975) Carbohydr. Res. 42: 267
73 Zwierzchowska-Nowakowska Z, Zamojski A (1974) Rocz. Chem. 48: 1019
74 Ferrier RJ, Ponpipom MM (1971) J. Chem. Soc. (C) 553
75 Ferrier RJ, Prasad N (1969) J. Chem. Soc. (C) 581
76 Hall RH, Jordaan A (1973) J. Chem. Soc. Perkin I 1059
77 Heyns K, Feldman J (1980) Starch 32: 40
78 Pearson RG, Songstad J (1967) J. Amer. Chem. Soc. 89: 1827
79 Guthrie RD, Irwine RW (1980) Carbohydr. Res. 82: 207
80 Guthrie RD, Irwine RW (1980) Carbohydr. Res. 82: 225
81 Ferrier RJ (1964) J. Chem. Soc. 5443
82 Ferrier RJ, Prasad N (1969) J. Chem. Soc. (C) 570
83 Brimacombe JS, Doner LW, Rollins AJ (1972) J. Chem. Soc. Perkin I 2977
84 Bhaté P, Horton D, Priebe W (1985) Carbohydr. Res. 144: 331
85 Bock K, Christiansen JK, Pedersen C (1971) Carbohydr. Res. 20: 73
86 Hadfield AF, Cunningham L, Sartorelli AC (1979) Carbohydr. Res. 72: 93
87 Garegg PJ, Norberg T (1975) Acta Chem. Scand. 29B: 507
88 Brimacombe JS, Doner LW, Rollins AJ, Al-Rhadi AK (1973) J. Chem. Soc. Perkin Trans. I 1295
89 Lemieux RU, Fraser-Reid B (1964) Can. J. Chem. 42: 532
90 Lemieux RU, Fraser-Reid B (1965) Can. J. Chem. 43: 1460
91 Igarashi K, Honma T, Imagawa T (1970) J. Org. Chem. 35: 610
92 Boullanger P, Descotes G (1976) Carbohydr. Res. 51: 55
93 Nakamura K, Tejima S, Akagi M (1964) Chem. Pharm. Bull. 12: 1302
94 Arita M, Ueda N, Matsushima Y (1972) Bull. Chem. Soc. Japan 45: 567
95 Horton D, Priebe W, Varela O (1985) Carbohydr. Res. 144: 305
96 Binkley RW, Bankaitis B (1982) J. Carbohydr. Chem. 1: 1
97 Lemieux RU, Levine S (1964) Can. J. Chem. 42: 1473
98 Monneret C, Choay P (1981) Carbohydr. Res. 96: 299
99 Tatsuta K, Fujimoto K, Kinoshita M, Umezawa S (1977) Carbohydr. Res. 54: 85
100 Umezawa S, Umezawa H, Tatsuta K (1979) Japan Tokkyo Koho 79 06,528. Chem. Abstr. (1979) 91: 57431
101 Umezawa S, Umezawa H, Tatsuta K (1978) Japan Kokai 78 23,918. Chem. Abstr. (1978) 89: 44147
102 Gnichtel H, Rebentisch D, Tompkins TC, Gross PH (1982) J. Org. Chem. 47: 2691
103 Thiem J, Karl H, Schwentner J (1978) Synthesis 696
104 Garegg PJ, Samuelsson B (1981) Carbohydr. Res. 92: 157
105 Horton D, Priebe W, Varela O (1984) Carbohydr. Res. 130: C-1
106 Horton D, Priebe W (1985) Carbohydr. Res. 136: 391
107 Zorbach WW, Payne TA (1960) J. Amer. Chem. Soc. 82: 4979
Zorbach WW, Bhat CC, Bhat KV (1968) Advan. Chem. Ser. 74: 1
108 Kochetkov NK, Chizov OS, Bochkov AF (1973) MTP International Reviews of Science. Butterworths, London (Org. Chem. Ser. 1 vol 7) p 148
Wulff G, Röhle G (1974) Angew. Chem. 86: 173
Igarashi K (1977) Advan. Carbohydr. Chem. Biochem. 34: 243
Paulsen H (1982) Angew. Chem. 94: 184
109 Garegg PJ, Köpper S, Ossowski P, Thiem J (1986) J. Carbohydr. Chem. 5: 59
110 Bock K, Lundt I, Pedersen C (1969) Acta Chem. Scand. 23: 2083 and references therein
111 Inglis GR, Schwarz JCP, McLaren L (1962) J. Chem. Soc. 1014
112 El Khadem HS, Swartz DL, Nelson JK, Berry LA (1977) Carbohydr. Res. 58: 230
113 Pelyvás I (1977) Definitive synthesis of L-ristosamine and preparation of D-ristosamine. Thesis, Lajos Kossuth University, Debrecen, Hungary
114 Pelyvás I, Sztaricskai F, Szilágyi L, Bognár R, Tamás J (1979) Carbohydr. Res. 68: 321
115 Vargha L, Kuszmann J (1963) Chem. Ber. 96: 2016

116 Maki T, Tejima S (1968) Chem. Pharm. Bull. 16: 2242
117 Manolopulos PT, Mednick M, Lichtin NN (1962) J. Amer. Chem. Soc. (1962) 84: 2203
118 Takiura K, Honda S (1973) Carbohydr. Res. 29: 477
119 Honda S, Kakehi K, Takai H, Takiura K (1973) Carbohydr. Res. 29: 488
120 Cheung TM, Horton D, Weckerle W (1977) Carbohydr. Res. 59: 276
121 Gibb VG, Hall LD (1978) Carbohydr. Res. 63: C-1
122 Sztaricskai F, Pelyvás I, Szilágyi L, Bognár R, Tamás J, Neszmélyi A (1978) Carbohydr. Res. 65: 193
123 Marsh JP, Mosher CW, Acton EM, Goodman L (1967) Chem. Commun. 973
124 Horton D, Tarelli JM, Wander JD (1972) Carbohydr. Res. 23: 440
125 Allgeier H (1968) Helv. Chim. Acta 51: 311
126 Takiura K, Honda S (1972) Carbohydr. Res. 23: 369
127 Michalska M, Borowieczka J (1983) J. Carbohydr. Chem. 2: 99
128 Borowieczka J, Michalska M (1979) Carbohydr. Res. 68: C-8
129 Williams NR (1970) Advan. Carbohydr. Chem. Biochem. 25: 109
130 Reichstein T, Weiss E (1962) Advan. Carbohydr. Chem. 17: 65
131 Elderfield RC (1945) Advan. Carbohydr. Chem. 1: 147
132 Fürst A, Plattner PA (1949) Helv. Chim. Acta 32: 275
133 Prins DA (1948) J. Amer. Chem. Soc. 70: 3955
134 Haunstein H, Reichstein T (1949) Helv. Chim. Acta 32: 22
135 Medgyes G, Kuszmann J (1981) Carbohydr. Res. 92: 225
136 Taki H, Yuki H, Takiura K (1975) Tetrahedron Lett. 3647
137 Gurjar MK, Yadav JS, Rao AVR (1983) Ind. J. Chem. 22B: 1139
138 Dwivedi SK, Khare A, Khare MP (1981) Carbohydr. Res. 91: 159
139 Kulkarni SU, Patil VD (1982) Heterocycles 18: 163
140 Buchanan JG (1973) MTP International Reviews of Science. Butterworths, London (Org. Chem. Ser. 1 vol 7) p 32
141 Umezawa S, Tsuchiya T, Hineno H (1970) Bull. Chem. Soc. Japan 43: 1212
142 Zobácová A, Hermánková V, Jary J (1970) Collection Czech. Chem. Commun. 35: 327
143 Zobácová A, Hermánková V, Jary J (1971) Collection Czech. Chem. Commun. 36: 1860
144 Strandtman M, Puchalski C, Shavel J (1968) J. Org. Chem. 33: 4015
145 Hedgley EJ, Mérész O, Overend WG (1967) J. Chem. Soc. (C) 888
146 Ekborg G, Svensson S (1978) Acta Chem. Scand. 27: 1437
147 Kiss L (1978) Acta Chim. Acad. Sci. Hung. 97: 345
148 Baer HH, Hanna HR (1982) Carbohydr. Res. 110: 19
149 Vis E, Carrer P (1954) Helv. Chim. Acta 37: 378
150 Baer HH, Mekarska-Falicki M (1985) Canad. J. Chem. 63: 3043
151 Showden JC, Fischer HOL (1947) J. Amer. Chem. Soc. 69: 1048
152 Williams EH, Szarek WA, Jones JKN (1969) Can. J. Chem. 47: 4467
153 Zorbach W (1964) J. Org. Chem. 29: 1970
154 Hanessian S (1966) Advan. Carbohydr. Chem. 21: 143
155 Grethe G, Mitt T, Williams TH, Uskokovic MR (1983) J. Org. Chem. 48: 5309
156 Wittig G, Geissler G (1953) Liebigs Ann. Chem. 580: 44
157 Zhdanov YuA, Alekseiev, YeE, Alekseieva VG (1972) Advan. Carbohydr. Chem. Biochem. 27: 227
158 Bestmann HJ, Angerer J (1969) Tetrahedron Lett. 3665
159 Hanessian S (1966) Carbohydr. Res. 2: 86
160 Hanessian S, Plessas NR (1969) J. Org. Chem. 34: 1035
161 Hanessian S, Plessas NR (1969) J. Org. Chem. 34: 1045
162 Hanessian S, Plessas NR (1969) J. Org. Chem. 34: 1053
163 Monneret C, Florent J-C, Gladieux N, Khuong-Kuu Q (1976) Carbohydr. Res. 50: 35
164 Thiem J, Elvers J (1978) Carbohydr. Res. 60: 63
165 Haga M, Chonan M, Tejima S (1971) Carbohydr. Res. 16: 486
166 Florent J-C, Monneret C (1980) Carbohydr. Res. 85: 243
167 Klemer A, Rodemeyer G (1974) Chem. Ber. 107: 2612
168 Klemer A, Thiemeyer H (1984) Liebigs Ann. Chem. 1094 and references therein
169 Clode DM, Horton D, Weckerle W (1976) Carbohydr. Res. 49: 305

170 Brimacombe JS, Hanna R, Saed MS, Tucker LCN (1982) Carbohydr. Res. 100: C-10
171 Brimacombe JS, Hanna R, Saed MS, Tucker LCN (1982) J. Chem. Soc. Perkin Trans. I 2583
172 Horton D, Weckerle W (1975) Carbohydr. Res. 44: 227
Horton D, Weckerle W (1977) US Patent 4, 024, 333 Chem. Abstr. (1977) 87: 68584
173 Morin C (1986) Chem. Lett. 1055
174 Horton D, Sorenson RJ, Weckerle W (1977) Carbohydr. Res. 58: 125
175 Horton D, Weckerle W (1976) Carbohydr. Res. 46: 227
176 Horton D, Nickol RG, Weckerle W, Winter-Mihály É (1979) Carbohydr. Res. 76: 269
177 Thang TT, Winternitz F, Olesker A, Lagrange A, Lukacs G (1979) J. Chem. Soc. Chem. Commun. 153
178 Crugnola A, Lombardi P, Gandolfi C, Arcamone F (1981) Gazz. Chim. Ital. 111: 395
179 Kimura Y, Matsumoto T, Suzuki M, Terashima S (1986) Bull. Chem. Soc. Japan 59: 663
180 Yoshimura J, Matsuzawa M, Funabashi M (1978) Bull. Chem. Soc. Japan 51: 2064
181 Brimacombe JS, Mengech AS, Saed MS (1979) Carbohydr. Res. 75: C-5
182 Brimacombe JS, Rahman KMM (1985) J. Chem. Soc. Perkin Trans. I 1073
183 Brimacombe JS, Mengech AS, Rahman KMM, Tucker LCN (1982) Carbohydr. Res. 110: 207
184 Arcamone F, Bargiotti A, Cassinelli G, Penco S, Hanessian S (1976) Carbohydr. Res. 46: C-3
185 Brimacombe JS, Hanna R, Tucker LCN (1982) Carbohydr. Res. 105: C-1
Brimacombe JS, Hanna R, Tucker LCN (1983) J. Chem. Soc. Perkin Trans. I 2277
186 Yoshimura J, Matsuzawa M, Sato KI, Nagasawa Y (1979) Carbohydr. Res. 76: 67
187 Brimacombe JS, Hanna R, Tucker LCN (1985) Carbohydr. Res. 136: 419
188 Pelyvás I, Sztaricskai F, Bognár R (1979) Carbohydr. Res. 76: 257
189 Pelyvás I, Whistler RL (1980) Carbohydr. Res. 84: C-5
190 Pelyvás I, Hasegawa A, Whistler RL (1986) Carbohydr. Res. 146: 193
191 Whistler RL (1983) US Patent 4, 413, 120. Chem. Abstr. (1984) 100: 86059
192 Yasumori T, Sato K, Hashimoto H, Yoshimura J (1984) Bull. Chem. Soc. Japan 57: 2538
193 Yoshimura J, Yasumori T, Kondo T, Sato KI (1984) Bull. Chem. Soc. Japan 57: 2535
194 Brimacombe JS, Rahman KMM (1985) J. Chem. Soc. Perkin Trans. I 1067
195 Kirithi L, Wickremesinghe G, Slessor KN (1980) Can. J. Chem. 58: 2628
196 Thiem J, Rasch D (1976) Synthesis 481
197 Szarek WA, Dmytraczenko A (1974) Synthesis 579
198 Lipták A, Imre J, Harangi J, Nánási P, Neszmélyi A (1982) Tetrahedron 38: 3721
199 Binkley RW, Goewey GS, Johnston JC (1984) J. Org. Chem. 49: 992
200 Jacobsen S, Pedersen C (1973) Acta Chem. Scand. 27: 2111
201 Lundt I, Pedersen C (1976) Acta Chem. Scand. 30B: 680
202 Shiue CY, Mac Gregor RR, Lade RE, Wan CN, Wolf AP (1979) Carbohydr. Res. 74: 323
203 Honda S, Kakehi K, Oguri S (1978) Carbohydr. Res. 64: 101
204 Callson B, Garegg PJ, Samuelsson B (1980) 10th International Symposium on Carbohydr. Chemistry, Sydney. Abstr. IL-7.
Garegg PJ, Samuelsson B (1980) J. Chem. Soc. Perkin Trans. I 2866
205 Collins PM, Munasinghe VRZ (1977) J. Chem. Soc. Chem. Commun. 927
206 Barton DHR, Subramanian R (1977) J. Chem. Soc. Perkin Trans. I 1718 and references therein
207 Haque MdE, Kikuchi T, Kanemitsu K (1986) Chem. Pharm. Bull. 34: 430
208 Bock K, Lundt I, Pedersen C (1981) Carbohydr. Res. 90: 7
Bock K, Lundt I, Pedersen C (1984) Carbohydr. Res. 130: 215
Bock K, Lundt I, Pedersen C (1986) Acta Chem. Scand. 40B: 740
209 Wong MYH, Gray GR (1980) Carbohydr. Res. 80: 87
210 Gurjar MK, Pawar SM (1987) Tetrahedron Lett. 28: 1327
211 Dornhagen J, Scharf HD (1986) J. Carbohydr. Chem. 5: 115
Dornhagen J, Klausner A, Runsink J, Scharf HD (1985) Liebigs Ann. Chem. 1838 and references therein
212 Pauls HW, Fraser-Reid B (1985) J. Carbohydr. Chem. 4: 1
213 Parker AJ (1962) Quart. Rev. (London) 16: 163
214 Haberfield P, Nudelman A, Bloom A, Romm R, Ginsberg H, Steinherz P (1968) Chem. Commun. 194

215 Wu MC, Anderson L, Slife CW, Jensen LJ (1974) J. Org. Chem. 39: 20
216 Miljkovic M, Glisin D, Gligorijevic M (1975) J. Org. Chem. 40: 1054
217 Richardson AC (1973) MTP International Reviews of Science. Butterworths, London (Org. Chem. Ser. 1 vol 7) p 106
218 Hill J, Hough L, Richardson AC (1968) Carbohydr. Res. 8: 7
219 Richardson AC (1969) Carbohyhr. Res. 10: 395
220 Ali Y, Richardson AC (1968) J. Chem. Soc. (C) 1764
221 Ali Y, Richardson AC (1969) J. Chem. Soc. (C) 320
222 Lieber E, Curtice JS, Rao CR (1966) Chem. Ind. (London) 586
223 Cheung TM, Horton D, Weckerle W (1979) Carbohydr. Res. 74: 93
224 Hanessian S, Desilets D, Rancourt G, Fortin R (1982) Can. J. Chem. 60: 2292
225 Sztaricskai F, Pelyvás I, Bognár R (1975) Tetrahedron Lett. 1111
226 Lee WW, Wu HY, Marsh JJ, Mosher CW, Acton EM, Goodman L, Henry DW (1975) J. Med. Chem. 18: 767
226a Monneret C, Gagnet R, Florent JC (1987) J. Carbohydr. Chem. 6: 221
227 Guthrie RD, Murphy D (1965) J. Chem. Soc. (C) 6956
228 Beynon PJ, Collins PM, Overend WG (1969) J. Chem. Soc. (C) 272
229 Richardson AC (1967) Carbohydr. Res. 4: 422
230 Richardson AC (1965) Chem. Commun. 1627
231 Boivin J, Pais M, Monneret C (1978) Carbohydr. Res. 64: 271
232 Cheung TM, Horton D, Sorenson RJ, Weckrele W (1978) Carbohydr. Res. 63: 77
233 Arndt RR, Boessenkool IK, Lourens GJ, Boeyens JCA (1981) S. Afr. J. Chem. 34: 1
234 Ikota N, Yoshino O, Koga K (1982) Chem. Pharm. Bull. 30: 1929
235 Acramone F, Bargiotti A, Cassinelli G, Redaelli S, Hanessian S, DiMarco A, Casazza AM, Dasdia T, Necco A, Regianni P, Supino R (1976) J. Med. Chem. 19: 733
236 Bargiotti A, Cassinelli G, Franchi G, Gioia B, Lazzari E, Redaelli S, Vigevani A, Arcamone F (1977) Carbohydr. Res. 58: 353
237 Kovar J, Dienstbierova W, Jary J (1967) Collection Czech. Chem. Commun. 32: 2498
238 Boivin J, Monneret C, Pais M (1980) Tetrahedron Lett. 21: 2413
239 Pelyvás I, Sztaricskai F, Bognár R (1977) Carbohydr. Res. 53: C-17
240 Baer HH, Georges FFZ (1977) Carbohydr. Res. 55: 253
241 Wu MC, Anderson L, Slife CW, Jensen JL (1974) J. Org. Chem. 39: 3014
242 Binkley RW, Ambrose MG (1984) J. Carbohydr. Chem. 3: 1
243 Pelyvás I, unpublished results
244 Malik A, Afza N, Roosz M, Voelter W (1984) J. Chem. Soc. Chem. Commun. 1530
Kowollik W, Malik A, Afza N, Voelter W (1986) J. Org. Chem. 50: 3325
245 Whistler RL, Doener LW (1970) J. Org. Chem. 35: 3562
246 Medgyes G, Pelczer I, Kuszmann J (1983) Carbohydr. Res. 111: 225
Kuszmann J, Medgyes G, Dvortsák P, Tory K (1986) Arzneim.-Forsch./Drug Res. 36: 786
247 El Khadem HS, Matsuura D (1982) Carbohydr. Res. 99: 143
248 Yamaguchi T, Kojima M (1977) Carbohydr. Res. 59: 343
249 Ohuri HH, Emoto S (1969) Carbohydr. Res. 10: 221
250 Maiti SN, Singh MP, Micetich RG (1986) Tetrahedron Lett. 27: 1423 and references therein
Knouz N, Vaultier M, Carrie R (1985) Bull. Soc. Chim. France 815
Mungall WS (1975) J. Org. Chem. 40: 1659
251 Garcia J, Vilarrasa J, Bordas X, Banaszek A (1986) Tetrahedron Lett. 27: 639
252 Kim KS, Vyas DM, Szarek WA (1979) Carbohydr. Res. 72: 25
253 Stoddart JF (1971) Stereochemistry of carbohydrates. Wiley-Interscience, New York, p 58
254 Guthrie RD, Liebmann JA (1974) Carbohydr. Res. 33: 355
255 Horton D (1968) The aminosugars. Academic Press, New York (vol 1A pp 27–31)
256 Foster AB, Inch TD, Lehmann J, Stacey M, Webber JM (1962) J. Chem. Soc. 2116
257 Baer HH, Chin Shet-Hing L (1973) Carbohydr. Res. 28: 390
258 Achmatowich, Jr. O, Bielski R (1977) Rocz. Chem. 51: 1389
259 Ozeki M, Mizoguchi T, Koga K, Yamada SI (1977) Chem. Pharm. Bull. 25: 2676
260 Koga K, Yamada SI, Yoh M, Mizoguchi T (1974) Carbohydr. Res. 36: C-9
261 Jary J, Kovár J, Capek K (1964) Collection Czech. Chem. Commun. 30: 1144

262 Capek K, Stanek, Jr. J, Capková J, Jary (1975) Collection Czech. Chem. Commun. 40: 3886
263 Florent JC, Monneret C, Khuong-Huu Q (1977) Carbohydr. Res. 56: 301
264 Martin A, Pais M, Monneret C (1983) Carbohydr. Res. 113: 189
265 Lee WW, Wu HY, Christensen JE, Goodman L, Henry DW (1975) J. Med. Chem. 18: 768
266 Gupta SK (1974) Carbohydr. Res. 37: 381
267 Capek K, Jary J (1966) Collection Czech. Chem. Commun. 31: 2558
268 Stevens CL, Glinski RP, Taylor KG, Blumbergs P, Gupta SK (1970) J. Amer. Chem. Soc. 92: 3160
269 Durette PL (1982) Carbohydr. Res. 100: C-27
270 Stewart AO, Williams RH (1984) Carbohydr. Res. 135: 167
271 Jary J, Capek K, Kovár J (1963) Collection Czech. Chem. Commun. 28: 2171
272 Georges M, Mackay D, Fraser-Reid B (1984) Carbohydr. Res. 130: 115
273 Csanádi J, Sztaricskai F, Batta Gy, Dinya Z, Bognár (1986) Carbohydr. Res. 147: 211
274 Carre MC, Houmounu JP, Caubere P (1985) Tetrahedron Lett. 26: 3107
275 Richardson AC (1973) MTP International Reviews of Science. Butterworths, London (Org. Chem. Ser. vol 7 p 132)
276 Buss BH, Hough L, Richardson AC (1965) J. Chem. Soc. 2736
277 Miyake T, Tsuchiya T, Takahashi Y, Umezawa S (1981) Carbohydr. Res. 89: 255
278 Ali Y, Richardson AC, Gibbs CF, Hough L (1968) Carbohydr. Res. 7: 225
279 Guthrie RD, Murphy D (1965) J. Chem. Soc. 3828
280 Guthrie RD, Williams GJ (1976) J. Chem. Soc. Perkin Trans. I 801
281 Butterworth RF, Hanessian S (1971) Synthesis 70
282 Szarek WA (1973) MTP International Reviews of Science. Butterworth, London (Org. Chem. Ser. 1 vol 7 p 72)
283 Szarek WA, Vyas MM (1976) MTP International Reviews of Science. Butterworths, London (Org. Chem. Ser. 2 vol 7 p 90)
284 Yoshimura J, Sato K, Hashimoto H (1977) Chem. Lett. 1327
285 Garegg PJ, Samuelsson B (1978) Carbohydr. Res. 67: 267
286 Herscovici J, Antonakis K (1980) J. Chem. Soc. Chem. Commun. 561
287 Andersson F, Samuelsson B (1984) Carbohydr. Res. 129: C-1
288 Tsuda Y, Matsuhira H, Kanemitsu (Yoshimoto) K (1985) Chem. Pharm. Bull. 33: 4095
289 Czernecki S, Georgioulis C, Stevens CL, Vijayakumaran K (1986) Synth. Commun. 16: 11
290 Roldan F, Gonzalez A, Palomo C (1986) Carbohydr. Res. 149: C-1
291 Pelyvás I, unpublished results
292 Rosenthal A, Catsoulacos P (1969) Can. J. Chem. 47: 2747
293 Tronchet JMJ (1970) Helv. Chim. Acta 53: 1489
294 Paulsen H, Bünsch H (1978) Chem. Ber. 111: 3484
295 Firouzabadi H, Ghaderi E (1978) Tetrahedron Lett. 839
Firouzabadi H, Mostafavipoor Z (1983) Bull. Chem. Soc. Japan 56: 914
296 Czernecki S, Vijayakumaran K, Ville G (1986) J. Org. Chem. 51: 5472
297 Thiem J, Elvers J (1978) Chem. Ber. 111: 3514
298 Synkar R, Zielinski J, Golik J, Borowski E (1983) J. Antibiotics 36: 1415
299 Baer HH, Fischer HOL (1959) J. Amer. Chem. Soc. 81: 5184, (1960) ibid; 82: 3709
300 Baer HH (1969) Advan. Carbohydr. Chem. Biochem. 24: 67
301 Lichtenthaler FW (1970) Fortschr. Chem. Forsch. 14: 556
Richardson AC (1976) MTP International Reviews of Science. Butterworths, London (Org. Chem. Ser. 2 vol 7 p 131)
302 Baer HH (1962) J. Amer. Chem. Soc. 84: 33
303 Baer HH, Kienzle F (1963) Can. J. Chem. 41: 1606
304 Baer HH (1979) J. Carbohydrates, Nucleosides, Nucleotides 6: 51
305 Capek K, Stanek, Jr. J, Jary J (1974) Collection Czech. Chem. Commun. 39: 1462
306 Sakakibara T, Minami T, Ishido Y, Sudoh R (1982) Chem. Lett. 67
307 Richardson AC (1962) J. Chem. Soc. 2758
308 Hochstein FA, Regna PP (1955) J. Amer. Chem. Soc. 77: 3353 and references therein
309 Baer HH, Capek K (1969) Can. J. Chem. 47: 99
310 Baer HH, Nielson T (1965) Can. J. Chem. 43: 840

311 Baer HH (1972) Methods in Carbohydr. Chem. 6: 241
312 Baer HH, Kienzle F (1967) Can. J. Chem. 45: 983
313 Baer HH, Chiu CW (1974) Can. J. Chem. 52: 111
314 Baer HH, Georges FFZ (1977) Can. J. Chem. 55: 1100
315 Baer HH, Hanna HR (1979) Carbohydr. Res. 77: C-12
316 Baer HH, Hanna HR (1980) Can. J. Chem. 58: 1751
317 Baer HH, Kienzle F (1965) Can. J. Chem. 43: 3074
318 Baer HH, Capek K, Cook MC (1969) Can. J. Chem. 47: 89
319 Baer HH, Rank W (1972) Can. J. Chem. 50: 1292
320 Baer HH, Chiu CW (1974) Can. J. Chem. 52: 122
321 Ganguly AK, Sarre OZ, McPhail AT, Onan KD (1977) J. Chem. Soc. Chem. Commun. 313 and references therein
322 Wright DE (1979) Tetrahedron 35: 1207
323 Mizsak SA, Hoeksema H, Pschgoda LM (1979) J. Antibiotics 32: 771
324 Hoeksema H, Mizsak SA, Baczynskij L (1979) J. Antibiotics 32: 773
325 Mallams AK, Puar MS, Rossmann RR, McPhail AT, Macfarlane RD, Stephens RL (1983) J. Chem. Soc. Perkin Trans. I 1497 and references therein
326 Ishii K, Nishimura Y, Kondo S, Umezawa H (1983) J. Antibiotics 36: 454
327 Kawai H, Hayakawa Y, Nakagawa M, Furihata K, Seito H, Otake N (1984) Tetrahedron Lett. 1941
Hütter K, Baader E, Frobel K, Zeeck A, Bauer K, Gau W, Kurz J, Schröder T, Wünsche C, Karl W, Wendisch D (1986) J. Antibiotics 39: 1193
328 Sharpless KB, Chong AO, Oshima K (1976) J. Org. Chem. 41: 177
329 Dyong I, Quan LC, Schulte G, Fraser-Reid B, Primeau JL (1977) Angew. Chem. Int. Ed. 16: 553
330 Dyong I, Schulte G, Quan LC, Friege M (1979) Carbohydr. Res. 68: 257
331 Dyong I, Norberg J (1979) Chem. Ber. 112: 1849
Dyong I, Norberg J, Quan LC (1979) Chem. Ber. 112: 1859
332 Schulte G, Meyer W, Starkloff A, Dyong I (1981) Chem. Ber. 114: 1809
333 Friege M, Friege H, Dyong I (1981) Chem. Ber. 114: 1822
334 Heyns K, Feldman J (1977) Tetrahedron Lett. 2789
335 Banaszek A, Zamojski A (1979) Pol. J. Chem. 53: 2029
336 Deboer A, Thanel CJ, Wilson GA (1972) Tetrahedron Lett. 5137
337 Haines AH (1972) Carbohydr. Res. 21: 99
338 Hanessian S, Bargiotti A, Laure M (1978) Tetrahedron Lett. 737
339 Radatus BK, Clarke IS (1980) Synthesis 47
340 Grynkiewicz G, Priebe W, Zamojski A (1979) Carbohydr. Res. 68: 33
341 Dave RD, Fraser-Reid B (1981) J. Chem. Soc. Chem. Commun. 1180
Fraser-Reid B, Kelly DR, Tulshian DB, Ravi PS (1983) J. Carbohydr. Chem. 2: 105
342 Yamazaki T, Sugiyama H, Yamaoka N, Matsuda K, Seto S (1976) Carbohydr. Res. 50: 279 and references therein
343 Current S, Sharpless KB (1978) Tetrahedron Lett. 5075
344 Pauls HW, Fraser-Reid B (1983) J. Org. Chem. 48: 1392
345 Bongini A, Cardillo G, Orena M, Sandri S, Tomasini C (1983) Tetrahedron 39: 3801
346 Pauls H, Fraser-Reid B (1983) J. Chem. Soc. Chem. Commun. 1031
347 Sammes PG, Thetford D (1985) J. Chem. Soc. Chem. Commun. 352
348 Cardillo G, Orena M, Sandri S, Tomasini C (1981) J. Org. Chem. 49: 3951
349 Pauls HW, Fraser-Reid B (1986) Carbohydr. Res. 150: 111
350 Mitsunobu O (1981) Synthesis 1
351 Springer D (1985) Synthesis of the saccharide components of anthracycline antibiotics. Thesis, University of Hamburg
352 Giuliano RM, Deisenroth TW, Frank WC (1986) J. Org. Chem. 51: 2304
353 Dyong I, Weigand J, Merten H (1981) Tetrahedron Lett. 22: 2965
354 Dyong I, Weigand J, Thiem J (1986) Liebigs Ann. Chem. 577
355 Baer HH, Hanna ZS (1980) Carbohydr. Res. 78: C-11
356 Baer HH, Hanna ZS (1981) Can. J. Chem. 59: 889
357 Baer HH, Hanna ZS (1981) Carbohydr. Res. 94: 33

358 Barluenga J, Jiménez C, Najera C, Yus M (1981) J. Chem. Soc. Chem. Commun. 1178
359 Heyns K, Lim M, Park JI (1976) Tetrahedron Lett. 1477
360 Heyns K, Feldmann J, Hadamczyk D, Schwentner J, Thiem J (1981) Chem. Ber. 114: 232
361 Heyns K, Hohlweg R (1978) Chem. Ber. 111: 1632
362 Boivin J, Pais M, Monneret C (1980) Carbohydr. Res. 79: 193
363 Boivin J, Montagnac A, Monneret C, Pais M (1980) Carbohydr. Res. 85: 223
364. Guthrie RD, Irwine RW, Jenkin ID (1980) Austr. J. Chem. 33: 2499
365 Thiem J, Springer D (1985) Carbohydr. Res. 136: 325
366 Lemieux RU, Bose RJ (1966) Can. J. Chem. 44: 1855
367 Chen SC (1974) Synthesis 691
368 Fraser-Reid B, Radatus B (1975) Can. J. Chem. 53: 2005
369 Boivin J (1981) Synthesis and glycosidation of 2,6-dideoxyhexoses, application to semisynthesis of daunorubicin analogues. Thesis, University of Paris, Orsay
370 Hall RH, Jordaan A, Lourens GJ (1973) J. Chem. Soc. Perkin Tans. I 38
371 Hall RH, Jordaan A, Villiers OG (1975) J. Chem. Soc. Perkin Trans. I 626
372 Shibahara S, Kondo S, Maeda H, Umezawa H, Ohno M (1972) J. Amer. Chem. Soc. 94: 4353
373 Tanaka S (1979) Japan Kokai Tokkyo Koho, 79 14,913 Chem. Abstr. (1979) 91: 5437
374 Brimacombe JS, Doner LW (1974) J. Chem. Soc. Perkin Trans. I 62
375 Daniels PJL, Mallams AK, Wright JJ (1973) Chem. Commun. 675
376 Lourens GJ, Jordaan A (1972) J. Heterocycl. Chem. 9: 975
377 Reid RJ, Mizsak SA, Reineke LM, Zurenko GE, Stern KF, Magerlein BJ (1981) J. Med. Chem. 24: 1487
378 Arcamone F, Cassinelli G (1977) US Patent 4,020,270
379 Fuchs EF, Horton D, Weckerle W, Winter B (1979) J. Antibiotics 32: 223
380 Thang TT, Imbach JL, Fizames C, Lavalie F, Pousinet G, Olesker A, Lukacs G (1985) Carbohydr. Res. 135: 241
381 Askin D, Angst C, Danishefsky SJ (1985) J. Org. Chem. 50: 5005
382 Guthrie RD, Johnson LF (1961) J. Chem. Soc. 4166
383 Patroni JJ, Stick RV (1985) Austr. J. Chem. 38: 947
384 Richardson AC (1964) J. Chem. Soc. 5364
385 Jacobsen F, Mols O (1981) Acta Chem. Scand. 35B: 521
386 Collins PM, Overend WG, Racz VM (1975) J. Chem. Soc. Chem. Commun. 181
Collins PM, Overend WG, Racz VM (1984) J. Chem. Soc. Perkin Trans. I 733
387 Chawala RK, McGonial WE (1974) J. Org. Chem. 39: 3821
388 Christensen JE, Goodman L (1961) J. Amer. Chem. Soc. 83: 3827
389 Meyer zu Reckendorf W, Bonner WA (1962) Chem. Ber. 95: 1917
390 Guthrie RD, Murphy D (1963) J. Chem. Soc. 5288
391 Buss DH, Hough L, Richardson AC (1963) J. Chem. Soc. 5295
392 Meyer zu Reckendorf W (1964) Chem. Ber. 97: 325
393 Gibbs CF, Hough L, Richardson AC (1965) Carbohydr. Res. 1: 290
394 Hough L, Penglis AAE, Richardson AC (1980) Carbohydr. Res. 83: 142
395 Gurjar MK, Patil VJ, Yadav JS, Rao AVR (1984) Carbohydr. Res. 129: 267
396 Baptistella LHB, Marsaioli AJ, Filho D-SJD, Oliveira GG, Oliveira AB, Dessinges A, Castillon S, Olesker A, Thang TT, Lukacs G (1985) Carbohydr. Res. 140: 51
397 Gurjar MK, Patil VJ, Yadav JS, Rao AVR (1984) Carbohydr. Res. 135: 174
398 Baer HH, Javorska-Sobiesiak A (1985) Carbohydr. Res. 140: 201
399 Picq D, Anker D (1985) J. Carbohydr. Chem. 4: 113
400 Castillon S, Dessinges A, Faghih R, Lukacs G, Olesker A, Thang TT (1985) J. Org. Chem. 50: 4913
401 Baptistella LHB, Marsaioli AJ, Imamura PM, Castillon S, Olesker A, Lukacs G (1986) Carbohydr. Res. 152: 310
402 Faghih R, Escribano FC, Castillon S, Garcia J, Lukacs G, Olesker A, Thang TT (1986) J. Org. Chem. 51: 4558
403 Hasegawa A, Goto M, Kiso M (1985) J. Carbohydr. Chem. 4: 627
404 Gunner SW, Overend WG, Williams NR (1964) Chem. Ind. (London) 1523
405 Brimacombe JS, Mengech AS (1980) J. Chem. Soc. Perkin Trans. I 2054

406 Ahmad HI, Brimacombe JS, Mengech AS, Tucker LCN (1981) Carbohydr. Res. 93: 288
407 Bourgeois JM (1974) Helv. Chim. Acta 57: 2533; (1975) ibid. 58: 363; (1976) ibid. 59: 2114
408 Howarth GB, Szarek WA, Jones JKN (1968) J. Chem. Soc. Chem. Commun. 62
409 Thang TT, Winternitz F, Lagrange A, Olesker A, Lukacs G (1980) Tetrahedron Lett. 21: 4495
410 Yoshimura J, Aquel A, Hong N, Sato KI, Hashimoto H (1986) Carbohydr. Res. 155: 236
411 Brimacombe JS, Miller JA, Zakir U (1975) Carbohydr. Res. 44: C-9
412 Brimacombe JS, Miller JA, Zakir U (1976) Carbohydr. Res. 49: 233
413 Brimacombe JS, Rahman KMM (1985) Carbohydr. Res. 140: 163
414 Brimacombe JS, Rahman KMM (1983) Carbohydr. Res. 113: C-6
415 Funabashi M, Hong NG, Komada H, Yoshimura J (1978) Carbohydr. Res. 67: 139
416 Brimacombe JS, Rahman KMM (1983) Carbohydr. Res. 123: C-19
417 Funaki K, Takeda K, Yoshii E (1982) Tetrahedron Lett. 23: 3069
418 Brimacombe JS, Saeed MS, Weakley JR (1980) J. Chem. Soc. Perkin Trans. I 2061
419 Giuliano RM, Deisenroth TW (1986) Carbohydr. Res. 158: 249
420 Dyong I, Schulte G (1981) Chem. Ber. 114: 1484
421 Giuliano RM (1984) Carbohydr. Res. 131: 341
422 Dyong I, Merten H, Thiem J (1986) Liebigs Ann. Chem. 600
423 Dyong I, Meyer W, Thiem J (1986) Liebigs Ann. Chem. 613
424 Picq D, Carret G, Anker D, Abou-Assali M (1985) Tetrahedron Lett. 26: 1863 and references therein
425 Picq D, Carret G, Anker D (1985) 3rd European Symposium on Carbohydrates and Glycoconjugates, 16–20 September, Grenoble Abstr. B.6–4P
426 Goodman L, Christensen JE (1963) J. Org. Chem. 28: 158
427 Ball DH, Parrish FW (1968) Advan. Carbohydr. Chem. 23: 233
428 Srivastava VK, Lerner LM (1978) Carbohydr. Res. 64: 263
429 Tipson RS (1953) Advan. Carbohydr. Chem. 8: 107
430 Weidmann H, Wolf N, Timpe W (1972) 24: 184
431 Barette EP, Goodman L (1984) J. Org. Chem. 49: 176
432 Hutchins RO, Kandasamy D, Dux F, Marynoff CA, Rotstein D, Goldsmith B, Burgoyne W, Cistone F, Dalessandro J, Puglis J (1978) J. Org. Chem. 43: 2259
433 Kuzuhara H, Sato K, Emoto S (1975) Carbohydr. Res. 43: 293
434 Hutchins RO, Kandasamy D, Marynoff CA, Masilamani D, Marynoff BE (1977) J. Org. Chem. 42: 82
435 Hutchins RO, Marynoff BE, Milewski CA (1971) J. Chem. Soc. Chem. Commun. 1097
436 Feather MS, Harris JF (1965) J. Org. Chem. 30: 153
437 Khan R, Hough L (1972) Carbohydr. Res. 24: 141
438 Khan R, Hough L (1972) Carbohydr. Res. 24: 147
439 Binkley RW, Hehmann DG (1978) J. Org. Chem. 43: 3244
Binkley RW, Ambrose MG, Hehmann DG (1980) J. Org. Chem. 45: 4387
Ambrose MG, Binkley RW (1983) J. Org. Chem. 48: 674
440 Marvell EN, Joncich HJ (1951) J. Amer. Chem. Soc. 73: 973
441 Rieche A, Schmitz E, Beyer E (1958) Chem. Ber. 91: 1935
442 Failla DL, Hullar TL, Siskin SB (1966) Chem. Commun. 716
443 Hanessian S (1968) Advan. Chem. Ser. 74: 159
444 Paulsen H (1971) Advan. Carbohydr. Chem. Biochem. 26: 127
445 Szarek WA (1973) Advan. Carbohydr. Chem. Biochem. 28: 225
446 Lipták A, Szurmai Z (1985) Magy. Kém. Lapja 40: 488
447 Pelyvás I, Sztaricskai F, Bognár R (1984) J. Chem. Soc. Chem. Commun. 104
448 László P, Pelyvás I, Sztaricskai F, Szilágyi L, Somogyi Á (1988) Carbohydr. Res. 175: 227
449 Iyengar BS, Kumar V, Remers WA (1986) Carbohydr. Res. 157: 95
450 Machado AS, Olesker A, Castillon S, Lukacs G (1985) J. Chem. Soc. Chem. Commun. 330
451 Redlich H, Roy W (1981) Liebigs Ann. Chem. 1215
452 Redlich H, Roy W (1981) Liebigs Ann. Chem. 1223
453 Gelas J (1981) Advan. Carbohydr. Chem. Biochem. 39: 71
454 Lipták A, Szurmai Z (1986) Magy. Kém. Lapja 41: 220

455 Lipták A, Szurmai Z (1986) Magy. Kém. Lapja 41: 355
456 Lipták A, Jodál I, Nánási P (1975) Carbohydr. Res. 44: 1
Fügedi P, Lipták A, Neszmélyi A (1980) Carbohydr. Res. 80: 233 and references therein
457 Fügedi P, Lipták A, Nánási P, Szejtli J (1982) Carbohydr. Res. 104: 55 and references therein
458 Garegg PJ, Hultberg M (1981) Carbohydr. Res. 93: C-10
459 Garegg PJ, Hultberg M, Wallin S (1982) Carbohydr. Res. 108: 97
460 Gorin PAJ, Finlayson AJ (1971) Carbohydr. Res. 18: 269
461 Gorin PAJ (1971) Carbohydr. Res. 18: 281
462 Garegg PJ (1984) Pure Appl. Chem. 56: 845 and references therein
463 Binkley RW, Goewey GS, Johnston JC (1984) J. Org. Chem. 49: 992
464 Johansson NR, Samuelsson B (1984) J. Chem. Soc. Perkin Trans. I 2371
465 Ponpipom MM, Hanessian S (1971) Carbohydr. Res. 18: 342
466 Hanessian S, Ponpipom MM, Lavallee P (1972) Carbohydr. Res. 24: 45
467 Boeckmann, Jr. RK, Ganem B (1974) Tetrahedron Lett. 913
468 Sharma M, Korytnyk K (1977) Tetrahedron Lett. 573
469 Anisuzzaman AKM, Whistler RL (1978) Carbohydr. Res. 61: 511
470 Garegg PJ, Samuelsson B (1979) J. Chem. Soc. Chem. Commun. 978
471 Durette PL (1980) Synthesis 1037
472 Garegg PJ, Johansson R, Samuelsson B (1984) Synthesis 168
473 Hanessian S, Leblanc Y, Lavallee P (1982) Tetrahedron Lett. 23: 4411
474 Hollingsworth RI, Hrabak EM, Dazzo FB (1986) Carbohydr. Res. 154: 103
475 Freudenberg K, Raschig K (1929) Chem. Ber. 62: 373
476 Helferich B, Himmen (1928) Chem. Ber. 61: 1825
Helferich B, Himmen (1929) Chem. Ber. 62: 2136
477 Blair MJ (1963) Methods in Carbohydr. Chem. 2: 415
478 Aspinall GO, Igarashi O, Krisnamurthy TN, Mitura W (1976) Can. J. Chem. 54: 1708
479 Chretien F (1985) 3rd European Symposium on Carbohydrates and Glycoconjugates, 16–20 September, Grenoble Abstr. B.1–23P
480 Ikeda D, Tsuchiya T, Umezawa S (1971) Bull. Chem. Soc. Japan 44: 2529
481 Abuaan MM, Brimacombe JS, Tucker LCN (1984) Carbohydr. Res. 132: 51
482 Gallagher TF, Horton D (1983) Carbohydr. Res. 116: 227
483 Meyer zu Reckendorf W (1964) Chem. Ber. 97: 1275 Meyer zu Reckendorf W, Wassiliadou-Micheli N (1970) Chem. Ber. 103: 37
Meyer zu Reckendorf W (1970) Chem. Ber. 103: 995
484 Monneret C, Conreur C, Khuong-Huu Q (1978) Carbohydr. Res. 65: 35
485 Pelyvás I, Sztaricskai F, Szilágyi L, Bognár R, Tamás J (1979) Carbohydr. Res. 76: 79
486 Mori M, Tejima S, Niwa T (1986) Chem. Pharm. Bull. 34: 4037
487 Cheung TM, Horton D, Weckerle W (1977) Carbohydr. Res. 58: 139
488 Boivin J, Pais M, Monneret C (1978) C. R. Acad. Sci. Ser. C, 286: 51
489 Descotes PG, Faure A, Martin JC (1974) J. Carbohydrates, Nucleosides, Nucleotides 1: 133
490 Wakamatsu T, Nakamura H, Naka E, Ban Y (1986) Tetrahedron Lett. 27: 3895
Chiba T, Tejima S (1979) Chem. Pharm. Bull. 27: 2838
Defaye J, Gadelle A, Wong CC (1981) Carbohydr. Res. 94: 131
491 Brockhaus M, Gorath W, Lehmann J (1976) Liebigs Ann. Chem. 89
492 Barata LES, Marsaioli AJ, Valente L, Olesker A, Lukacs G, Thang TT (1981) Carbohydr. Res. 90: 326
493 Hough L, Khan R, Otter BA (1968) Advan. Chem. Ser. 74: 120
494 Lehmann J (1965) Angew. Chem. Intn. Edn. 4: 874
495 Houghes NA, Tyson PD (1977) Carbohydr. Res. 57: 317 and references therein
496 Wovkulich PM, Uskokovic MR (1981) J. Amer. Chem. Soc. 103: 3956
497 Wovkulich PM, Uskokovic MR (1985) Tetrahedron 41: 3455
498 Fronza G, Fuganti C, Grasselli P (1980) J. Chem. Soc. Chem. Commun. 442
499 Fronza G, Fuganti C, Grasselli P, Pedrocchi-Fantoni G (1981) Tetrahedron Lett. 22: 5073
500 Fronza G, Fuganti C, Grasselli P, Pedrocchi-Fantoni G (1983) J. Carbohydr. Chem. 2: 225
501 Brown HC, Krisnamurthy S (1972) J. Amer. Chem. Soc. 94: 7159

Krisnamurthy S (1974) Aldrichimica Acta 7: 55
Wigfield DC, Feiner S (1978) Can. J. Chem. 56: 789
502 Arcamone F, Penco S, Vigevani A, Readelli S, Franchi G, Di Marco A, Casazza AM, Dasdia T, Formelli F, Necco A, Soranzo C (1975) J. Med. Chem. 18: 703
503 Bargiotti A, Cassinelli G, Arcamone F (1977) Ger. Offen. P. 2,752,115. Chem. Abstr. (1978) 89: 180312
504 Fronza G, Fuganti C, Grasselli P (1980) Tetrahedron Lett. 21: 2999
505 Suarto A, Penco S, Vigevani A, Arcamone F (1981) Carbohydr. Res. 98: C-1
506 Behrens CH, Sharpless KB (1983) Aldrichimica Acta 16: 67
507 Dyong I, Bendlin H (1978) Chem. Ber. 111: 1677
508 Jensen JE, Torsell K (1978) Acta Chem. Scand. 32B: 457
509 Hirama M, Shigemoto T, Ho S (1985) Tetrahedron Lett. 26: 4137
510 Hirama M, Shigemoto T, Yamazaki I, Ho S (1985) Tetrahedron Lett. 26: 4133
511 Hauser FM, Ellenberger SR, Glusker JP, Smart CJ, Carell HL (1986) J. Org. Chem. 51: 50
512 Overman LE (1976) J. Amer. Chem. Soc. 98: 2901
513 Zamojski A, Banaszek A, Grynkiewicz G (1982) Advan. Carbohydr. Chem. Biochem. 40: 1
514 Herczegh P, Zsély M, unpublished results
515 Wong CM, Ho TL, Nieczura WP (1975) Can. J. Chem. 53: 3144
516 Hung HK, Lam HY, Niemczura W, Wang MC, Wong CM (1978) Can. J. Chem. 56: 638
517 Hauser FM, Rhee RP (1981) J. Org. Chem. 46: 227
518 Hauser FM, Rhee RP, Ellenberger SR (1984) J. Org. Chem. 49: 2236
519 Manhas MS, Hedge VR, Wagle DR, Bose AK (1985) J. Chem. Soc. Perkin Trans. I 2045
520 Grethe G, Sereno J, Williams TH, Uskokovic MR (1983) J. Org. Chem. 48: 5315
521 Fuganti C, Grasselli P (1977) Chem. Ind. (London) 983
522 Fuganti C, Grasselli P, Marinoni G (1979) Tetrahedron Lett. 1161
523 Fronza G, Fuganti C, Grasselli P, Marinoni G (1979) Tetrahedron Lett. 3883
524 Mayer H, Seebach D (1975) Liebigs Ann. Chem. 2261
525 Musich JA, Rapoport H (1978) J. Amer. Chem. Soc. 100: 4865
526 Dyong I, Weimann R (1978) Angew. Chem. Intn. Edn. 17: 682
527 Dyong I, Friege H, zu Höne T (1982) Chem. Ber. 115: 256
528 Dyong I, Weimann R (1980) Chem. Ber. 113: 2666
529 Dyong I, Weimann R (1980) Chem. Ber. 113: 1592
530 Sharpless KB, Hori T, Truesdale LK, Dietrich CO (1976) J. Amer. Chem. Soc. 98: 269
531 Sharpless KB, Hori T (1976) J. Org. Chem. 41: 176
532 Dyong I, Friege H (1979) Chem. Ber. 112: 3273
533 Dyong I, Friege H, Luftmann H, Merten H (1981) Chem. Ber. 114: 2669
534 DeShong P, Leginus JM (1983) J. Amer. Chem. Soc. 105: 1686
535 DeShong P, Dicken CM, Leginus JM, Whittle RR (1984) J. Amer. Chem. Soc. 106: 5598
536 Kita Y, Itoh F, Tamura O, Ke YY, Tamura Y (1987) Tetrahedron Lett. 28: 1431
537 Iwataki I, Nakamura Y, Takahashi K, Matsumoto K (1979) Bull. Chem. Soc. Japan 52: 2731
538 Mukaiyama T, Goto Y, Shoda S (1983) Chem. Lett. 671
539 Iida H, Yamazaki N, Kibayashi C (1986) J. Org. Chem. 51: 4245
540 Hiyama T, Nishide K, Kobayashi K (1984) Chem. Lett. 361
541 Hiyama T, Nishide K, Kobayashi K (1984) Tetrahedron Lett. 25: 569
542 Tietze LF, Voß E (1986) Tetrahedron Lett. 27: 6181
543 Danishefsky SJ, Maring CJ (1985) J. Amer. Chem. Soc. 107: 1269
544 Fuganti C, Grasselli P, Pedrocchi-Fantoni G (1981) Tetrahedron Lett. 22: 4017
545 Fuganti C, Grasselli P, Pedrocchi-Fantoni G (1983) J. Org. Chem. 48: 909
546 Fronza G, Fuganti C, Grasselli P (1982) J. Chem. Soc. Perkin Trans. I 885
547 Fronza G, Fuganti C, Grasselli P, Majori L, Pedrocchi-Fantoni G, Spreafico T (1982) J. Org. Chem. 47: 3289
548 Mukaiyama T, Yamada T, Suzuki K (1983) Chem. Lett. 5
549 Warm A, Vogel P (1985) Tetrahedron Lett. 26: 5127
550 Warm A, Vogel P (1986) J. Org. Chem. 51: 5348
551 Hirama M, Nishizaki I, Shigemoto T, Ito S (1986) J. Chem. Soc. Chem. Commun. 393
552 Hanessian S, Kloss J (1985) Tetrahedron Lett. 26: 1261

553 Brandänge S, Lindquist B (1985) Acta Chem. Scand. 39B: 589
554 Suami T, Tadano KI, Suga A, Ueno Y (1984) J. Carbohydr. Chem. 3: 429
555 Kawai A, Hamada Y, Shioiri T (1984) Heterocycles 21: 776
556 Hamada Y, Kawai A, Shioiri T (1984) Tetrahedron Lett. 25: 5409
557 Hamada Y, Kawai A, Shioiri T (1985) Chem. Pharm. Bull. 33: 5601
558 Hamada Y, Kawai A, Shioiri T (1984) Tetrahedron Lett. 25: 5413
559 Kawai A, Hamada Y, Shioiri T (1986) Heterocycles 24: 248
560 Heathcock CH, Montgomery SH (1983) Tetrahedron Lett. 24: 4637
561 Hanessian S (1983) Total synthesis of natural products: the 'chiron' approach. Pergamon Press, Oxford
562 Pelyvás I, Hasegawa A, Whistler RL, unpublished results
563 Sallam MAE, Whistler RL, Cassady JM (1985) Can. J. Chem. 63: 2697
564 Lazzari E, Vigevani A, Arcamone F (1977) Carbohydr. Res. 56: 35
565 Acton EM, Ryan KJ, Smith T (1981) Carbohydr. Res. 97: 235
566 Acton EM, Ryan KJ, Tracey M (1984) Tetrahedron Lett. 25: 5743
567 Acton EM, Ryan KJ, Tracey M, Arora SK (1986) Tetrahedron Lett. 27: 4245
568 Mosher CW, Wu HY, Fujiwara AN, Acton EM (1982) J. Med. Chem. 25: 18 and references therein
569 Acton EM, Mosher CW, Gruber JM (1982) Approaches to more effective anthracyclines by analog synthesis and evaluation. In: El Khadem HS (ed) Anthracycline antibiotics. Academic Press, New York, p 119
570 Acton EM, Tong GL, Smith TH, Taylor DL, Streeter DG, Peters JH, Gordon GR, Filippi JA, Wolgemuth RL, Giuliani FC, Penco S (1986) J. Med. Chem. 29: 2120
571 Fujiwara AN, Acton EM, Henry DW (1974) US Patent 3,940,383. Chem. Abstr. (1976) 84: 180557
572 Roger P, Choay P, Monneret C, Fournier JP (1985) PCT Intn. Appl. WO 85 01,050. Chem. Abstr. (1985) 103: 142312
573 Martin JH, Ellestad GA, Lovell FM, Morton GO, Hargreaves RT, Perkinson NA, Baker JL, Gamble J, McGahren J (1981) Carbohydr. Res. 89: 237
574 Giuliani FC, Kaplan O (1980) Cancer Res. 40: 4682
575 Arcamone F, Penco S, Redaelli S, Hanessian S (1976) J. Med. Chem. 19: 1424
576 Suarto A, Penco S, Giuliani FC, Arcamone F (1985) Ger. Offen. DE 3,428,945. Chem. Abstr. (1985) 102: 221148
577 Florent JC, Gesson JP, Jacquesy JC, Mondon M, Monneret C (1986) Fr. Demande FR 2,565,982. Chem. Abstr. (1986) 105: 172984
578 Sztaricskai F, Pelyvás I, Bognár R, Tamás J (1983) Acta Chim. Acad. Sci. Hung. 112: 275
579 Chmielewski M (1980) Polish J. Chem. 54: 1197
580 Konoval A, Jurczak J, Zamojski A (1968) Rocz. Chem. 42: 2045
581 Chmielewski M, Jurczak J, Zamojski A (1978) Tetrahedron 24: 2977
582 Bargiotti A, Cassinelli G, Penco S, Vigevani A, Arcamone F (1982) Carbohydr. Res. 100: 273
583 Bargiotti A, Cassinelli G, Arcamone F, Di Marco A (1980) Ger. Offen. 2,942,818. Chem. Abstr. (1980) 93: 204982
584 Arcamone F, Cassinelli G, Penco S (1982) Recent developments in the chemistry of doxorubicin-related anthracycline glycosides. In: El Khadem HS (ed) Anthracycline antibiotics. Academic Press, New York, p 59
585 Toyama Chem. Co. (1984) Japan Kokai Tokkyo Koho JP 58 174,376. Chem. Abstr. (1984) 100: 156934 Toyama Chem. Co. (1984) Japan Kokai Tokkyo Koho JP 58 174,377. Chem. Abstr. (1984) 100: 156938
586 Melillo DG, Shinkai I, Liu T, Ryan K, Sletzinger M (1980) Tetrahedron Lett. 21: 2738
Melillo DG, Liu T, Ryan K, Sletzinger M, Shinkai I (1981) Tetrahedron Lett. 22: 913
587 El Khadem HS, Matsuura D, Swartz DL, Cermak R (1982) Anthracycline analogs modified in the sugar moiety. In: El Khadem HS (ed) Anthracycline antibiotics. Academic Press, New York, p 253
588 Welch JT, Svahn BM, Eswarakrishnan S, Hutchinson JP, Zubieta J (1984) Carbohydr. Res. 132: 221
589 Horton D, Weckerle W, Winter B (1979) Carbohydr. Res. 70: 59
590 Pelyvás I, Bognár R (1980) Acta Chim. Acad. Sci. Hung. 105: 141

591 Menyhárt M, Sztaricskai F, Bognár R (1983) Acta Chim. Acad. Sci. Hung. 113: 459
592 Sztaricskai F, Menyhárt M, Bognár R, Neszmélyi A (1982) Carbohydr. Res. 100: C-14
593 Kondo S, Shibahara S, Takahashi S, Maeda K, Umezawa H, Ohno M (1971) J. Amer. Chem. Soc. 93: 6305
594 Cooper DJ, Jaret RS, Reimann H (1971) J. Chem. Soc. Chem. Commun. 285
595 Ferrier RJ (1979) J. Chem. Soc. Perkin Trans I 1455
596 Mádi-Puskás M, Pelyvás I, Bognár R (1985) J. Carbohydr. Chem. 4: 323
597 Blattner R, Ferrier RJ, Haines SR (1985) J. Chem. Soc. Perkin Trans. I 2413
598 Barton DHR, Géro SD, Augy S, Quiclet-Sire B (1986) J. Chem. Soc. Chem. Commun. 1399
599 Binkley RW, Sivik MR (1986) J. Org. Chem. 51: 2621
600 Binkley RW, Sivik MR (1986) J. Carbohydr. Chem. 5: 647 and references therein
601 Arcamone F, Di Marco A, Lazzari E (1977) Ger. Offen. 2,633,913. Chem. Abstr. (1977) 86: 171789
602 Paulsen H, Stoye D (1969) Chem. Ber. 102: 834
603 Guthrie RD, Wells DD (1972) Carbohydr. Res. 24: 11
604 Arcamone F, Bargiotti A, Di Marco A, Penco S (1976) Ger. Offen. 2,618,822. Chem. Abstr. (1977) 86: 140416
605 Bargiotti A, Cassinelli G, Arcamone F (1978) Ger. Offen. 2,743,675. Chem. Abstr. (1978) 89: 44144
606 Welch JT, Svahn BM (1985) J. Carbohydr. Chem. 4: 421
607 Soc. Farmaceutici Italia SpA (1976) Brit Patent 1,457,660. Chem. Abstr. (1977) 87: 23686
608 Cassinelli G, Ruggieri D, Arcamone F (1979) J. Med. Chem. 22: 121
609 Cassinelli G, Arcamone F, Di Marco A (1978) Ger. Offen. 2,757,102. Chem. Abstr. (1978) 89: 197891
610 Sato KI, Matsuzawa M, Ajisaka K, Yoshimura J (1980) Bull. Chem. Soc. Japan 53: 189
611 Ganguly AK, Sarre OZ, McPhail AT, Ohan KD (1977) J. Chem. Soc. Chem. Commun. 313
612 Alexander J, Khanna I, Mitscher LA, Veysoglu T, Wielogorski Z, Wolgemuth RL (1984) J. Med. Chem. 27: 1343
613 Martin A, Pais M, Monneret C (1983) J. Chem. Soc. Chem. Commun. 305
614 Dyong I, Hermann R (1986) Liebigs Ann. Chem. 545
615 Sasajima K, Hayashi M, Kakiya Y (1986) Japan Kokai Tokkyo Koho JP 61 60,693. Chem. Abstr. 105: 97874
616 Whistler RL (1980) US Patent 4,181,795
617 Smith TH, Fujiwara AN, Lee WW, Wu HY, Henry DW (1977) J. Org. Chem. 42: 3653
618 Acton EM, Fujiwara AN, Henry DW (1974) J. Med. Chem. 17: 659
619 Vigevani A, Gioia B, Cassinelli G (1974) Crabohydr. Res. 32: 321
620 El Khadem HS, Liav A, Swartz DL (1979) Carbohydr. Res. 74: 345
621 Penco S, Arcamone F (1977) Ger. Offen. 2,642,837
622 El Khadem HS, Liav A (1979) Carbohydr. Res. 74: 199
623 Smith RM, Johnson AW, Guthrie RD (1972) J. Chem. Soc. Chem. Commun. 361
624 Arcamone F, Bargiotti A, Di Marco A, Penco S (1976) Ger. Offen. 2,618,882. Chem. Abstr. (1977) 86: 140416
625 Mallams AK (1973) J. Chem. Soc. Perkin Trans. I 1369
626 Dyong I, Bendlin H (1979) Chem. Ber. 112: 717
627 Cheung TM, Horton D, Priebe W, Turner WR, Weckerle W (1985) J. Antibiotics 38: 683
628 Boivin J, Monneret C, Pais M (1978) Tetrahedron Lett. 1111
629 Jordaan A, Lourens GJ (1971) J. Chem. Soc. Chem. Commun. 581

References to "Note Added in Proof"

630 Debono M, Merkel KE, Molloy RM, Barnhart M, Presti E, Hunt AH, Hamill RL (1984) J. Antibiotics 37: 85
631 Hunt AH, Debono M, Merkel KE, Barnhart M (1984) J. Org. Chem. 49: 635
632 Takatsu T, Takahashi S, Najima M, Haneishi T, Nakamura T, Kuwano H, Kinoshita T (1987) J. Antibiotics 40: 933 and references cited therein.
633 Sakakibara T, Yatabe T, Yoshino K, Ishido Y (1987) Chem. Lett. 7
634 Malik A, Kazmi NH, Khan AQ, Ahmad Z (1987) J. Chem. Soc. Chem. Commun. 1073

635 Florent JC, Monneret C (1987) J. Chem. Soc. Chem. Commun. 1171
636 Dessinges A, Cabrera Escribano F, Lukacs G, Olesker A, Thang TT (1987) J. Org. Chem. 52: 1633
637 Giuliano RM, Deisenroth TW (1987) J. Carbohydr. Chem. 6: 295
638 Klemer A, Wilbers H (1987) Liebigs Ann. Chem. 815
639 Holzapfel CW, Koekemoer JM, Verdoorn GH (1987) S. Afr. J. Chem. 40: 87
640 Hauser FM, Ellenberger SR, Rhee RP (1987) J. Org. Chem. 52: 5041
641 Roush WR, Straub JA, Brown RJ (1987) J. Org. Chem. 52: 5127
642 Hiyama T, Kobayashi K, Nishide K (1987) Bull. Chem. Soc. Jpn. 60: 2127
643 Hirama M, Shigemoto T, Ito S (1987) J. Org. Chem. 52: 3342
644 Banfi L, Cardani S, Potenza D, Scolastico C (1987) Tetrahedron 43: 2317
645 Trost BM, Sudhakar AR (1987) J. Amer. Chem. Soc. 109: 3792
646 Kametani T, Suzuki Y, Ban C, Kanada K, Honda T (1987) Heterocycles 26: 1789
647 Ha DC, Hart DJ (1987) Tetrahedron Lett. 28: 4489
648 Czerwinski A, König WA, Martelli S, Sowinski P, Bontemps-Gracz M, Mörhinger C, Kolodziejczyk P, Borowski E (1987) J. Antibiotics 40: 1067
649 Umezawa H, Nakajima S, Kawai H, Komeshima N, Yoshimoto H, Urata T, Odagawa A, Otsuki N, Tatsuta K, Otake N, Takeuchi T (1987) J. Antibiotics 40: 1058
650 Bargiotti A, Pirangelo Z, Penco S, Fernando G (1986) Belg. BE 904, 431. Chem. Abstr. (1987) 107: 154685
651 Binkley RW, Abdulaziz MA (1987) J. Org. Chem. 52: 4713

Subject Index*

* When searching for substituted derivatives of 3-amino-2,3-dideoxy- and -2,3,6-trideoxyhexoses, please consult also with Tables 19–21 (in the "Appendix") and Tables 22 and 23 (in the "Note Added in Proof" section).

P

R

S

GPSR Compliance
The European Union's (EU) General Product Safety Regulation (GPSR) is a set of rules that requires consumer products to be safe and our obligations to ensure this.

If you have any concerns about our products, you can contact us on

ProductSafety@springernature.com

In case Publisher is established outside the EU, the EU authorized representative is:

Springer Nature Customer Service Center GmbH
Europaplatz 3
69115 Heidelberg, Germany

www.ingramcontent.com/pod-product-compliance
Lightning Source LLC
Chambersburg PA
CBHW061205240926
51846CB00006B/1119

* 9 7 8 3 6 4 2 7 3 4 0 1 4 *